MODELING AND CHARACTERIZATION OF RF AND MICROWAVE POWER FETS

This is a book about the compact modeling of RF power FETs. In it, you will find descriptions of characterization and measurement techniques, analysis methods, and the simulator implementation, model verification, and validation procedures that are needed to produce a transistor model that can be used with confidence by the circuit designer. Written by semiconductor industry professionals with many years' device modeling experience in LDMOS and III–V technologies, this is the first book to address the modeling requirements specific to high-power RF transistors.

A technology-independent approach is described, addressing thermal effects, scaling issues, nonlinear modeling, and in-package matching networks. These are illustrated using the current market-leading high-power RF technology, LDMOS, as well as with III–V power devices. This book is a comprehensive exposition of FET modeling, and is a must-have resource for seasoned professionals and new graduates in the RF and microwave power amplifier design and modeling community.

All three authors work in the RF Division at Freescale Semiconductor, Inc., in Tempe Arizona. PETER H. AAEN is Modeling Group Manager; JAIME A. PLÁ is Design Organization Manager; and JOHN WOOD is Senior Technical Contributor responsible for RF CAD and Modeling, and a Fellow of the IEEE.

The Cambridge RF and Microwave Engineering Series

Series Editor

Steve C. Cripps, Hywave Associates

Peter Aaen, Jaime Plá and John Wood, *Modeling and Characterization of RF and Microwave Power FETs*

Dominique Schreurs *et al.*, *RF Power Amplifier Behavioral Modeling*

Sorin Voinigescu and Timothy Dickson, *High-Frequency Integrated Circuits*

J. Stephenson Kenney, *RF Power Amplifier Design and Linearization*

Allen Podell and Sudipto Chakraborty, *Practical Radio Design Techniques*

Paul Young, *RF and Microwave Networks: Measurement and Analysis*

Dominique Schreurs, *Microwave Techniques for Microelectronics*

MODELING AND CHARACTERIZATION OF RF AND MICROWAVE POWER FETS

PETER H. AAEN
Freescale Semiconductor, Inc

JAIME A. PLÁ
Freescale Semiconductor, Inc

JOHN WOOD
Freescale Semiconductor, Inc

CAMBRIDGE
UNIVERSITY PRESS

CAMBRIDGE UNIVERSITY PRESS
Cambridge, New York, Melbourne, Madrid, Cape Town,
Singapore, São Paulo, Delhi, Tokyo, Mexico City

Cambridge University Press
The Edinburgh Building, Cambridge CB2 8RU, UK

Published in the United States of America by Cambridge University Press, New York

www.cambridge.org
Information on this title: www.cambridge.org/9780521336178

First published 2007
First paperback edition 2011

A catalogue record for this publication is available from the British Library

ISBN 978-0-521-87066-5 Hardback
ISBN 978-0-521-33617-8 Paperback

To our families: Ljubica & Luka; Sandra, Andrea & Gaby; Gayle, Diane & Audrey.

Contents

Preface

This is a book about the modeling of RF power transistors, in particular, field effect transistors, or FETs. In it, we shall describe characterization and measurement techniques, analysis and synthesis methods, the model implementation in the simulator, and the verification and validation techniques that are needed to produce a transistor model that can be used with confidence by the circuit designer.

The demand for accurate transistor models for RF and microwave circuit design has increased as a result of the more stringent requirements that are placed upon power amplifier and transmitter designs by the customers and regulating agencies. Modern power amplifiers for communications systems are tightly specified in terms of the required linearity performance, multi-channel capability, bandwidth, and so forth. At the same time, there is a quest for higher-efficiency operation, to be realized perhaps by inherently nonlinear modes of operation, such as Class D, E, F, and others. These demands are generally conflicting, and the designer is faced with a multi-dimensional compromise. The traditional high-frequency design approaches of 'cut-and-try' are simply not appropriate for the design of RF power amplifiers for complex signal communications systems, and the designer must turn to computer-aided design (CAD) techniques and circuit simulation to optimize his design to meet the specifications. This increased use of CAD methods for RF power amplifier design places a greater reliance on the availability of accurate transistor models for simulation.

Transistor models have been used extensively in the design of analogue circuits, from lower-frequency multi-function circuits with hundreds to thousands of transistors, to higher-frequency microwave and millimeter-wave circuits with a relatively low transistor density. Simulation-based design is essential in these arenas: the circuits are virtually impossible to tune after manufacture, and so the design must be close to perfect first time, meaning that accurate models are essential.

The combination of high powers and high frequencies in RF and microwave power amplifiers brings together a unique set of challenges for the modeling engineer. The transistors themselves are physically large, and may be a significant fraction of a wavelength, even at microwave frequencies, for power transistors; the electrical behaviour in this distributed environment must be captured in the model. Even with the trend towards higher efficiency modes of operation, the device will generate a lot of heat, which must be dissipated effectively; the thermal effects on the transistor's electrical behaviour need to be characterized and modeled accurately to enable high-power designs.

Our goal is to produce compact models of the power FETs that can be used in the RF circuit simulator. The compact models will be designed to preserve the dynamics of the transistor, while being simple to develop and extract. We shall adopt a technology-independent approach to creating the compact model, based on observations of the transistor's electrical behaviour: the models will be derived directly from electrical and thermal measurements, and so careful characterization is necessary. We shall address the thermal effects on the device, scaling issues, nonlinear modeling of the active transistor, and the modeling of the internal package and matching networks. These modeling techniques are illustrated using LDMOS FETs, as this is the current market-leading high-power technology for RF cellular infrastructure applications, as well as with GaAs power devices. This is the first book to address the modeling requirements specific to high-power RF transistors. That said, the methodology that we outline can be applied to almost any FET modeling application.

We shall introduce the book by reviewing some of the historical developments in both applications and device technology that have resulted in transistor devices capable of delivering tens to hundreds of watts at RF and microwave frequencies. We shall also review the basics of FET operation, although we have left the detailed physical principles to the semiconductor device texts, and introduce the concepts behind compact modeling. This provides a basis for our analysis and construction of the transistor model in later parts of the book.

Accurate measurements form the foundation of the model. We describe how calibration and fixturing are used to ensure that repeatable and accurate measurements of the high-power transistor are made. A range of DC and RF measurement techniques and principles is outlined, for both model extraction and validation measurements. The analysis and construction of the compact model are then described. We partition the transistor into passive and active components, and address the modeling of these elements in detail.

The passive components comprise the transistor package and the in-package bondwire and capacitor components that provide the internal matching network of the transistor. This matching network is used to control the impedances presented at the terminals of the packaged transistor. The strict specifications imposed by the regulatory authorities require careful design of these matching networks. Consequently, these networks must be equally carefully modeled, to provide an accurate description of the transistor for circuit design.

The active transistor can then be accessed by de-embedding the package elements. We shall construct a large-signal model of the FET from DC and RF measurements, using a charge conservative approach for highest accuracy. The development of the charge conservative model is described in some detail, to give some insight into how the model works and its advantages for large-signal nonlinear applications.

The thermal environment is, of course, very important for power transistors: as the device heats up, its electrical properties change and so we spend some time describing a number of modern techniques for measuring the static and dynamic thermal properties. The thermal description is then used to construct a self-consistent electro-thermal model for the power FET.

At this point we consider how the model can be constructed in the circuit simulator, using function approximation or data fitting techniques, and how we can verify that the model is working correctly. Finally, we compare the model predictions with high-power RF measurements, using loadpull and large-signal network analyzer instruments, to validate the model accuracy and provide the circuit designers with confidence in its use.

Acknowledgments

The production of this book has not been carried out in isolation. We have had a great deal of support from our friends and colleagues, and we would like to take this opportunity to acknowledge their help.

We would like to thank some of our colleagues in the RF Division of Freescale Semiconductor Inc. for providing us with their expertise and assistance with some of the techniques that we have described and used. In particular, we thank Jeff Crowder for his help with the thermal measurements and simulations, and his feedback on thermal modeling; Daren Bridges, Dan Lamey, Michael Guyonnet and Daniel Chan for their inputs on nonlinear modeling, thermal models, and measurement techniques; and Chris Dragon and David Burdeaux for insights in LDMOS technology.

We would like to acknowledge all of our colleagues for providing a supportive and rewarding environment in which to work and learn, and also the many individuals in the technical community, with whom the exchange of ideas has provided us with an almost infinite supply of encouragement, knowledge, and fresh ideas. In particular, we would like to acknowledge the help and encouragement given to (two of) us by Aryeh Platzker (Raytheon), Wayne Struble (TriQuint Semiconductor), and David Root (Agilent Technologies), early in our modeling careers, which has shaped our experience and approach to transistor modeling.

We also thank Steve Cripps for his encouragement and support of this project; the staff at Cambridge University Press, in particular Julie Lancashire, for their responsiveness and assistance in getting the manuscript ready; and Andrea Plá for her assistance with figures.

To our family members we would like to express our deepest gratitude for their support and understanding during the entire process of bringing this book from an idea to its full fruition.

Peter H. Aaen
Jaime A. Plá
John Wood

1

RF and Microwave Power Transistors

1.1 Introduction

While wireless communications standards may come and go with developments in the latest digital coding technology, or the liberation of new fragments of the electromagnetic spectrum, the common denominator among the various communications systems is the power amplifier. Over the last few decades, the transition from vacuum tubes and other forms of amplification to solid state devices has been almost complete, especially at power levels less than 1 kW. Nowadays, at the heart of the power amplifier, we find the power transistor.

In the world of RF wireless communications, the base-stations and long range transmitters use silicon LDMOS (laterally-diffused MOS) high power transistors almost exclusively. In addition to modern cellular communication systems, LDMOS devices are also used in a wide range of applications requiring radio-frequency power amplification: HF, VHF, and UHF communications systems; pulsed radar; industrial, scientific and medical (ISM) applications; avionics; and most recently in WiMAX™ communication systems. The frequency range of these applications is from a few megahertz and up to 4 GHz.

While LDMOS technology is pre-eminent in high-power RF and lower frequency microwave applications, a wide variety of compound (III–V) semiconductors are used as effective power amplification devices, especially for application frequencies above about 5 GHz, and also for lower power applications such as cellular handsets, Bluetooth™ and other wireless local area networks (WLANs), which generally demand output powers of about 1 watt or below. The most commonly used compound semiconductor material for RF and microwave applications is gallium arsenide (GaAs), which is used as the basis for field effect transistor devices such as metal-semiconductor

1

FETs (MESFETs), heterojunction transistors such as high electron mobility transistors (HEMTs), and heterojunction bipolar transistors (HBTs). The basics of operation of these devices will be outlined later in this chapter. Gallium arsenide FETs and pseudomorphic HEMTs (PHEMTs) are used for low-power handset power amplifiers, as well as in some high-power cellular base-station applications, and they are also used for wide bandwidth power transistors up to the millimeter-wave regime. Recent compound semiconductor technology developments have led to the introduction of gallium nitride (GaN)-based HEMT devices, which boast very high power densities, and, depending on the substrate material used, can also have very low thermal resistance, making these devices well suited for high-power amplification. Gallium nitride FETs also have high transition frequencies, similar to some GaAs FET technologies, and so they also have the potential for high-power amplification while operating at microwave and millimeter-wave frequencies.

Even as the complexity of wireless systems continues to increase, the pressure to reduce the design cycle time and the time-to-market is challenging the designer to evaluate and develop alternative design techniques. Traditional empirical design methods based on experience and measurements are being replaced with computer-aided design (CAD) approaches. In a CAD-based design flow, the need for accurate and validated models that are properly implemented in the design tools is of paramount importance.

In this book, we shall focus on the device modeling issues peculiar to high power RF applications. Here we should state that we shall use the term 'RF' to include the RF, microwave and millimeter-wave frequency bands, in which the field effect transistor is widely used for power amplification. The use of transistors at high powers brings a unique set of problems that must be overcome in the development and deployment of a transistor model that can be used in the successful design of high power RF amplifiers. The power transistors are of considerable physical size, as can been readily appreciated from Fig. 1.1. This figure shows a transistor capable of delivering 140 watts of power at 1-dB compression, at 2.1 GHz. The complexity of the transistor in terms of the number of constituent components is a major factor to be considered in how the device is modeled. This complexity is manifest in a number of issues that must be addressed in a successful realization of a model, including electromagnetic interactions between matching networks, packaging considerations, thermal management, and the self-consistent integration of the thermal model with the electrical model of the device. The power transistor has very large total gate-width, generally much too large

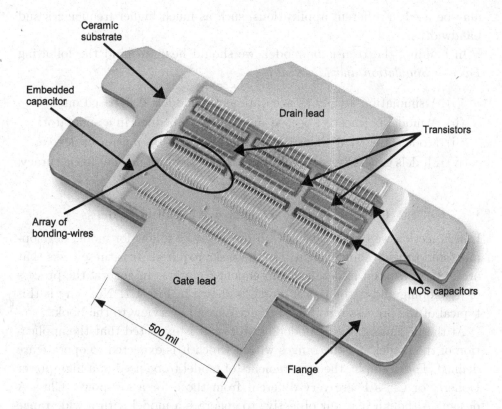

Fig. 1.1. A view of a high-power LDMOS transistor, with the lid removed to show the complexity of the internal matching networks and the LDMOS die [1]. © 2006 IEEE. Reprinted with permission.

to measure and characterize directly, so the question of scaling must be addressed. How these problems are overcome depends upon the overall device modeling strategy: the interaction between measurement techniques and function approximations, and between accuracy and implementation in the CAD tool. Our goal is to describe how these compromises can be addressed successfully.

Although the focus of this book is primarily on the high-power LDMOS FET, as this is the current market-leading technology, we shall develop, describe, and apply many measurement, analysis, and model synthesis techniques that can be more widely applied to FET modeling in general. For example, the nonlinear analysis and modeling techniques can be applied to other material technologies, such as GaAs or GaN power devices, which may be used in different applications, such as much higher frequencies and bandwidths.

In building the transistor model, we should bear in mind the following *Laws of Simulation and Modeling*†:

(i) A simulation is only as accurate as the models it is based on;

(ii) A model is (mostly) useless unless it is embedded in a simulator;

(iii) Models are, by definition, inaccurate; it's just a matter of degree;

(iv) Models generally trade off complexity (simulation time) for accuracy.

1.2 Outline of the Transistor Modeling Process

In this section we present the overall processes we follow for model development and provide some insight into the tasks required. It is these tasks that are to be presented in detail in subsequent chapters. In general the process of developing and extracting a model is shown in Fig. 1.2. Not only is this typical of the process but it is also serves as an overview to the book.

At the beginning of the modeling process it is expected that the application of the model and the ranges within which it is expected to operate are defined. For example, the requirements of model to be used in a high-power Doherty or Class F are very different from those for a low-power Class A design. Although it is our objective to generate a model with a wide range of applicability, the model must function in its intended application. From our experience, this step of defining the model is very time-consuming and is often overlooked. A clear and complete definition of the scope and deliverables of a project should be received before moving into the execution phase; this is fundamental to project management methodology. The time taken to investigate and document the model specifications is well invested, as it forces the modeling engineer and the customer to agree upon a set of objectives and validation criteria for the model. The type of model that is required is determined from these discussions. Without completing this task, which is simple in concept, but difficult to achieve in practice, there is a high probability that a model will not match the desired application or expectations of the customer.

Once the model topology has been finalized, data need to be gathered either through measurement or simulation. Typically, the frequency range, impedance ranges, and power levels are used for the specification of nonlinear models. For linear devices the frequency range and parametric variations

---

† These 'laws' have been attributed to both Colin MacAndrew and Mike Golio; we are unsure who has the precedence.

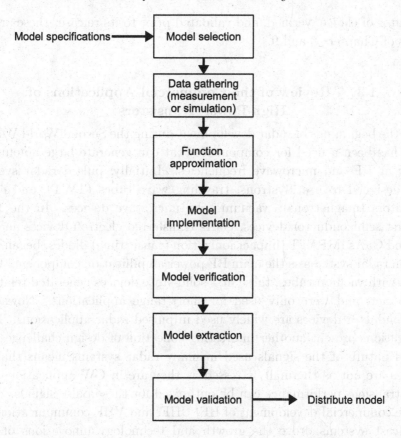

Fig. 1.2. A flow chart illustrating the distinct processes required to generate a model. Each step is covered in detail in the following chapters.

of the geometry are often specified. In Chapter 3, several measurement techniques for the extraction of linear and nonlinear models are presented. In Chapter 5, we discuss the measurements required for thermal sub-circuit generation. Electromagnetic and thermal simulation techniques are also covered in Chapters 4 and 5, respectively.

The model is generated by transforming the measured data into a format that is suitable for the implementation of the model in the chosen topology. As an example, for large-signal transistor models, the nonlinear charges and currents for the transistor model need to be extracted (see Chapter 6), then a function approximation technique is applied to generate a mathematical representation of the data that is suitable for use within a circuit simulator. Several practical approaches to function approximation are described in Chapter 7. The transistor model must then be implemented in the circuit simulator of choice, verified, and validated prior to its release: these are the topics of Chapters 8 and 9.

1.3 A Review of the Commercial Applications of High-Power Transistors

Since the beginnings of radar development during the Second World War [2], there has been a need for components that can generate large amounts of power at RF and microwave frequencies. Initially, pulsed radar systems were designed to use klystrons, traveling wave tubes (TWT), and cavity resonators (magnetrons): vacuum tube microwave devices. In the 1960s the first semiconductor devices, GaAs transferred electron devices and silicon and GaAs IMPATT (impact ionization transit time) diodes, began to be used in radar systems as the main RF power amplification component. Compared with vacuum tubes, these first solid-state devices generated relatively low power, and were only used for short-range applications. Nowadays, semiconductor devices are widely used in pulsed radar applications. These applications present another interesting collection of design challenges: the pulsed nature of the signals used in many radar systems means that the devices are not as thermally stressed as they are in CW applications, and their transient performance can be optimized for these pulse signals.

The commercial development of HF, UHF, and VHF communication and broadcast systems drove the growth and technology innovations of high power transistors, power amplifiers and communication systems further during the 1980s. Paging and cellular communication systems then brought a new wave of products and technologies operating at higher frequencies. The first generation of cellular systems, based on analogue modulation, were introduced in the 1980s and were limited to use within the 900 MHz frequency bands. During the 1990s, the second-generation ('2G') cellular systems revolutionized the communication industry by integrating voice and data through the introduction of digital modulation. These second-generation systems also shifted to higher frequencies in their quest for available spectrum and higher data rates (900 MHz, 1.9 GHz, and 2.1 GHz). In the early years of this century, third-generation ('3G') cellular systems were becoming available; these more complex systems are distinguished by even higher data rates, resulting in more stringent requirements on the efficiency and linearity of the power amplifier.

A broad category of applications that requires the generation and amplification of high powers is the industrial, scientific, and medical (ISM) market. Examples of such applications are plasma generators for etching, surface finishing, coating, magnetic resonance imaging (MRI), industrial lighting, microwave heating and drying, and so forth. In this market, there is a growing need for higher power levels, and for some applications many devices are assembled in parallel to obtain power amplifiers that can deliver several kilowatts. A new trend is to offer high performance RF transistors that are designed to operate at 50 V, thereby increasing the power density per device, and hence reducing the number of devices per power amplifier. A further benefit of using the higher voltage transistors is that the high power levels can be obtained with higher terminal impedances [3], making it easier to design amplifiers of broad frequency bandwidth.

A relatively new application in which RF high-power semiconductor devices are being used instead of tube RF amplification devices is digital television or video broadcasting (DVB). This application spectrum covers 480 MHz to 880 MHz, and because of its large bandwidth, presents a different set of challenges to the RF transistor designer. In this application, many transistors are used in a large combining network to achieve the required amount of RF power. To maximize the amount of power output per device, the devices used in digital broadcast applications operate at 32 V DC instead of the more common 28 V DC that is used in other RF power applications, such as wireless infrastructure base-stations.

Another recent application area for high-power RF transistors is in WiMAX communication systems. These systems represent a new opportunity for LDMOS power transistors as they are pushing the frequency of operation higher than is found in the traditional wireless communication systems. WiMAX systems are being designed and built to operate at several frequency bands for different regions of the world, with the majority of the development centered around 2.7 GHz and the 3.4 to 3.8 GHz bands. These higher frequencies also bring a new set of challenges to designers, as more distributed phenomena must be considered for optimum transistor and circuit design. In addition, these systems are designed with bandwidths that are much wider than traditional wireless communication systems, further increasing the complexity of the design.

1.4 Silicon Device Technology Development

Silicon metal-oxide-semiconductor field effect transistors (MOSFETs) are the most common devices in use today in the high-power RF market. They offer several advantages over bipolar devices in RF power applications as they tend to exhibit lower intermodulation distortion, and do not suffer from thermal runaway [4]. With the requirement for ever more linear transistors,

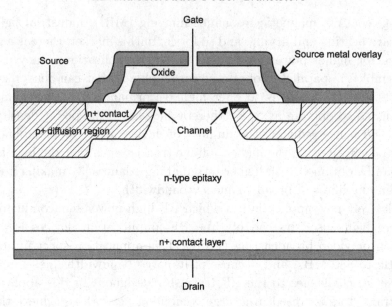

Fig. 1.3. Cross-section of a VDMOS transistor.

the BJT was relegated in the early 1990s. The cost and performance trade-off of the MOS devices also provides a clear advantage over many III–V technologies at frequencies below 4 GHz. The silicon LDMOS FET is the predominant technology among all the silicon high power FET transistors. In this section we will review the technology progression that led to the development of the silicon LDMOS transistor.

The development of the power MOSFET was marked by the introduction of a structure that allowed the current to flow through the substrate vertically. Figure 1.3 shows a cross-section of a vertically-diffused MOS (VDMOS) transistor, which is characterized by a structure that has the source and drain terminals on opposite sides of the silicon wafer. The VDMOS FET allows a more efficient use of the space on the wafer since it enables a higher concentration of the active cells. The VDMOS structure is an enhancement-mode device, which requires only a single bias polarity.

The vast majority of RF transistors are used in the common source configuration and, therefore, the VDMOS device structure has a major disadvantage in that its drain electrode is located in the back side of the wafer. As a consequence the source of the transistor, which is at the top surface of the silicon wafer, needs to be connected to the grounded terminal of the package. This results in a package construction that is very complex, requiring the use of an insulating material to isolate the back side of the wafer (the drain terminal of the transistor) from the grounded metal carrier of the package. Beryllium oxide (BeO) was commonly used as the insulating material in the VDMOS RF packages, which was expensive and highly toxic. Wire bonds were used to provide the source connection from the top of the wafer to the grounded part of the package, introducing an inductive path to ground, severely limiting the gain and the frequency response of the transistors. As a result the commercial success of VDMOS transistor was limited to applications less than 1 GHz.

Another high-power transistor technology that enjoyed commercial acceptance in the 1980s and early 1990s is the high-power bipolar junction transistor (BJT). The state-of-the-art sub-micron, high-power RF BJT technology achieved power levels around 60 W in single-ended common emitter Class AB devices at 2 GHz [5]. The high-power RF BJT suffered from the same drawback as the VDMOS device as the back side of the wafer is the collector terminal, which has to be insulated from the grounded package, and the emitter contact is at the top of the silicon wafer, requiring bondwires from the top of the wafer to the grounding bar of the package. These bondwires introduced a high feedback inductance limiting the gain of the devices. In addition, the mutual coupling between the collector and emitter bondwire arrays made the design of the matching network more challenging. The ruggedness of this sub-micron BJT was quite limited, and special care was needed in the design of bias network to ensure the thermal stability of the device, and prevent thermal runaway.

In the late 1980s and early 1990s, researchers at Motorola's Semiconductor Product Sector, now Freescale Semiconductor, developed a laterally-diffused MOS field effect transistor, or LDMOS FET, to address the inherent limitations of the VDMOS FET and silicon BJT transistor [6–9]. Figure 1.4 shows the LDMOS FET structure, with the source, gate, and drain contacts at the top surface of the wafer. It is desirable to have the source terminal at the back of the die because it can be easily grounded to an electrically and thermally conductive heatsink. The LDMOS FET also has a very low resistance and inductance connection from the source terminal at the top of the wafer to the back side of the wafer, which simplifies the design of the package and eliminates the need for source bondwires.

The LDMOS structure has a number of important advantages over the VDMOS structure. The LDMOS structure has significantly lower parasitic capacitances, by virtue of its structure; this feature results in an extended high-frequency response compared with the vertical structure. The gate-to-source capacitance, C_{gs}, and the drain-to-source capacitance, C_{ds}, have inherently lower values compared with the VDMOS structure. Another

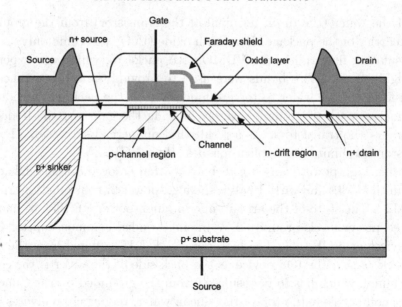

Fig. 1.4. Cross-section of an LDMOS transistor.

important advantage is that the LDMOS FETs can be used in RF integrated circuits, since they have the gate and drain terminals located on the same side of the wafer, enabling the use of microstrip transmission lines, integrated capacitors, resistors and inductors, and so forth, in the design of integrated power amplifiers and circuits.

The commercial predecessor of the LDMOS device was the sub-micron RF BJT, so it is interesting to compare the main performance differences between these two technologies. The LDMOS transistor is more reliable in the field owing to its improved ruggedness over the BJT transistor. For example, it is common for LDMOS transistors to withstand a 10:1 VSWR when driven well past their rated output power without resulting in any damage, whereas the BJT can only accept 3:1 VSWR at rated power before damage to the transistor is observed. Also, LDMOS FETs have a significant gain advantage over the BJTs. One reason for this is that the LDMOS source is connected directly to the package ground plane, avoiding the inductive feedback of the grounding bondwires.

The power saturation characteristic of an LDMOS device is quite different from that of a BJT device. The LDMOS FET has a soft saturation compared with the more abrupt power saturation of the bipolar transistor. This gives the LDMOS FET an advantage when the peak-to-average ratio (PAR) of the stimulus signal is high, avoiding the sudden clipping of the peaks in the signal and, as a result, improving the distortion characteristics of the LDMOS device.

1.5 Compound Semiconductor (III–V) Device Technology Development

The III–V compound semiconductor materials have been studied since the early 1960s because their electronic band structure and material properties offered practical device possibilities that promised significant improvements over what could be obtained in silicon, germanium, and other elemental semiconductors. The prototypical III–V compound semiconductor that is used for RF and microwave applications is gallium arsenide (GaAs); examples of such microwave devices include: transferred electron diodes (TEDs), impact ionization transit time (IMPATT) diodes, metal-semiconductor field effect transistors (MESFETs), and bipolar junction transistors (BJTs). Gallium arsenide as a material and in devices has been the subject of much development for microwave and optical applications.

The family of III–V compound semiconductors includes aluminium phosphide (AlP) and indium antimonide (InSb), and also cross-periodic compounds such as indium phosphide (InP) and aluminium antimonide (AlSb). Alloy semiconductors can be created by substituting some of the Group III or V atoms with others from the same group: gallium arsenide phosphide (GaAsP) has probably enjoyed the most commercial success of all III–V semiconductors, as a light-emitting diode (LED) material; aluminium gallium arsenide (AlGaAs) allows the band-gap of the material to be changed as the proportion of aluminium is changed, without introducing strain or stress into the lattice structure – this has enabled the development of heterojunction devices such as the high electron mobility transistor (HEMT) and the heterojunction bipolar transistor (HBT), which will be described later.

The earliest commercial III–V semiconductor microwave devices were the transferred electron diodes and IMPATT diodes, which were used primarily as oscillators, covering the frequency range from about 1–100 GHz, and from a few dBm to several watts for the highest power IMPATT diodes. These solid state signal sources displaced vacuum tube-based circuits, because of their smaller size, lower cost and higher reliability. These devices were also used in reflection amplifier configurations, over a similar frequency and power regime, although transistor-based circuits have all but replaced these applications.

The first GaAs-based transistors were bipolar junction transistors, developed in the early 1970s for RF applications. Silicon RF BJTs dominated the power transistor market up to about 3 GHz at this time, and while the GaAs transistors were being developed for higher frequency applications than their silicon counterparts, they suffered from low gain and high noise. The performance limitation in GaAs BJTs was attributable to the low hole mobility in the p-type base region of the transistor. Even at the high base doping levels that were used to minimize the base contact resistance, this resulted in a relatively high base resistance, which led to the poor noise performance. The high base doping also compromised the emitter efficiency of the BJT, resulting in low forward gain, and high leakage current across the base-emitter junction. Gallium arsenide BJTs were not a commercial success.

With the advent of modern epitaxial growth techniques for controlling the content, purity, and layer thickness of the semiconductor layers, such as molecular beam epitaxy (MBE) and metal-organic chemical vapour deposition (MOCVD), the fabrication of heterojunction bipolar transistors (HBTs) became a reality. The HBT uses the difference in the energy bandgap between the emitter and base materials; this band-gap difference is used either to produce an enhanced flow of majority carriers from emitter to base, or to introduce a potential barrier to the flow of minority carriers from base to emitter, or both. The HBT principle was patented by Shockley in 1951 [10], well before the semiconductor technology was able to build such a device. The fundamental analysis of the HBT was published by Kroemer in 1957, [11]. The first HBTs in III–V semiconductors used a wide band-gap n-type AlGaAs emitter, with narrower band-gap p-type GaAs for the base region; the collector was n-type GaAs. The potential barrier to hole transport across the emitter-base junction enabled the base region to be very heavily doped p-type, reducing its bulk resistance and metal contact resistance considerably. Epitaxial growth techniques also enabled the base region to be grown to a consistent and very small thickness, enabling a high base transport factor and hence high operating frequencies. Modern HBTs use InGaP base regions to make the band offsets more pronounced, and to improve the hole mobility. The current density in the emitter region is usually the limiting factor in the size of HBTs; the individual emitter areas are kept relatively small – a few μm^2 – to minimize the power dissipation and temperature rise at the base-emitter junction. Even multi-finger emitter structures are presently limited to relatively small absolute emitter currents even though the current density can be high. This limits these devices to relatively low powers, in the context of RF power amplification applications, although AlGaAs (emitter)–InGaP (base) HBTs are used for the power amplifiers in the majority of hand-held cellular telephones.

In the early 1970s, a number of laboratories worldwide began to develop FET device technology using gallium arsenide: the GaAs metal-semiconductor FET, or MESFET. Unlike the dominant silicon MOS technology which uses the insulating silicon dioxide between the gate terminal and the FET channel, the GaAs MESFET uses a Schottky barrier diode for the gate connection. In normal operation this diode is reverse biased, so there is little gate current, and the FET conducting channel is defined by the depletion width of the Schottky barrier in the GaAs material – the device operates in a similar manner to the junction FET (JFET); the principles of operation of the MESFET will be outlined in more detail later in this chapter. In contrast to the bipolar device, the current flow in the MESFET is parallel to the semiconductor surface, and to obtain a high transition frequency, f_T, the channel length must be short: there is an approximate inverse relationship between the channel length and f_T. There is also a direct proportionality between the carrier (electron) velocity and the transition frequency; GaAs has a much higher electron mobility than silicon, which is one of the reasons for the interest in GaAs and similar III–V semiconductors for microwave applications. The channel is essentially the region under the gate, and is hence defined by the length of the gate metal contact. To obtain f_T in the microwave frequency range requires a gate-length of a micron or less. This was a significant technological challenge at the time. Silicon MOS technology used self-alignment of the source and drain contacts to the channel region, avoiding any critical layer-to-layer alignment issues, and at that time the critical dimension in silicon MOS was of the order of two to three microns. In contrast, the fabrication of the GaAs MESFET required the definition of the gate metal, and the source and drain connections by separate photolithographic steps: this required not only the definition of sub-micron features, but also layer-to-layer alignment of features to sub-micron resolution. Nevertheless, these and other fabrication challenges were overcome and the GaAs MESFET began to see some commercial successes as a low-noise and medium-power microwave transistor.

Contemporary reviews of GaAs MESFET technology and applications can be found in [12, 13]. The GaAs MESFET was used initially in low-noise and small-signal microwave amplifiers, and was the pre-eminent solid-state device for such commercial applications in the 1980s. One example was its application in direct broadcasting of TV over satellite links, at around 12 GHz: a GaAs MESFET was used in the low-noise amplifier block (LNB), ahead of the first mixer. MESFETs also saw applications as power amplifiers, generally at low output powers, below one watt at microwave frequencies.

After the demonstration of the growth of modulation-doped AlGaAs/GaAs superlattice structures using MBE by Dingle *et al.* in 1978 [14], in which the wide band-gap material (AlGaAs) was heavily doped, and the narrow band-gap material was undoped, enabling this material to exhibit a high electron mobility, this technology was applied to the construction of FET devices in the early 1980s. These FETs used a heavily-doped n-type AlGaAs layer grown close to the heterojunction with the undoped GaAs; the freed electrons would 'spill over' into the thin undoped GaAs region, where they would occupy a thin sheet close to the heterojunction, and have a high mobility. These heterojunction FET devices were able to demonstrate higher transition frequencies, and hence operating frequencies, than MESFETs of the same gate-length (or channel length). Generally, heterojunction FETs also showed higher gain and lower noise figures than MESFETs, and have supplanted the GaAs MESFET in many applications. In the early years of the development of these heterojunction FETs, the various proponents of the device coined their own taxonomy: the modulation-doped FET (MODFET), from Cornell University; the two-dimensional electron gas FET (TEGFET), from Thomson-CSF; and the high electron mobility transistor (HEMT), generally used by the physics community. This last term, HEMT, is the one that is common currency today (despite the fact that it is the high saturated electron velocity that gives this device its microwave and millimeter-wave performance, not the electron mobility).

Further developments in the MBE growth technology enabled the stable growth of thin semiconductor layers of mismatched crystal lattice dimension compared with the substrate material. The prototype example of this is the growth of indium gallium arsenide (InGaAs) on GaAs substrates. Provided that the fraction of indium is kept reasonably low, below about 25%, and is of the order of a tenth of a micron in thickness, the strain due to the mismatch in crystal lattice size can be accommodated in the epitaxial layer structure. The advantage obtained from this technique is yet higher electron mobility and saturated velocity in the InGaAs channel, and hence higher frequency of operation of the transistor. In all other respects, the construction or fabrication of the HEMT is, in principle, the same as the MESFET. Because of the mismatch in the semiconductor layer lattice constants, these devices are known as *pseudomorphic* HEMTs, or PHEMTs. Epitaxial growth and fabrication technology for PHEMTs is nowadays well established, and these devices are used in the lowest noise applications, and also for small-signal and power amplifiers.

Nowadays, MESFETs and PHEMTs are available commercially as discrete packaged transistors for medium power applications from RF to microwave frequencies. Plastic packages are commonplace at lower frequencies, and ceramic packages are used at higher frequencies, for lower parasitics. These devices are used in broadband microwave amplifiers for instrumentation applications: for example, the Agilent 83020A 2–26 GHz bandwidth, 1 W amplifier, and 83050A 2–50 GHz bandwidth, 100 mW amplifier are distributed amplifiers built using MESFET and PHEMT technologies, respectively, and probably represent the yardstick for broadband medium-power amplifiers. GaAs FET and PHEMT transistors have also been used in high power base-station amplifiers for cellular wireless communications. Commercial transistors are available with powers comparable to LDMOS, [15], and very high powers from a single die have been recorded, for example [16], at the time, a world record of about 300 watts.

The PHEMT is also used at much lower powers in the cellular frequency bands, in the power amplifier of mobile telephones. In this application, the threshold voltage of the PHEMT is adjusted by controlling the growth and fabrication conditions, so that the transistor is in the 'off' state when no gate voltage is applied; in other words, no drain current flows in this condition. This is known as 'normally-off' or *enhancement-mode* PHEMT, or E-PHEMT. Such a device consumes no DC power as there is no current flow, and this helps prolong the battery life in the hand-held device. When switched on, the E-PHEMT power amplifier is biased into Class AB to deliver the RF output signal. Typically, these power amplifiers are rated at 1–2 W, and are used in GSM and CDMA applications.

An advantage claimed for III–V MESFETs and PHEMTs is that the output conductance of the transistor is relatively small, and approximately independent of the drain voltage [17]. In contrast, LDMOS transistors have a higher output capacitance than comparably-rated GaAs PHEMT power transistors. This parameter is important for Doherty amplifier applications. In the Doherty amplifier configuration, an auxiliary or peaking amplifier is used to modulate the load seen by the main amplifier, to improve the overall efficiency of the power amplifier [18]. The auxiliary amplifier is switched off at lower input powers, and so it should present an open circuit to the main amplifier's output. The ability of the transistor to present an open circuit at RF is compromised by its output capacitance. Even so, LDMOS transistors have been used successfully in Doherty configuration for RF power amplifiers [19–21].

Since the early 1990s, there has been an upsurge of interest in wide band-gap semiconductors for power amplifiers. The advantage of a wide band-gap is the increase in breakdown voltage between the FET gate and drain; this allows higher operating drain voltages and RF voltage amplitudes and so higher output powers can be realized. A collateral benefit of the higher RF drain voltage is that there is a higher impedance at the output of the transistor, which is easier to match to in the amplifier circuit. The first wide band-gap semiconductor to be studied for RF power transistors was silicon carbide, SiC. This material has an indirect band-gap of about 3.2 eV in the 4H-SiC form, and also has a very high thermal conductivity, which means that the transistor can operate at a high specific dissipated power while the junction temperature can be kept to a practical value for device reliability [22]. Good laboratory results have been obtained for SiC MESFETs, but the high cost of substrates and difficult processing have so far prevented commercial exploitation.

More recently, gallium nitride has received considerable attention as a wide band-gap semiconductor for power FET applications. This material has similar properties to the prototype III–V semiconductor, GaAs, but has a wide band-gap coupled with high electron mobility and saturated velocity. The GaN FET device is a HEMT structure, with aluminium nitride used as the barrier layer, and gallium nitride as the conducting channel. Gallium nitride can be grown epitaxially onto silicon or silicon carbide substrates, although as a result of the stresses between the GaN and the substrate, the GaN layer can have a high density of dislocations and defects. This picture has improved in recent years as the semiconductor growth and processing technology has been focused on this problem. The GaN-on-SiC technology is very expensive, and targeted primarily at military applications. The GaN-on-silicon technology is less expensive, but has inferior thermal properties; this technology is being focused on commercial power amplifier applications at communications frequencies that are not attainable by LDMOS devices. Very high specific power densities of around 10 W/mm have been quoted for GaN HEMTs, significantly higher than that of GaAs or LDMOS power transistors, which are about 1 W/mm. Such a high power density is not such a clear benefit, even if the power amplifier operates under high efficiency conditions, because the dissipated power or heat still has to be extracted from the transistor package to maintain a reasonable junction temperature at the device. In commercial applications, efforts are made to reduce the power density for good thermal management. At the time of writing, several semiconductor companies are poised to announce commercial availability of GaN-based power transistors, directed at cellular and WiMAX base-station applications.

1.6 The Basics of FET Operation

The field effect transistor is a charge-controlled device. By this we mean that by applying a voltage to the 'control' terminal – the gate – we can control the amount of charge in the region of the semiconductor beneath the gate. This region is confined laterally and vertically by the structure of the transistor, so the charge is effectively confined to a channel that connects the drain and source terminals. By controlling the amount of charge in the channel, we can control the flow of current between the source and drain terminals. Thus, the input voltage controls the output current; this is also known as a *transconductance* device:

$$
\begin{aligned}
I_{\text{drain}} &\propto V_{\text{gate}} \\
&= g_{\text{m}} V_{\text{gate}}
\end{aligned}
\tag{1.1}
$$

where the parameter g_{m} is the transconductance. The subscript 'm' remains as a legacy of the old-fashioned term, *mutual* conductance.

We shall investigate this charge control mechanism, as it applies to the various field effect transistor devices that are used for power transistors, to determine the drain current. We shall highlight only the main results for each device; the details of the basic semiconductor physics and fundamentals of operation of these transistors can be found in, for example [13, 23–26].

1.6.1 Basic Theory of Operation of the MOS Transistor

The essential feature of the MOSFET is the MOS capacitor. This is a two-terminal device, comprising a metal contact that is separated from the semiconductor by a thin insulating layer. In practical terms, the semiconductor is silicon, and the thin insulator is silicon dioxide, which can be grown directly onto the silicon with very few defects at the silicon–silicon dioxide interface. The thickness of the oxide can be carefully controlled in the fabrication process, and the high quality of the interface permits mobile charges to flow near the silicon surface, without being trapped by the defects. It is the high structural quality of the silicon-silicon dioxide interface that sets this metallurgical system apart: many attempts at building metal-insulator-semiconductor (MIS) FET devices have failed because of the inability to control the defect densities at the interface, and in the insulator, at the levels required to permit charge to accumulate and move freely at the interface.

The charge density at the surface of the silicon, beneath the oxide, is controlled by the application of a voltage to the metal gate. The applied voltage is dropped across the oxide and the semiconductor, and the voltage at the surface of the semiconductor causes bending of the electron energy

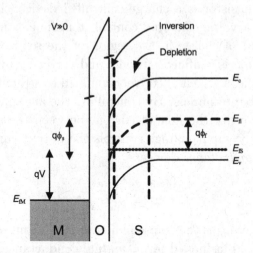

Fig. 1.5. Electron energy-band diagram for an idealized MOS capacitor in inversion.

bands in the silicon. The MOS devices used at high frequencies are typically n-channel transistors; electrons have a higher mobility than holes in silicon, and so can operate at higher frequencies. The electron channel is created at the oxide-silicon interface, in an *inversion* layer, by applying a positive voltage to the metal.

A positive voltage applied to the metal will cause the energy bands in the silicon to bend downwards, creating a depletion region at the silicon surface. We can also think of this condition arising from the electrostatic force between the positive charge on the metal due to the applied voltage and the positively-charged holes, repelling the holes from the semiconductor surface and hence creating the depletion region. If the positive voltage on the metal is increased, the energy-band bending will become greater, and at some voltage the Fermi energy of the silicon at the surface will be nearer to the conduction band than the valence band: the p-type semiconductor 'looks' n-type at the surface. This electron-rich layer is the inversion layer. In practice, the onset of *strong inversion* occurs when the electron density in the inversion layer is the same as the bulk hole density in the substrate. The voltage that is applied at this point is the *threshold voltage*. The electron energy-band structure in inversion is shown in Fig. 1.5.

In (strong) inversion, the positive charge on the metal, Q_M, is balanced by the negative charge in the semiconductor, Q_s, which comprises the depletion region charge, Q_{dep}, and the charge due to the electrons in the inversion layer, Q_n:

$$Q_M = -Q_s = -Q_{dep} - Q_n \qquad (1.2)$$

At the onset of strong inversion, while the surface electron density is the same as the bulk hole density, the total surface charge is small compared with the charge in the depletion region. The positive charge on the metal is, therefore, approximately equal to the negative charge in the depletion region, and the threshold voltage is given by

$$V_T = V_{FB} - \frac{Q_{dep}}{C_{ox}} + 2\phi_f \qquad (1.3)$$

where the surface potential at threshold, $2\phi_f$, is related to the doping density of the p-type silicon, and V_{FB} accounts for any defect charge, and differences between the metal and silicon work functions. C_{ox} is the parallel-plate capacitance of the oxide layer.

An expression for the charge control by the applied voltage can be developed from the expressions above, and accounting for the applied voltage drop across the oxide and depletion regions, to yield

$$Q_n = -C_{ox}\left(V - V_T - (\phi_s - 2\phi_f)\right) \qquad (1.4)$$

The charge in the inversion layer is controlled by the amount of applied voltage above threshold, less the excess band-bending at the surface, denoted by ϕ_s. This defines the capacitance of the device.

The MOS capacitance–voltage relationship is shown in Fig. 1.6. At extreme negative applied voltages we attract the majority carrier holes to the surface, and we have effectively a parallel-plate capacitance, C_{ox}. As the applied voltage passes through zero volts, we obtain the flat-band condition for this idealized MOS structure. As the voltage becomes positive, we begin to create the depletion region in the semiconductor, as described earlier. The capacitance is a series combination of the oxide capacitance and the semiconductor depletion capacitance. As the applied voltage increases, the depletion region widens, and the total capacitance falls, until we reach the minimum capacitance at the threshold voltage: the depletion region is at its maximum width. Above threshold, the depletion width does not increase significantly with further applied voltage. The charge density in the inversion layer at the surface grows, and screens the depletion region from the applied voltage. The increase in the positive charge on the metal is balanced by an increase in the inversion layer charge, and above threshold the MOS capacitor behaves like a parallel-plate capacitor C_{ox}. This characteristic is true at low frequencies. At high frequencies, we see that the capacitance in

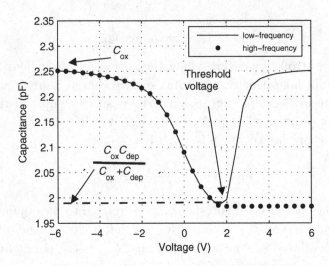

Fig. 1.6. Capacitance–voltage relationship for an idealized MOS capacitor.

the inversion regime remains at or close to the minimum value, with the depletion capacitance corresponding to its value at threshold. The reason for this is that the minority carrier electrons which comprise the inversion layer arise from the generation–recombination (G–R) processes at work in the semiconductor. These G–R processes have relatively long time constants, and so they cannot respond to the high-frequency excitation.

The MOS transistor is created when the inversion layer of a MOS capacitor is used as the conducting channel between the source and drain regions. The MOS capacitor metal electrode forms the gate contact. A schematic of an n-channel MOSFET is shown in Fig. 1.7, indicating the source and drain regions, and their terminals, the inversion layer and the depletion region under the MOS capacitor, and the gate terminal.

In the MOS transistor, the gate capacitance will follow the 'low-frequency' curve in Fig. 1.6, for all frequencies. This is because in the FET, the source and drain contacts are abundant sources of minority carriers – electrons – and as long as the gate bias is above the threshold voltage, the inversion channel will be filled with electrons, and the gate capacitance will look, to first order, like the parallel-plate capacitor. As noted above, in the MOS capacitor, the only sources of electrons in the inversion layer are thermal generation processes, which have time constants that are too great to respond to high-frequency excitation.

In the MOS capacitor, the potential beneath the oxide is well defined and the inversion layer is at equipotential. In the MOS transistor, we also

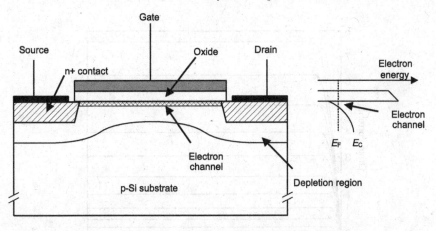

Fig. 1.7. Schematic view of an n-channel MOSFET, showing the inversion layer which forms the conducting channel between the source and drain, and the depletion layer, under the gate.

apply a potential difference between the drain and source terminals, causing current to flow in the inversion layer or channel. The applied voltages at the gate and drain are V_G and V_D, with the source at zero volts. There is now also a voltage drop along the length of the inversion layer, and each point in the channel experiences a different voltage or potential. The drain current in the transistor is calculated by expressing the surface potential and hence the channel charge in eq. 1.4 as a function of the local potential. The drain current can then be found by integration along the channel, to give

$$I_D = \frac{W C_{ox} \mu_s}{L} \left[(V_G - V_T) V_D - \frac{1}{2} V_D^2 \right] \tag{1.5}$$

where μ_s is the *surface* or *channel* electron mobility, and W and L are the width and length of the gate, respectively. We stress that this mobility is different from the mobility of electrons in bulk silicon; generally the surface mobility is much lower. This is thought to be due to additional electron scattering mechanisms present at the oxide-silicon interface, and to quantum confinement effects induced by the narrowness of the inversion channel.

The expression in eq. 1.5 is the 'simple charge control model' and is based on the gradual channel approximation (GCA). This model is applicable at low drain voltages; this is often called the *linear* or *triode* region of the FET drain current characteristics. As the drain voltage is increased, the maximum value of the drain current is reached at the *saturation* voltage:

$$V_D = V_G - V_T = V_{sat} \tag{1.6}$$

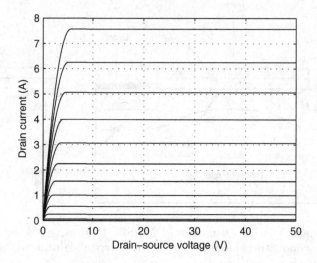

Fig. 1.8. Drain current characteristics for a MOSFET calculated using the simple charge control model, showing the 'linear' or 'triode' region and saturation.

This is the onset of *saturation* in the FET. In saturation, the drain current is independent of the drain voltage, and has a value given by substituting the expression in eq. 1.6 into the drain current equation, eq. 1.5:

$$I_\mathrm{D} = \frac{W C_\mathrm{ox} \mu_\mathrm{s}}{2L} \left(V_\mathrm{G} - V_\mathrm{T}\right)^2 \tag{1.7}$$

The MOSFET drain current characteristics for the simplified charge control model are shown in Fig. 1.8. While qualitatively useful in terms of understanding the origins of MOSFET behaviour, and estimation of some device parameters, a brief examination of this model reveals its drawbacks: at the onset of saturation, the charge in the channel falls to zero. This is a somewhat unphysical result, and is generally side-stepped by calling this condition 'pinch-off' and invoking an infinitesimally-thin channel for the drain current to pass through.

As MOS technology developed, and transistor gate-lengths became shorter, the electric fields along the channel became larger, and a more enlightened view of electron transport in the channel led to the conclusion that the saturation in the drain current characteristic is result of the electrons in the channel reaching a limiting velocity at high electric fields. This limiting velocity is the *saturation velocity*. At the onset of drain current saturation, the potential drop along the channel is just sufficient to cause electron velocity saturation. A number of empirical expressions for the electron channel velocity as a function of the applied electric field have been reported [27], generally concerned with smoothing the transition from ohmic (linear) to limiting behaviour.

The practical MOSFET has drain current characteristics that differ from these ideal curves. We will outline below a few effects on the drain current curves that occur largely as a result of high electric fields, which are consequences of the short gate-lengths and transport physics in aggressively-scaled MOSFETs. We shall follow up this discussion with more detailed descriptions and models in Chapter 6 where we present the foundations of our nonlinear FET model.

The ideal drain current characteristics shown in Fig. 1.8 display ideal current source behavior in saturation. Practical devices have a finite output conductance, both at DC and RF; this is attributable to the effect known as *channel-length modulation*. As the drain voltage is increased in saturation, the electric field in the channel must increase, and the point in the channel where the electrons reach the saturation velocity moves slightly closer to the source. At this point the channel voltage will be slightly smaller, and hence the band-bending at the surface ϕ_s will be slightly larger, admitting a higher current density. An increasing channel current with drain-source voltage in saturation produces a positive slope, that is, a finite output resistance.

At high drain voltages, the MOSFET suffers a *breakdown* caused by the high voltage across the oxide at the drain end of the gate, resulting in high gate-drain current flow. Another effect arising from the high electric field in this region is *hot-carrier injection*. The high fields and small distances in the MOSFET give rise to non-equilibrium transport effects on the electrons in the channel: some electrons are accelerated to much higher velocities than the average, and hence have sufficient (kinetic) energy to cross the potential barrier at the oxide-silicon interface, and become trapped in the oxide. The resulting space charge causes shifts in threshold in the transistor, and hence changes to the operating conditions and device performance.

These high-field effects are mitigated in the laterally-diffused (LD) MOS structures developed for power MOSFETs. The use of a lightly-doped n-type region at the drain end of the gate moves the heavily-doped drain contact region away from the high field region, and has a number of benefits. The lightly-doped semiconductor can support a high voltage, enabling the high RF voltage swing required for a high-power device. The electric field in saturation at the drain edge of the gate is reduced, thereby reducing the hot-carrier injection and increasing the gate breakdown voltage.

1.6.2 Basic Theory of Operation of the GaAs MESFET

The gallium arsenide MESFET is a member of the junction FET family. In this case, the gate contact is a metal-semiconductor Schottky barrier junction instead of the more common p-n junction diode that is found in classical silicon JFETs (in fact, silicon MESFETs can also be made, but so far have not found a significant commercial application). The gate diode is normally operated in reverse-bias, using the bias-dependent depletion region depth to define the channel region between the source and drain contacts. In reverse bias, there is also very little gate current flow, so the device presents a high input resistance at DC.

A simple schematic cross-section of the GaAs MESFET structure is shown in Fig. 1.9. The channel region is formed in a layer of n-type GaAs that is grown epitaxially onto a semi-insulating GaAs substrate. The electron mobility in GaAs is much greater than the hole mobility, so only n-channel MESFETs are practical microwave devices. An undoped GaAs 'buffer layer' is often grown between the substrate and the active layer to minimize the incorporation of defects and unwanted impurities at the lower interface of the channel. The source and drain contacts are made to highly-doped GaAs to reduce the contact resistance and the resistance of the semiconductor that is in series with the channel region. The height of the conducting channel between the source and drain is determined by the depth of the depletion region under the gate contact: the voltage on the reverse-biased gate contact controls the current flow between the drain and source. For a given gate voltage and channel opening, the device will behave as a (nonlinear) resistor at low values of drain-source voltage. As the drain voltage is increased, at some point the drain current becomes independent of any further increase in the drain voltage, and the MESFET is in saturation. The drain voltage at the onset of saturation is known as V_{dsSAT}. When sufficient voltage is applied to the gate to cause the depletion region to reach completely across the conducting channel to the buffer layer, the flow of current is cut off. The gate voltage required for this condition is known as the *pinch-off voltage*, V_{PO}. Strictly speaking, this definition applies at zero drain voltage, but the channel can still be pinched-off for non-zero drain voltage, by the application of a suitable gate voltage.

The transistor action in the GaAs MESFET is in principle the same as the silicon MOSFET: a change in the channel conductance controlled by the gate voltage. The difference is that in the MESFET it is the physical dimension of the channel that is changed, whereas in the MOSFET it is the channel charge and hence conductivity that is controlled by the applied gate

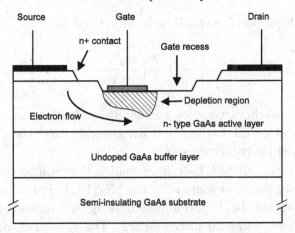

Fig. 1.9. Schematic cross-section of a GaAs MESFET.

voltage. The drain current characteristics can be developed from some basic semiconductor physics. We shall use the channel charge control approach, for consistency of the analysis with the MOSFET drain current character- istics derived earlier. The charge Q at a point x in the channel under the depletion region is given by

$$Q(x) = qn_0 W D \left(1 - \frac{D_{\text{dep}}(x)}{D} \right) \qquad (1.8)$$

where D is the total thickness of the active channel, and D_{dep} is the depletion layer thickness; n_0 is the equilibrium electron density in the active channel. The depletion region depth is controlled by the reverse bias voltage on the gate, giving the charge control. The current in the channel can be found by integrating along the length of the channel from drain to source, and after some simplifying assumptions [23], yields

$$I_D = G_0 V_P \left[\frac{V_D}{V_P} - \frac{2}{3} \left(\frac{V_D - V_G + \phi_{bi}}{V_P} \right)^{\frac{3}{2}} + \frac{2}{3} \left(\frac{\phi_{bi} - V_G}{V_P} \right)^{\frac{3}{2}} \right] \qquad (1.9)$$

where G_0 is the open-channel conductance, obtained from eq. 1.8 when the depletion depth is zero, and ϕ_{bi} is the built-in potential of the Schottky contact [25]. This relation is only valid up to the point of pinch-off, or saturation, where $V_D - V_G = V_P$. In saturation we assume that the drain current remains essentially constant, independent of any further increase in drain voltage. Equation 1.9 can then be simplified to

$$I_D = G_0 V_P \left[\frac{V_G}{V_P} + \frac{2}{3} \left(\frac{\phi_{bi} - V_G}{V_P} \right)^{\frac{3}{2}} + \frac{1}{3} \right] \qquad (1.10)$$

The drain current characteristics are qualitatively the same as those obtained for the MOSFET earlier, and demonstrate that the charge-control approach can yield satisfactory results.

In saturation, the MESFET drain current is determined by velocity saturation effects, in a similar manner to the MOSFET. From Fig. 1.9, we can see that the shape of the depletion region along the channel reflects the local gate-to-channel potential in the device. The gate-to-channel voltage is highest at the drain edge of the gate, and so the depletion region is widest at this point. The electric field in the channel is highest at the drain edge of the gate and, consequently, the electron velocity is highest at this point. The drain voltage at which the channel electrons reach the limiting velocity therefore defines the onset of saturation in the GaAs MESFET; this is the *saturation voltage*.

Increasing the drain voltage beyond the saturation voltage requires that the channel narrows even more around the drain end of the gate, as the local channel voltage is higher, and hence the depletion region is deeper. The point in the channel corresponding to the electrons reaching the saturated velocity moves towards the source. There is a region of the channel under the drain end of the gate where the channel height is smaller than the height at velocity saturation. To maintain current continuity in this region, the electron density must increase above its equilibrium value, and there is an accumulation of charge. Beyond the drain edge of the gate, the channel widens sharply; the electric field here is still above the critical value, so the electrons continue to travel at their saturated drift velocity in this region, and to maintain current continuity, the charge density in this region falls below the equilibrium value: a partial depletion of charge in the channel occurs. The MESFET in saturation is characterized by a charge dipole at the drain edge of the gate, and most of the applied drain voltage above saturation falls across this dipole.

The MESFET suffers from similar high electric field limitations as the MOSFET, *viz*, finite output conductance, and gate-drain breakdown effects. The output conductance arises because as the drain voltage increases, the point at which the electrons in the channel reach velocity saturation moves nearer to the source. At this point the channel voltage will be slightly smaller, the depletion depth smaller, and hence the channel will be slightly larger, admitting a higher current density. An increasing channel current with drain-source voltage in saturation produces a positive slope, that is, a finite output resistance.

The highest voltage in the MESFET is at the drain edge of the gate, and this is the usual location for the breakdown. In high power MESFETs, the region between the gate and drain contacts is made quite wide compared with typical general-purpose devices, up to several microns, and etched to a small thickness: this allows a high voltage to be supported without the creation of high-field domains associated with the *Gunn Effect*, which can lead to instability and oscillations [28].

1.6.3 Basic Theory of Operation of the High Electron Mobility Transistor

The demonstration of high electron mobility in modulation-doped super-lattice structures, comprising thin alternating layers of n-type AlGaAs and undoped GaAs, by Dingle *et al.* [14] in 1978 led to a race to find a commercial device application for this phenomenon. Within about a year, several laboratories around the world announced a field effect transistor based on a simplified version of the superlattice: the high electron mobility transistor (HEMT) using a single heterojunction. The structure and associated electron energy bands in the channel are shown in Fig. 1.10. The thin electron charge sheet in the GaAs forms the channel between the source and drain contacts. The electron density in the charge sheet is controlled by the gate voltage, which controls the band-bending in the structure and hence charge density. The electronic band structure of the channel and means of charge control are very similar to the MOSFET (see Fig. 1.5). The ability of this structure to control the charge at the heterojunction interface, and hence produce a current flow parallel to this interface, is a statement of the crystalline purity and defect control provided by the MBE semiconductor growth technique, which was beginning to show its capabilities at this time [29]. Nowadays, virtually all high quality commercial grade heterojunction FETs are grown exclusively using MBE.

Higher performance HEMT structures were achieved through the synergetic development of the MBE technique and an understanding of the electron transport behaviour in the confined channel. By sandwiching a layer of very narrow-gap semiconductor between layers of high band-gap material, good charge confinement was obtained, and high transconductance and transition frequency f_T were observed. The narrow-gap material, indium gallium arsenide (InGaAs), has a slightly different crystal lattice constant

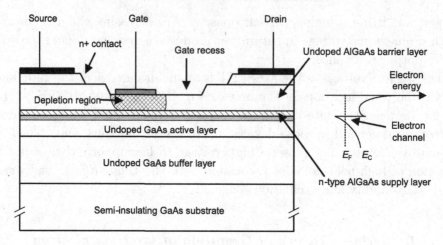

Fig. 1.10. Schematic diagram of AlGaAs–GaAs HEMT structure and its electron energy-band diagram.

from the GaAs substrate, but for low concentrations of indium in the alloy, typically 25% or below, the strain in the structure can be accommodated without generating the crystal defects that would destroy the device performance. In fact, the strain is beneficial, as it causes the electron mobility in the InGaAs layer to increase slightly. This structure is the basis of the pseudomorphic HEMT or PHEMT. The schematic structure of the PHEMT and the associated electron energy-band structure are shown in Fig. 1.11. While the HEMT has the MOS-like advantage of a confined charge sheet, with its direct charge control and a well defined threshold of conduction, it also has the MESFET's drawback of requiring the drain and source contacts to be place at some distance from the channel region, introducing an access resistance that lowers the extrinsic gain, increases noise, and dissipates power. In spite of this, the PHEMT has all but supplanted the GaAs MESFET for commercial applications, offering lower noise, higher gain, and high power capability.

Early attempts at modeling the HEMT focused on analytical models of the charge control mechanism in the channel, for example, [30]. The objective of these analyses was to establish a simple relationship between the epitaxial layer structure of the device and the charge confinement in the channel. The quantum confinement of the charge in the channel, and the difficulty in expressing this simply in analytical form, led to adoption of numerical modeling schemes involving self-consistent Poisson and Schrödinger equation solvers (SPS) to model the charge confinement [31] and also couple this with the electron transport to yield a full, physically-based model [32].

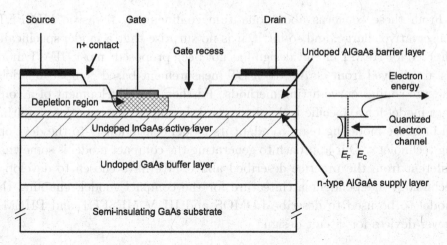

Fig. 1.11. Schematic diagram of AlGaAs–InGaAs–GaAs PHEMT structure and its electron energy-band diagram.

While such models are useful for optimizing the predicted device performance by adjusting the epitaxial layer structure, they are not suitable for circuit design, as they are difficult to implement or embed in the circuit simulator software, and can take a very long time to converge. Since the basic principle of operation of the HEMT is the control of the channel charge by the gate voltage, existing FET compact models can be used, by adjusting various of the model parameters to suit the HEMT. The basic I_d–V_{ds} characteristic of the HEMT is similar to that of the other charge-controlled field effect transistors, so adapting the *hyperbolic tangent* model of the GaAs MESFET is straightforward. The charge model for the HEMT is closer to that of the MOSFET than the MESFET, and, in principle, some accommodation should be made for this, although as we shall see in Chapter 6, the charge models are rarely implemented in a correct charge-conservative manner in many of the available compact models.

A further discouragement to developing a simple analytic model for PHEMTs used as power transistors is the fact that these devices often take advantage of the parasitic MESFET that exists in the upper barrier layer to increase the current density. The power capability is related to the current and voltage that the device can support, so increasing the current in this way is a method of increasing the power capability of the HEMT. This perhaps goes against the grain of the purist, who might try to engineer the band structure for maximum quantum confinement of the charge, but in a practical power transistor there are more prosaic concerns.

From these concerns about quantum confinement, parasitic MESFET, conservative charge, and so forth, it is no surprise that a model specifically directed for HEMT circuit design has not been proposed: most HEMT models are derived from GaAs MESFET measurement-based compact models, and use similar curve-fitting methods. Further, the development of a compact model for a specific technology can take many man-months, requiring an in-house modeling team to adapt its model continuously to the developing technology. Our approach to generating the compact model is somewhat different from the practice described above. We have chosen to develop a technology-independent architecture for the compact model, enabling the model to be used to describe LDMOS and III–V MESFET and PHEMT power devices for circuit design.

1.6.4 FET Figures of Merit

The figures of merit for a transistor or technology are used as simple indicators of performance and for comparison. They are generally measured qualities derived from S-parameters and related to a small-signal equivalent circuit model of the transistor, or features of the I_d–V_{ds} characteristics, for example. Small-signal figures of merit include the transition frequency, f_T, the maximum frequency of oscillation, f_{max}, and the transconductance or small-signal gain g_m. The figures of merit that can be derived from the DC characteristics include the maximum breakdown voltage, and the channel on-resistance. Large-signal figures of merit of the transistor can be measured using loadpull techniques to determine the power output and gain, efficiency, distortion behaviour, and so forth, under conditions that replicate the operating conditions of the transistor in a power amplifier.

A generic charge-control FET I_d–V_{ds} characteristic is shown in Fig. 1.12, indicating regions of non-zero output conductance, and gate-drain breakdown current. In Fig. 1.13 we show an intrinsic small-signal model. The capacitances correspond to the small-signal gate-source and gate-drain capacitances, found by measurement or by differentiation of the charge expression, and the transconductance and output conductance are found from differentiation of the drain current expression with respect to V_{gs} and V_{ds}, respectively. The series resistors in the gate-source and gate-drain branches are included to accommodate the gate charging current paths.

The (intrinsic) transition frequency for the transistor is defined as that frequency for which the current gain is unity. We can find this by applying a short circuit to the output of the small-signal model in Fig. 1.13. Ignoring

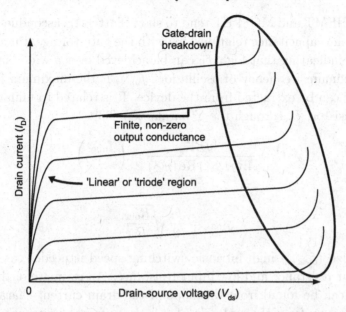

Fig. 1.12. The prototype drain current characteristics for a charge-controlled field effect transistor, showing the 'linear' or 'triode' region and saturation region.

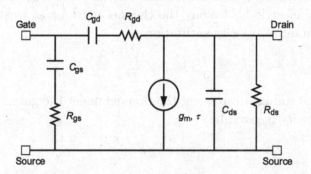

Fig. 1.13. A simple and generic small-signal model of the field effect transistor

the feedback capacitance, the current gain is

$$A_i = \frac{g_m v_{gs}}{C_{gs} \left(dv_{gs}/dt \right)} = \frac{g_m}{j\omega C_{gs}} \tag{1.11}$$

for a sinusoidal excitation; and hence the transition frequency is

$$f_T = \frac{g_m}{2\pi C_{gs}} \tag{1.12}$$

This is a measure of how quickly the transistor can transfer the charge in its channel onto the gate of another transistor of the same size, acting as the load. The HEMT and MOSFET tend to show 'flatter' transconductance and input channel capacitance relationships with the gate voltage, than does the MESFET, indicating that high f_T can be achieved over a wide bias range.

The maximum frequency of oscillation, f_{max}, is the maximum frequency that power can be extracted from the device. It is related to unilateral gain of the transistor, U, through the Y-parameters [33]

$$U = \frac{|y_{21}|^2}{4\mathrm{Re}\,(y_{11})\,\mathrm{Re}\,(y_{22})} = \left(\frac{f_{max}}{f}\right)^2 \tag{1.13}$$

and

$$f_{max} = \frac{f_T}{2}\sqrt{\frac{R_{out}}{R_{in}}} \tag{1.14}$$

To maximize f_{max}, a high intrinsic switching speed is needed, as well as a high output resistance and low input resistance. An estimate of the output resistance can be found from the slope of the drain current characteristics in saturation, or from the real part of y_{22}, the output admittance.

In the FETs we have described so far, the drain current saturation is reached when the electrons in the channel reach the saturated velocity. At this point the electric field along the channel must be at least the critical field for the onset of velocity saturation:

$$E_{crit} = \frac{V_i}{L} \tag{1.15}$$

where V_i is the voltage drop along the channel under the gate. The electron saturated velocity is given by

$$v_{sat} = \mu E_{crit} \tag{1.16}$$

Hence, the channel voltage at the onset of saturation is

$$V_i = \frac{v_{sat} L}{\mu_0} \tag{1.17}$$

By substituting this voltage for the saturation voltage into the MOSFET or MESFET drain current and gate charge expressions, we can obtain the following result (derived in Appendix 1.1)

$$f_T = \frac{g_m}{2\pi C_{gs}} = \frac{v_{sat}}{L} \tag{1.18}$$

This indicates that a shorter gate-length in a given technology, or choosing a material technology with high saturation velocity, results in higher frequency capability.

The transconductance of the MOSFET in the saturation region is given by

$$g_{m(sat)} = \frac{dI_D}{dV_G} = \frac{WC_{ox}\mu_s}{L}(V_G - V_T) \tag{1.19}$$

This ideal equation does not account for the presence of *access resistances*, which connect the device terminals to the channel region beneath the gate, and degrade the measured transconductance through

$$g_{m,meas} = \frac{g_m}{1 + g_m(R_s + R_g)} \tag{1.20}$$

The output conductance in saturation for the ideal model is zero – the drain current is perfectly flat in saturation – but in practice there is a positive slope to the high-frequency characteristic. The output conductance should be as low as possible, although this parameter is difficult to control, and is affected by the temperature of the device. In the 'linear' or 'triode' region of the FET, the slope of the drain current characteristic is

$$g_{0(linear)} = \frac{dI_D}{dV_{DS}} \simeq \frac{WC_{ox}\mu_s}{L}(V_G - V_T) \tag{1.21}$$

Again, the simple expression for MOSFET drain current has been used. The reciprocal of this is the *on-resistance* of the transistor, $R_{ds(on)}$. The on-resistance can be minimized by using a material system with a high carrier mobility, so III–V FETs have the advantage over MOSFETs in this regard. Another indicator of high $R_{ds(on)}$ is a large knee voltage, which also limits the available linear swing of the output voltage in a power amplifier [18].

The large-signal figures of merit for a power transistor include power gain, saturated power and maximum linear power; drain efficiency; and linearity, usually measured in terms of intermodulation distortion (IMD), and adjacent channel power (ACP) or error vector magnitude (EVM) for digitally-modulated signals. The maximum power is determined by the maximum current and voltage swings available. The maximum voltage swing is limited by the gate-to-drain breakdown voltage, and, as indicated earlier, both MOSFET and III–V FET technologies use doping and etching techniques to maximize the breakdown voltage. Gallium nitride heterojunction FETs have the advantage of being made from a wide band-gap material, which naturally has a large breakdown voltage; with careful processing methods, modern LDMOS devices can achieve breakdown voltages in excess of 100 V.

The maximum drain current in III–V FETs depends on the maximum forward current permissible in the gate Schottky diode; exceeding this value can have catastrophic consequences. In contrast, LDMOS FETs have an insulating gate dielectric, and can support a very high drain current, although other limitations such as the maximum permissible power dissipation in the device will come into play well before any practical limitation on maximum drain current.

The power dissipation is also limited by the thermal parameters of the device and the heatsink. The power density in the die is limited by how much heat can be conducted away through the die and heatsink. Gallium arsenide substrates have poor thermal conductivity, and consequently modest power density, in the context of RF power transistors. Silicon is a better thermal conductor, and can realize power densities similar to the GaAs FETs despite generally lower gains and worse on-resistance. Gallium nitride devices have potentially the best power density: silicon carbide substrates have high thermal conductivity, and devices fabricated on SiC have very high power density. Gallium nitride FETs built on silicon substrates have a lower power density, but it is still higher than LDMOS or GaAs FETs, because of the higher breakdown voltage and also because the saturated velocity of the electrons in the channel is higher, yielding higher current density.

Power FET efficiency is largely a circuit-controlled phenomenon, but some device features can have an influence, notably the knee voltage. Through the knee voltage, the drain efficiency is related to the on-resistance of the FET. A small knee voltage will allow a greater RF voltage swing for a given drain bias, and therefore the drain efficiency is improved. LDMOS transistors have a larger knee voltage than GaAs, and should, therefore, suffer by comparison in terms of efficiency, though for high-power base-station applications this is not significant. The reduction in efficiency is particularly noticeable under low drain bias conditions. The 'sharper' knee of GaAs FETs is why these devices are for used many low-voltage handset PA applications.

The device linearity is a measure of the drive-up gain compression characteristic. This is a reflection of the linearity of the drain current–gate voltage characteristic, or the transconductance. The HEMT devices are noted for a relatively broad and flat transconductance over a range of gate voltages, whereas MESFETs exhibit more of a peak.

The figures of merit can be used with confidence when comparing devices manufactured using the same technology. More care should be exercised when using such figures to compare different technologies, such as LDMOS and GaN HEMTs, for example, as some of the figures may not be realizable in practice, resulting in some necessary modification to the device technology that may outweigh the apparent advantage.

1.6.5 Variation of the FET Physical Parameters with Temperature

The basic principles of operation of the FETs that have been described so far have assumed a constant temperature; no explicit variations with temperature of the physical parameters of the transistor have been considered. As we are concerned with the modeling of power transistors, we can expect the thermal properties of the semiconductor, the package, and the device physics to play a role in the transistor's behaviour under large-signal, high-power conditions. We need to be aware of how the transistor's characteristics vary with temperature, so that these effects can be accommodated in the compact model.

When measured under controlled, isothermal conditions, the transistor drain current shows an approximately linear variation with temperature at a constant drain-source voltage [34, 35], which can be expressed as

$$I_d(T) = \frac{I_{d0}}{1 + c(T - T_0)} \tag{1.22}$$

where I_{d0} is the drain current measured at the reference temperature T_0, and c is the drain current static thermal coefficient. If we now try and relate this empirical result to the simple drain current expressions outlined earlier, we can see that there are several material and device parameters that contribute to this temperature dependence. These parameters include the threshold, pinch-off and breakdown voltages, the mobility and saturation velocity of the charge carriers in the channel, changes in oxide thickness due to thermal expansion, and band-gap changes that affect the band-bending in MOSFETs and HEMTs. Before proceeding with a detailed analysis, we shall neglect, as second-order effects, the thermal expansion and band-gap changes, as their thermal coefficients are very small. This leaves the voltage parameters and electron transport in the channel as the first-order influences.

The threshold and breakdown voltages have a positive, linear temperature dependence, and the thermal coefficients can be extracted directly from isothermal drain current measurements over temperature. The electron mobility in the channel, and the saturation velocity have a more complicated temperature dependence, which is governed by the electron scattering mechanisms in the channel. The dominant electron scattering mechanisms include scattering by the ionized dopants in the semiconductor, and the interaction between the electron momentum and the thermal vibrations of the semiconductor crystal, known as *phonons*. These two mechanisms have temperature coefficients of opposite sign. While the impurity density can affect the mobility significantly, it is scattering by phonons that dominates

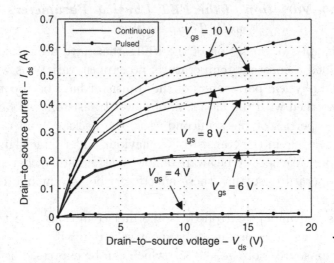

Fig. 1.14. A comparison between pulsed and DC I_d–V_{ds} characteristics, indicating the reduction in current with temperature. In this case, the increase in temperature in the transistor is due to self-heating.

the electron transport in the channel at room temperature and above, conditions typical of power amplifier applications. The temperature coefficient for phonon-limited mobility is negative at these temperatures, and is approximately –1 in lightly-doped GaAs, –1.5 in lightly-doped silicon, and up to –2.5 in LDMOS channels [36]. The electron scattering in the channel also affects the limiting velocity of the electrons. The velocity saturation is a result of inelastic scattering of the electrons by high energy phonons. As the temperature increases, the phonon energy increases, and the limiting velocity decreases.

The decrease in the electron velocity in the channel, due to thermal effects on the mobility and saturated velocity, can be observed in Fig. 1.14. The graph is a comparison of I_d–V_{ds} characteristics obtained by DC and pulsed I–V measurements. The measurements made at DC exhibit a noticeable reduction in the current in the saturation region, owing to the reduced electron velocity in the channel of the transistor [37].

Another important transistor parameter that exhibits a significant variation with temperature is the threshold voltage in LDMOS and GaAs PHEMT devices, and the pinch-off voltage in MESFETs. Changes in the Fermi energy-level with temperature reduce the magnitude of the threshold voltage as temperature increases.

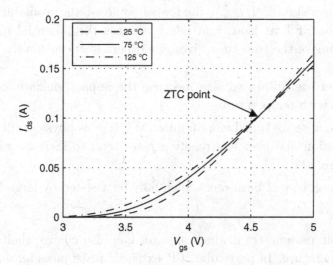

Fig. 1.15. The pulsed I_d–V_{gs} transfer characteristic measured from different heatsink temperatures, indicating the zero temperature coefficient (ZTC) point.

We shall examine the drain current characteristics in an LDMOS transistor to illustrate these thermal effects. The simple expression for the saturated drain current in an LDMOS transistor, eq. 1.7 can be expressed in the following way when temperature dependence of the electron mobility and threshold voltage are included:

$$I_d = \mu(T)C_{ox}(\frac{W}{2L})(V_{gs} - V_T(T))^2 \tag{1.23}$$

For small values of gate voltage and correspondingly small currents, the threshold voltage term dominates, and the decrease in threshold voltage with increasing temperature results in a positive temperature coefficient for saturation current, that is, the saturation current increases with temperature. For large gate voltages, the negative temperature coefficient of the mobility term dominates, and the saturation current decreases with temperature. For the special case dubbed the 'zero-temperature coefficient' (ZTC) bias condition, the saturation current does not change with temperature, as shown in Fig. 1.15. The ZTC point arises from the cancellation of the opposite thermal dependences of the threshold voltage and the electron mobility.

A large on-resistance, $R_{ds(on)}$, dissipates some of the available output power from the FET as heat, and hence reduces the potential efficiency. This self-heating of the transistor gives rise other effects such as:

(i) A higher breakdown voltage, because the impact ionization rate decreases with temperature;

(ii) A reduced mean time before failure (MTBF), as physical effects such as diffusion and chemical reaction rates tend to increase with temperature; and

(iii) The possibility of burn-out, which may be assisted by large RF voltages.

The extrinsic parameters in the transistor can also change their properties with temperature. In particular, the extrinsic resistances have a linear, positive temperature coefficient of resistance in the temperature range of interest. This means that at higher temperatures, a slightly higher proportion of the applied terminal voltage will be dropped across these components, thereby reducing the drain current further.

It is critical that the FET model incorporates the temperature dependence of the device electrical characteristics, in order to describe accurately the electrical behaviors that result from the self-heating, the most obvious being a reduction in the RF output power.

Because LDMOS devices do not experience frequency-dispersive effects, pulsed I–V measurements can be used to produce isothermal I_d–V_{ds} data, by changing only the base-plate temperature. After the drain current characteristics curves have been collected at several temperatures (typically 25°C, 75°C, 125°C, and 175°C), the temperature-mapping equations for each of the temperature-varying electrical parameters can be developed, either by optimizing algorithms or by direct extraction of the model parameters at each of the measured heatsink temperatures.

1.7 Packages

High-power RF and microwave semiconductor transistors are generally enclosed in air-cavity or over-molded plastic packages. These packages protect the internal circuitry from the external environment, and they aid in the removal of heat generated by the active area of the transistor. In addition, these packages also serve as components of the low-loss matching network used to transform the impedance of the transistor to a higher value at the

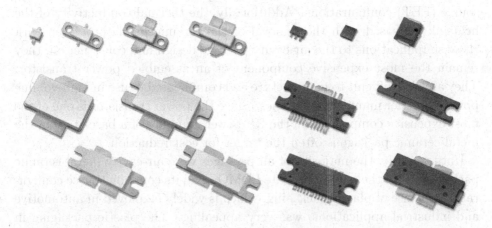

Fig. 1.16. Photographs of packages used within cellular base-stations. Several examples of air-cavity and over-molded plastic packages are shown on the left and right halves of the figure (courtesy of Freescale Semiconductor Inc.).

edge of the package. Transistors used for wireless infrastructure applications generate some of the largest heat fluxes amongst all semiconductor devices and must be carefully designed [38]. Stringent electrical and thermal-mechanical design practices are required to ensure the package does not degrade the electrical performance of the transistor and to enable the dissipation of the substantial heat-flux [39].

The design and use of packages for high-power semiconductor transistors has been primarily driven by the dominant transistor technology of the era. Throughout the evolution of the transistor, air-cavity metal-ceramic packages have been widely employed for GaAs, silicon bipolar, VDMOS, and LDMOS technologies.

As previously mentioned, bipolar and VDMOS transistors require connections to the back side of the die; this significantly complicates the packaging because insulating ceramics needed to be placed between the transistor and the package flange; unfortunately these materials were often toxic and expensive. Rows of vias within the ceramic permitted connections to the package flange, for electrical contact.

With the advent of LDMOS transistors constructed on high-conductivity silicon and GaAs transistors, employing through vias, the insulating ceramics were removed because the source contact of the FETs could be bonded directly to the package flange. This eliminated the requirement for a backside connection and since the sources were already attached to the flange, no grounding wires were necessary in common collector (BJT) or common source (FET) configurations. Additionally, the thermal conductivity of the heatsink increased with the removal of the ceramic insulator. Even with these simplifications to the air-cavity package design and construction, they remain the most expensive component of an assembled power transistor. They are also difficult to manufacture and complicated to use in high-volume power amplifier manufacturing lines. Since the power transistor is one of the most expensive components in the RF power amplifier of a base-station, the metal-ceramic package is often the target for cost reduction.

Until recently, the majority of all packages were air-cavity metal-ceramic packages. With the adoption of the LDMOS and its common source configuration, the use of plastic packaging, which is widely employed in automotive and industrial applications, was very appealing. The plastic packaging in use at the time was restricted to low frequency and lower temperature applications. Since the 1990s, research has been conducted in the development of new low-loss dielectric compounds able to work at RF frequencies and withstand the high junction temperatures of the transistors. This has culminated in the development of a high-frequency plastic packaging technology suitable for use in high-volume RF power transistor manufacturing with significant cost benefits compared with conventional metal-ceramic packaging. Research and development continues for cost reduction initiatives in the face of the continuing pressure to reduce the overall costs of the packaged power transistor.

Photographs of typical air-cavity and over-molded plastic packages are shown in Fig. 1.16. The internal components of the transistor within the air-cavity package are not encapsulated while those of the plastic package have been encapsulated, or over-molded, by the plastic material. The leads of each package are separated from the flange by either the ceramic or the plastic dielectrics as illustrated in Figs. 1.17(a) and 1.17(b). The majority of high-power transistor packages generally have two or four leads, although new multi-stage high-power RFICs, incorporating higher functionality, require more leads, as shown in Fig. 1.16. The packages are designed for the leads to rest on top of the microstrip transmission lines on the PCB. The back side of the flange contacts the heatsink of the power amplifier forming a conductive electrical connection to the bottom conductor of the microstrip and a conductive thermal connection to the heatsink, which enables heat to flow away from the packaged transistor. Alternatively, the leads of the packages may be bent such that the package is surface-mountable. In this case, vias through the PCB contact the package flange provide for thermal transfer and electrical grounding, and the bent leads contact the metal bond-pads on top of the the board.

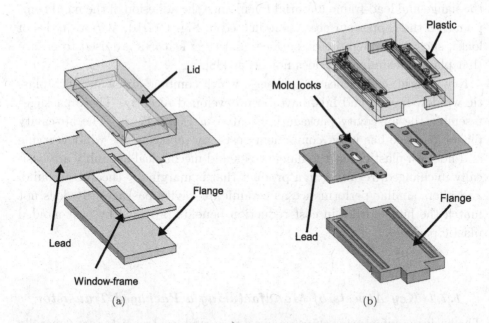

Fig. 1.17. Illustrations of typical packages showing their constituent components. Shown in (a) is a metal-ceramic air-cavity package and in (b) an over-molded plastic package.

The construction of both air-cavity and over-molded plastic packages is presented through exploded views outlining the internal package details as shown in Fig. 1.17. The air-cavity metal-ceramic package, shown in Fig. 1.17(a) consists of a metal flange, ceramic window-frame, gate and drain leads, and ceramic lid. The leads are attached to the ceramic window-frame, forming a short microstrip transmission line with the flange. A ceramic lid is attached using epoxy to seal the package. The gate and drain leads are manufactured from a nickel-iron alloy ('Alloy-42') and then gold plated. The flanges are constructed from proprietary copper-tungsten (CuW) or copper laminates designed for optimal thermal transfer and to minimize the stresses generated between the die and flange that arise from differences in the coefficients of thermal expansion.

Over-molded packages are similarly constructed as illustrated in Fig. 1.17(b), with the flange and package leads made from high-grade copper. The over-molded plastic provides mechanical rigidity and environmental protection. Notable differences in the design of the flange and lead-frame are the incorporation of the mold-locks. These are features incorporated into the flange and lead-frame material to enhance the adhesion of the mold compound to the copper surfaces, as indicated in Fig. 1.17(b). Various types of locks, such as through-holes, surface, and edge features are used to ensure that plastic delamination does not occur [38].

New hybrid approaches to packages, which combine air-cavity with plastic window-frames and lids, have been developed recently. These packages resemble the air-cavity package illustrated in Fig. 1.17(a). The air-cavity plastic package has lower component costs, by replacing the window-frame and lid with plastic, but the flange costs and mechanical assembly are basically unchanged, resulting in a product that is marginally cheaper to build, and offers similar performance as ceramic air-cavity packages. It does not match the huge strides in cost reduction benefits provided by over-molded plastic packages.

1.7.1 Key Aspects of Manufacturing a Packaged Transistor

The basic manufacturing steps required to assemble a transistor are generally the same regardless of the package type. Each transistor typically consists of transistor dies, pre- and post-matching capacitors, and arrays of bondwires. The first step in manufacturing is to attach the die to the package flange.

Achieving a uniform die attachment is very important for thermal management of high-power transistors, as the package flange is the main path for the heat-flux to be removed from the transistor. Any area beneath the die that has not been electrically connected to the flange is commonly referred to as a void. Since the void is a region of very low thermal conductivity, it results in an overall increase in the thermal resistance of the die-to-package interface. The impact of voids is magnified for RFIC applications where the heat generating areas of the die are much smaller compared with the total die area. A void beneath one of the smaller transistors can seriously degrade the amplifier performance.

The quality of a die-attach can be assessed through the use of acoustic micro-imaging, where a focused beam of ultra-high frequency pulses is scanned over the back side of the package flange [40]. As the pulses of ultrasonic energy travel though the flange they reflect at material interfaces, in a manner analogous to the operation of a radar system. Ultrasound will not pass through air spaces caused by delamination, cracks, or voids present in the sample. The return time of the pulse is a function of the distance from the interface to the transducer. By selecting different return times the interfaces at different depths can be examined and a picture or the intensity

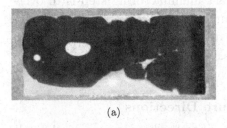
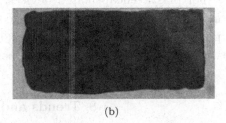

(a)   (b)

Fig. 1.18. Images from a scanning acoustic microscope, illustrating a poor die-attach with voiding (light spots) beneath the die in (a). An example of a good die-attach is shown in (b). Note the uniformity and lack of voids.

of the reflected energy is plotted as the acoustic source is scanned over the sample.

Examples of a poor- and good-quality die-attach are provided in Fig. 1.18. The light spots in Fig. 1.18(a) indicate large voids beneath the die. These voids will increase the thermal resistance of the die and limit the performance of the transistor. An example of an acceptable die-attach is provided in Fig. 1.18(b). Notice the lack of voids and the rectangular shape of the die-attach layer.

Once the transistor and matching capacitors have been attached to the flange, the package, capacitors, and die are connected together by bondwires, in an automated process. Bondwires are the most common inductive matching element within packaged high-power transistors. They are available in several diameters, ranging from 1 mil to 2.4 mil, and in various metals, including gold and aluminium alloys, for low resistance and high-temperature capability.

The inductance of the bondwire arrays within the package is controlled by: the number of wires; the three-dimensional profile; the spacing between adjacent wires (and arrays); and the wire diameter. These are all adjustable design parameters. Repeatability and controllability of these parameters are crucial to the production of transistors with tight performance distributions. At high frequencies, even small variations can have a serious effect on the performance of the device, and so the maintenance of tight mechanical and assembly tolerances is critical.

The other component in the in-package matching network is the MOS capacitor. The value of the capacitance is defined by the thickness of the silicon dioxide dielectric layer, its dielectric constant, and the size of the metal plate. The bottom plate of the capacitor is a heavily-doped bulk silicon substrate.

For plastic-encapsulated devices, the last manufacturing step is to over-mold the assembly with a plastic compound. In the case of air-cavity components, a ceramic or plastic lid is bonded to the flange and window-frame with epoxy to seal the transistor package.

1.8 Trends and Future Directions

To predict trends and the future directions of technology is always a difficult proposition. The rate of change of the RF and microwave semiconductor industry over the last couple of decades has been phenomenal, and there are no signs that technological innovation in the field will slow down in the future. Therefore, as engineers we need to understand how the proposed solutions to today's modeling problems will work on tomorrow's challenges.

A clear trend observed over the last decade is the constant upward mobility of the spectrum allocated for communication systems. Based on this observation, the modeling engineer needs to be prepared to face the challenge of higher power levels at higher frequencies. With this trend, the importance of coupled circuit-level and electromagnetic simulation will increase, to enable higher levels of simulation accuracy as distributed effects become more important at higher frequencies.

The persistence of Moore's law means that the circuits themselves are becoming more complex, with multi-stage integrated power amplifiers becoming more common, and including extra functionality such as thermal tracking and bias control. The transistor models must then be scalable and operate accurately over a range of dynamic bias conditions. The packaging requirements must follow the circuit complexity, and multi-lead packages are required to provide the DC, RF connections and other functions such as feedback or power control. The complexity of the package structure requires detailed electromagnetic simulation, coupled with the circuit and thermal simulations.

An important trend that will affect the modeling engineer's outlook is the blurring of the lines between traditional RF and microwave circuit design and system-level design, which typically involves aspects of digital signal processing. The modeling engineer needs to become more knowledgeable of system aspects, as his interaction with the system and circuit designers is likely to increase. To satisfy the simulation needs of a coupled circuit-level and system-level design approach, the transistor-level models and the transistor- or circuit-level behavioural models will need to converge seamlessly with one another. As computer processing power and advances in simulation technology continue to increase, we envision a time in which circuit-level simulations of the whole system being driven by digitally-modulated signals will be possible with simulation times that are comparable with today's circuit-level simulation times.

Historically, in the design of the high-power RF transistor, the thermal and electrical parts of the design activity have been decoupled. To extract the best performance from the device, the thermal and electrical designs must be combined to obtain the best compromise between the design objectives. This goal can only be accomplished through the use of coupled thermal and electrical simulations, including electromagnetic interactions, in the design of the packaged transistors and integrated circuits. In this book we will present modeling approaches that combine electrical and thermal compact models into a single electro-thermal model of a transistor that can be used to predict dynamic electro-thermal behaviour, which is a necessary first step toward the goal of a coupled electro-thermal simulation environment. A further extension of this approach is to couple electrical (electromagnetic) and thermal models with a physical simulation of the electronic processes within the semiconductor device, to achieve a self-consistent model at a physical level [41]. Currently, while such simulations can provide invaluable insight into the physical design of the transistor, they are very slow. As computer processing power increases, and the development of fast algorithms to solve the difference-equation matrices progresses, such modeling approaches may become more commonplace in the design of the transistor itself. For the circuit design using this component, the compact model will still be the primary choice.

A complementary point of convergence is the use of compact model techniques with 'Technology CAD' (TCAD) approaches to the design of semiconductor devices based on large signal figures of merit. This capability has already been demonstrated [42], and we see an acceleration of this practice as the complexity of the designs continues to increase.

On the model implementation and verification front, we envisage that the adoption of an open architecture in the CAD tools will bring a much higher level of model portability. This will be a benefit for modeling organizations that need to support multiple CAD packages from different vendors, in terms of the simplification of the model implementation.

The technology-independent modeling approach that we are presenting in this book is be able to accommodate new FET device, structure, and technology trends, provided the physics of operation of the device does not change significantly.

Appendix 1.1 – Derivation of f_T in the MESFET

To allow a simple analysis of the saturation behaviour under high electric field conditions, we shall consider that the channel current for a given gate and drain bias is limited by the minimum channel height, that is, the height of the channel at the drain edge of the gate. This corresponds to the depletion depth at that point. The channel current is

$$I_{channel} = G_{channel} V_i$$
$$= G_0 \left(1 - \frac{D_{dep}}{D}\right) V_i \qquad (1.24)$$

where D_{dep} is the depletion layer depth at the drain edge of the gate, where it is largest, and V_i is the voltage drop in the channel along the length of the gate, in the x-direction.

At the onset of saturation, the electric field in the channel is the critical field

$$E_{crit} = \frac{V_i}{L} \qquad (1.25)$$

and the electron saturated velocity is given by

$$v_{sat} = \mu_0 E_{crit} \qquad (1.26)$$

Hence, the channel voltage at the onset of saturation is

$$V_i = \frac{v_{sat} L}{\mu_0} \qquad (1.27)$$

The (intrinsic) transconductance can then be determined from the derivative of the channel current with respect to the gate voltage

$$G_m = \frac{dI_{channel}}{dV_G}\bigg|_{V_i = constant}$$

$$= -\frac{G_0}{D} \frac{v_{sat} L}{\mu} \frac{dD_{dep}}{dV_G} \qquad (1.28)$$

$$= \sqrt{\frac{q n_0 \varepsilon_{GaAs}}{2}} \frac{1}{(\phi_{bi} - V_G)^{1/2}} v_{sat} W$$

where ϕ_{bi} is the built-in potential of the Schottky gate metal-semiconductor contact, and ϵ_{GaAs} is the permittivity of the GaAs channel.

The input capacitance of the MESFET is approximately equal to the derivative of the charge in the depletion region under the gate, with respect to the gate voltage. The charge in the depletion region is

$$Q_{\text{dep}} = \int\limits_0^{D_{\text{dep}}} qn(z)dz \tag{1.29}$$

$$\approx WLqn_0 D_{\text{dep}}$$

approximating the true shape of the depletion region by a box. The input capacitance is then given by

$$C_{\text{in}} = \left. \frac{dQ_{\text{dep}}}{dV_{\text{G}}} \right|_{V_i=\text{constant}}$$

$$= qn_0 WL \frac{dD_{\text{dep}}}{dV_{\text{G}}} \tag{1.30}$$

$$= \sqrt{\frac{qn_0 \varepsilon_{\text{GaAs}}}{2}} \frac{1}{(\phi_{\text{bi}} - V_{\text{G}})^{1/2}} WL$$

Hence, the transition frequency is

$$f_{\text{T}} = \frac{g_{\text{m}}}{2\pi C_{\text{gs}}} = \frac{v_{\text{sat}}}{L} \tag{1.31}$$

References

[1] P. H. Aaen, J. A. Plá, and C. A. Balanis, "Modeling techniques suitable for CAD based design of internal matching networks of high-power RF/microwave transistors," *IEEE Trans. Microwave Theory Tech.*, 54, no. 7, 3052–3059, July 2006.

[2] R. Buderi, *The Invention that Changed the World*. New York, NY: Simon & Schuster, 1996.

[3] J. Norling, "A look at the world's leading radio frequency power supplier and a review of its strategies and markets," in *Freescale Technology Forum*, Orlando, FL, July 2006.

[4] J.-J. Bouny, "Advantages of LDMOS in high power linear amplification," *Microwave Engineering Europe*, 37–40, Apr. 1996.

[5] M. Shaw and A. Wood, "Characterization of a 2 GHz submicron bipolar 60 watt power transistor with single tone, multi-tone, and CDMA signals," in *47th ARFTG Conference Digest*, San Francisco, CA, June 1996, 26–31.

[6] R. B. Davies, R. J. Johnsen, and F. Y. Robb, "Semiconductor device having low source inductance," U.S. Patent 5,155,563, Oct. 13, 1992.

[7] A. Wood, C. Dragon, and W. Burger, "High performance silicon LDMOS technology for 2 GHz RF power amplifier applications," in *Int. Electron Devices Mtg. Tech. Dig.*, San Francisco, CA, Dec. 1996, 87–90.

[8] A. Wood and W. Brakensiek, "Applications of RF LDMOS power transistors for 2.2 GHz wideband-CDMA," in *Proc. IEEE Radio and Wireless Conference, (RAWCON) 98*, Colorado Springs, CO, Aug. 1998, 309–312.

[9] W. Burger, H. Brech, D. Burdeaux, C. Dragon, G. Formicone, M. Honan, B. Pryor, and X. Ren, "RF-LDMOS: a device technology for high power RF infrastructure applications," in *Compound Semiconductor Integrated Circuit Symp. Dig.*, Monterey, CA, Oct. 2004, 189–192.

[10] W. Shockley, U.S. Patent 2,569,347, June 26, 1951.

[11] H. Kroemer, "Theory of a wide-gap emitter for transistors," *Proc. IEEE*, 45, no. 11, 1535–57, Nov. 1957.

[12] J. V. DiLorenzo, *GaAs FET Principles and Technology*.   Norwood, MA: Artech House, 1982.

[13] B. Turner, "GaAs MESFETs," in *Gallium Arsenide Materials, Devices, and Circuits*, M. J. Howes and D. V. Morgan, Eds.   Chichester UK: John Wiley & Sons, 1985, ch. 10.

[14] R. Dingle, H. L. Stoermer, A. C. Gossard, and W. Wiegmann, "Electron mobilities in modulation-doped semiconductor heterojunction superlattices," *Appl. Phys. Lett.*, 33, no. 7, 665–667, 1978.

[15] M. F. O'Keefe, J. S. Atherton, W. Bösch, P. Burgess, N. I. Cameron, and C. M. Snowden, "GaAs pHEMT-based technology for microwave applications in a volume MMIC production environment on 150-mm wafers," *IEEE Trans. Semiconduct. Manufact.*, 16, no. 3, 376–383, Aug. 2003.

[16] K. Ebihara, K. Inoue, H. Haematsu, K. Yamaki, H. Takahashi, and J. Fukaya, "An ultra-broad-band 300W GaAs power FET for W-CDMA base stations," in *IEEE MTT-S Int. Microwave Symp. Dig.*, Phoenix, AZ, May 2001, 649–652.

[17] R. G. Ranson, "Three steps to more efficient 3G base station power amplifiers," in *IEEE MTT-S Int. Microwave Symp. Workshop 'Advances in high-efficiency power device and circuit technologies'*, Long Beach, CA, June 2005.

[18] S. C. Cripps, *RF Power Amplifiers for Wireless Communications*, 2nd edn. Norwood, MA: Artech House, 2006.

[19] W. Bella and J.-C. Nanan, "2-stage 200 W Doherty amplifier for WCDMA applications," in *IEEE Topical Power Amplifier Workshop Dig.*, San Diego, CA, Jan. 2006.

[20] J. Cha, J. Kim, B. Kim, J. S. Lee, and S. H. Kim, "Highly efficient power amplifier for CDMA base stations using Doherty configuration," in *IEEE MTT-S Int. Microwave Symp. Dig.*, Fort Worth, TX, June 2004, 533–536.

[21] K.-J. Cho and S. P. Stapleton, "Design of a 30 W WCDMA Doherty amplifier," in *IEEE Radio and Wireless Symposium Workshop 'High Power RF Devices and Amplifiers'*, San Diego, CA, Jan. 2006.

[22] R. J. Trew, "SiC and GaN transistors – is there one winner for microwave power applications?" *Proc. IEEE*, 90, no. 6, 1032–47, June 2002.

[23] B. G. Streetman, *Solid State Electronic Devices*. Englewood Cliffs: Prentice-Hall International, Inc, 1990.

[24] J. M. Golio, *Microwave MESFETs and HEMTs*. Norwood, MA: Artech House, 1991.

[25] S. M. Sze, *Physics of Semiconductor Devices*. New York, NY: John Wiley & Sons, 1969.

[26] P. H. Ladbrooke, *MMIC Design: GaAs FETs and HEMTs*. Norwood, MA: Artech House, 1989.

[27] T. Ytterdahl, Y. Cheng, and T. A. Fjeldly, *Device Modeling for Analog and RF CMOS Circuit Design*. Chichester, UK: John Wiley & Sons, 2003.

[28] M. J. Howes, "Transferred electron devices," in *Gallium Arsenide Materials, Devices, and Circuits*, M. J. Howes and D. V. Morgan, Eds. Chichester, UK: John Wiley & Sons, 1985, ch. 8.

[29] A. Y. Cho and J. R. Arthur, "Molecular beam epitaxy," *Progress in Solid State Chemistry*, 10, 157–191, 1975.

[30] D. Delagebeaudeuf and N. T. Linh, "Metal-(n) AlGaAs-GaAs two-dimensional electron gas FET," *IEEE Trans. Electron Devices*, 29, no. 6, 955–960, June 1982.

[31] C. G. Morton and J. Wood, "MODFET versus MESFET: the capacitance argument," *IEEE Trans. Electron Devices*, 41, no. 8, 1477–80, Aug. 1994.

[32] R. Drury and C. M. Snowden, "A quasi-two-dimensional HEMT model for microwave CAD applications," *IEEE Trans. Electron Devices*, 42, no. 6, 1026–32, June 1995.

[33] G. D. Vendelin, A. M. Pavio, and U. L. Rohde, *Microwave Circuit Design*. New York, NY: John Wiley & Sons, 1990.

[34] R. Anholt, *Electrical and Thermal Characterization of MESFETs, HEMTs, and HBTs*. Norwood, MA: Artech House, 1995.

[35] P. C. Canfield, S. C. F. Lam, and D. J. Allstot, "Modeling of frequency and temperature effects in GaAs MESFETs," *IEEE J. Solid State Circuits*, 25, no. 1, 299–306, Feb. 1990.

[36] G. Dolny, G. Nostrand, and K. Hill, "Characterization and modeling of the temperature dependence of lateral DMOS transistors for high-temperature applications of power ICs," in *Int. Electron Devices Mtg. Tech. Dig.*, San Francisco, CA, Dec. 1990, 789–792.

[37] J. Plá, "Characterization and modeling of high power RF semiconductor devices under constant and pulsed excitations," in *Proc. Fifth Annual Wireless Symposium*, San Jose, CA, Feb. 1997, 467–472.

[38] D. Abdo, F. Danaher, A. Elliot, and M. Mahalingam, "Continuous operation at 200°C device junction temperature: the final frontier for RF power semiconductor plastic packaging," in *Proc. 54th Electronic Components and*

Technology Conf., Las Vegas, NV, June 2004, 437–443.

[39] Z. Radivojevic, K. Andersson, L. Bogod, M. Mahalingam, J. Rantala, and J. Wright, "Novel materials for improved quality of RF-PA in base-station applications," *IEEE Trans. Comp., Packag., Manufact. Technol. A*, 28, no. 4, 644–649, Dec. 2005.

[40] J. E. Semmens and L. W. Kessler, "Acoustic micro imaging in the Fourier domain for evaluation of advanced packaging," in *Proc. 7th Annual Pan Pacific Microelectronics Symp.*, Maui, Hawaii, Dec. 2002, 233–237.

[41] D. Denis, C. M. Snowden, and I. C. Hunter, "Coupled electrothermal, electromagnetic, and physical modeling of microwave power FETs," *IEEE Trans. Microwave Theory Tech.*, 54, no. 6, 2465–70, June 2006.

[42] G. H. Loechelt and P. A. Blakey, "A computational load-pull system for evaluating RF and microwave power amplifier technologies," in *IEEE MTT-S Int. Microwave Symp. Dig.*, Boston, MA, June 2000, 465–468.

2

An Introduction to the Compact Modeling of High-Power FETs

2.1 Introduction

In this chapter we set the stage for the detailed discussion of the model analysis, extraction and construction choices that are described in subsequent chapters. So far, we have presented a background outlining how field effect transistors (FETs) have been developed for and used in RF and microwave power amplifiers. This has covered a high level introduction to how the FET-based transistors are structured and fabricated, in both silicon LDMOS and III–V semiconductor technologies, and an outline of how these field-effect transistors operate electrically. With this background in place, we can now discuss in greater detail some of the modeling issues that need to be considered carefully in order to construct an accurate transistor model that can be used in the design of RF power amplifiers.

Our aim is to build models for the transistors that can be used in circuit simulators for the design of power amplifiers and power amplifier integrated circuits. These models are known as *compact* models. To achieve this objective, the models must be able to reproduce with acceptable fidelity the measured electrical and thermal properties of the transistors, and to simulate them quickly, with robust convergence.

Another common modeling objective is to be able to inform the physical device design: in other words indicate which of the material and structural properties of a given transistor affect its electrical performance. The accuracy with which any model can achieve this depends on the level of abstraction of the model in the first place. As this aspect is not always appreciated, we shall discuss here some different modeling approaches, commonly found in the literature. We shall refer to Fig. 2.1, which describes a hierarchy of modeling approaches.

52

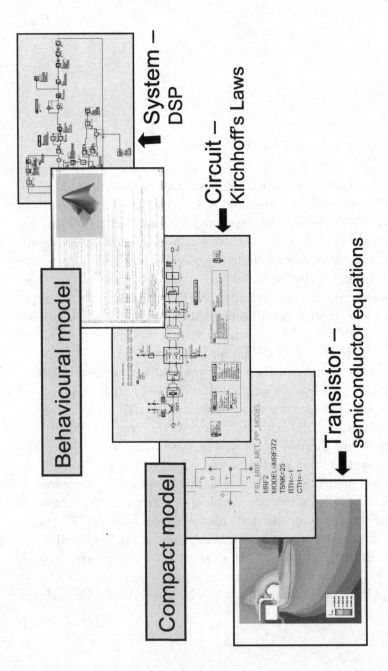

System – DSP

Circuit – Kirchhoff's Laws

Behavioural model

Compact model

Transistor – semiconductor equations

FSL_MRF_MET_PP_MODEL
MRF2
MODEL=MRF372
TSNK=25
RTH=1
CTH=1

Fig. 2.1. A general modeling hierarchy that we may apply to the level of abstraction of the design.

Finally, we shall discuss 'memory effects.' This term has become popular recently as a means of describing well known electrical phenomena such as frequency response and hysteresis, as well as nonlinear effects such as intermodulation distortions that appear out-of-band. This latter can be a particular problem for power amplifiers: even-order distortion mechanisms generate products at the baseband frequency. This low frequency energy can affect the DC bias of the transistor – observed as *self-biasing* – and alter the gain and other properties of the transistor and amplifier. As this happens at a timescale that is much longer than the RF period, the term *memory* is coined to describe it. Clearly, this is a real physical phenomenon that must be captured in a successful model of a power transistor. We shall outline in more detail how memory effects arise and can be measured, and hence how we can construct our model to accommodate this behaviour.

2.2 Physical Modeling

The 'physics of operation' of a given transistor are best described using a physical model simulation, in which the geometry, topography, and the material properties of the semiconductors, metals, and insulators that form the transistor are captured in the model description in the simulator. In physical modeling, the nonlinear partial differential equations (PDEs) that describe charge distribution, charge transport, current continuity, and so forth, in the transistor structure are solved in the simulator. As the transistor geometries become smaller, quantum-mechanical (QM) effects also need to be incorporated into this solution. In the simulation, the transistor geometry is discretized in two or three dimensions, and the solution of the PDEs and QM equations is performed for each cell or node in the structure. This requires complex solution techniques to be used, such as finite-difference [1, 2] and finite-element methods [3, 4]. These numerical methods are also found in commercial software for physical transistor modeling, for example: ISE [5] and Silvaco [6]. This class of device modeling is also known as 'TCAD' – Technology CAD.

Despite the many recent mathematical developments in the solution techniques for large systems of nonlinear PDEs, such as parallelization of the problem, using harmonic balance methods instead of traditional time-stepping [7], or modern advanced matrix mathematics [8], and the advances in computer technology such as increases in processor speed and available memory, this is still a huge problem to solve. The solution of these nonlinear PDEs takes a long time, and the accuracy of the solution depends on how well the physical properties and dimensions of the device are estimated, on the approximations used in the fundamental semiconductor equations, and on the numerical techniques applied in the solution of the system of equations.

This means that physical modeling is generally unsuitable for circuit design. However, these modeling techniques have been applied successfully to the technology development cycle for new generations of transistors. Physical model simulations can be used to generate DC *I–V* characteristics and bias-dependent S-parameter data directly from the simulation. These data can then be used to assess how the physical and geometrical design of the transistor need to be adjusted to improve the device RF performance. Even 'computational loadpull' simulations can be performed using the physical device model loaded by a tunable impedance [9] to assess the large signal performance of the transistor design and technology, and to inform the device designers of how the nuances in the physical structure of the transistor can influence the RF behaviour. Such TCAD simulations are performed at huge computational cost, though it could be argued that this is time better spent, in terms of economy and resources, than building up an array of process lot variations and measuring the electrical performance of each transistor.

A further development of the physical modeling approach is known as *global modeling*. Here, the physical model simulation describing the semiconductor equations and device geometry is coupled with a thermal model describing the heat transfer in the transistor, and an electromagnetic simulation of the device geometry, metallizations, and substrate [10]. This brings all of the physical principles of operation of the device together in one simulation. Such an approach can provide valuable feedback to the device design team, especially for the physically large structures typical of RF power transistors. For example, the thermal distribution in the transistor under RF drive can be understood and related to the metallization, feeding structures and manifolds, and device structural features such as source-to-drain spacing. Global modeling enables the complete design space of the transistor to be investigated to arrive at the optimal semiconductor and structural arrangement to meet a given design brief.

2.3 Compact Models

While physical modeling provides a viable route for detailed device design and technology development and optimization, it is generally impractical for circuit design. Aside from the computational overhead involved, it is usually expected that the transistor or integrated circuit (IC) process has been determined ('frozen') by the time circuit design begins, and therefore we can devise a more appropriate model that can be used in the circuit simulator for the design of the IC or discrete transistor product. Such models generally fall under the rubric *compact*, and these models are the main focus in this book. In fact, in many new device technology developments, compact models can be created from the physical model simulations, and hence the circuit or IC design can begin well before the process is finally frozen. The compact models can be updated as the technology develops, so that the circuit design is always in step with the latest technology improvements, enabling a much reduced new product introduction cycle. The question remains, what should these compact models look like?

2.3.1 Measurement-Based Equivalent Circuit Models

Generally, compact models are 'equivalent circuit' representations of the transistor. The electrical measurements that are performed during the characterization of the transistor can be mapped directly onto a network of circuit components to mimic this electrical behaviour. The values of the equivalent circuit parameters are extracted directly from the DC I–V and S-parameter measurements [11]. For example, at a given DC bias point and RF frequency, we can measure the S-parameters, which can then be readily converted into generalized two-port Y-parameters using standard conversion relationships. After de-embedding the extrinsic components, this yields an equivalent circuit containing conductance and susceptance components that can be readily incorporated into the circuit simulator. An example is shown in Fig. 2.2, where the admittance elements in the gate-source and gate-drain branches were transformed into series R–C networks. The values of the equivalent circuit components should, generally speaking, be independent of frequency.

This approach works well for small-signal models; the extraction of the equivalent circuit parameter values can be carried out over a range of bias voltages (V_{gs}, V_{ds}), and the values can be stored in a table indexed by the bias, using interpolation to find the required component values for the given bias voltages, to produce a bias-dependent linear transistor model [12]. The equivalent circuit parameter values can be fitted with parameterized functions of the bias voltages, and the model extraction consists of finding these function parameters for each of the circuit elements in the model.

Large-signal models can also be implemented in the simulator using equivalent circuit components, but now the component parameter values are dependent on the large-signal voltages. The example circuit topology shown in Fig. 2.3 is the Motorola electro-thermal (MET) model [13].

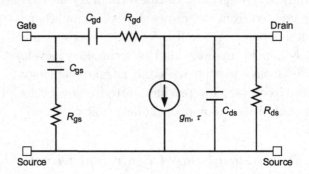

Fig. 2.2. Small-signal equivalent circuit model of the intrinsic transistor; this model is derived from Y-parameters, with some admittance-to-impedance transformation in the gate-source and gate-drain branches.

For a compact model that can be used for small- and large-signal applications, the functional dependences of the equivalent circuit parameter values on the instantaneous terminal voltages are required. We cannot simply use the DC bias voltages as indices and expect the dynamic behaviour of the transistor to be predicted correctly. The small-signal model is a representation of the first moment of the Taylor series expansion of the terminal admittances about the DC bias point, that is, the slopes defined by the infinitesimal voltage swings around the quiescent condition. In the large-signal model, we need to include the responses to large excursions of the terminal voltages. This can be done by integrating the voltage-dependent admittances over the voltage space, subject to some fundamental physical constraints. The construction of such a large-signal compact model is described in detail in Chapter 6.

We can then apply mathematical function fitting techniques to these derived large-signal equivalent circuit parameters, to obtain a nonlinear functional description for the large-signal model, as outlined in Chapter 7. An example of this mathematical function-fitting approach is the description of the FET I_d–V_{ds} characteristics using a hyperbolic tangent ($\tanh(x)$) curve. The hyperbolic tangent itself has no physical meaning in the context of FET operation, but describes the basic shape of the curve from quasi-linear through saturation regions reasonably well. This function is used in several examples of FET compact models [13–16] and is generally modified

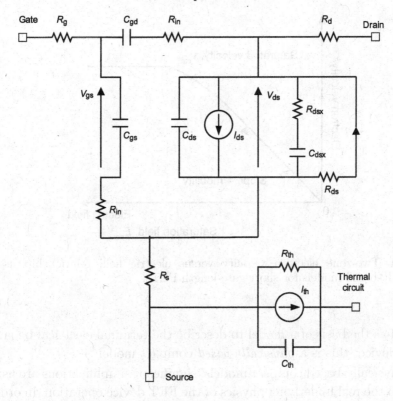

Fig. 2.3. An example of a large-signal equivalent circuit (compact) model, from Ref [13]. © 1999 IEEE. Reprinted with permission.

with other parameters to describe the sharpness of the 'knee' region, output conductance, and so forth.

Instead of fitting these derived parameters with some function approximation, we can simply store them in table form, indexed by the instantaneous terminal voltages $(V_{gs}(t), V_{ds}(t))$ [17], to produce a 'table model'.

2.3.2 Physically-Based Equivalent Circuit Models

The main alternative to the measurement-based approach to compact modeling, described above, is a physically-based approach in which the fundamental device physics is used as a basis for a set of 'phenomenological' equations that describe the terminal behaviour of the transistor in terms of macroscopic physical qualities or parameters, such as the thickness of the active semiconductor layer, gate length, active layer doping, electron mobility, gate oxide thickness, and so forth. We have outlined the physical structure and principles of operation of the FET in Chapter 1, and these equations can

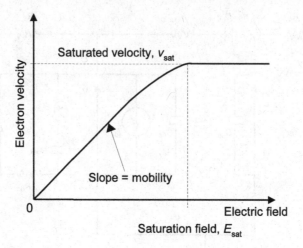

Fig. 2.4. Two-zone electron velocity versus electric field relationship used in physically-based models for short gate-length FETs.

be used as the basis of a model to describe the terminal electrical behaviour of the device: this is a *physically-based* compact model.

In physically-based compact models, significant simplifications are usually made to the real underlying physics of the FET device operation, in order to construct a model that can run quickly enough in the circuit simulator to be of use in circuit design. Classic examples of such an approach are found in Shockley's original drain current model of the silicon JFET [18], and in the extension of this for MESFETs by Pucel [19], in which the electron velocity saturation is modeled by a two-zone velocity field relationship, shown in Fig. 2.4; similar long-channel and velocity saturation models exist for MOS-FETs [20]. In these models the transistor I–V characteristics are calculated using the phenomenological equations. Thus, the influence of the material and device parameters on the terminal I–V characteristics can, in principle, be determined. In a similar manner, the gate current and depletion capacitances associated with the Schottky gate contact in MESFETs and HEMTs are often modeled using simple one-dimensional expressions for the rectifying diode. As we shall see in Chapter 6, this simple one-dimensional description of the charge storage as a two-terminal capacitance neglects charge conservation principles and results in an incorrect description of the bias-voltage dependence of the measured FET capacitances.

In contrast, physically-based compact models for bipolar transistors are relatively straightforward: their basic physics of operation can be described by back-to-back diodes and controlled current sources found in Ebers–Moll and Gummel–Poon models. These models are implemented in simulators and can be used successfully in circuit design. Extensions of these basic models to accommodate second-order effects and transport effects in III–V semiconductor heterojunction bipolar transistors (HBT) have been made, resulting in the VBIC model for silicon-based transistors [21], and the *AgilentHBT* model for III–V HBTs [22].

For any given FET transistor device, we will need to perform measurements to determine the values of the parameters that are used in the equations describing the device physics. This process is known as parameter or model extraction. To extract the parameters for a physically-based compact model, we will generally measure the I–V and C–V or S-parameters of the transistor, and fit the simplified equations by adjusting the parameters in these equations. The curve-fitting is often done automatically in a modeling tool such as *Agilent-EEsof* IC-CAP™. As we try to improve the model by accounting for higher-order physical phenomena into the equations, we create more parameters that need to be extracted from the measurements. We therefore need to make more measurements to illuminate these higher-order effects. The higher-order parameters are usually determined after the first-order model is created, often by using optimization methods. This can lead to a complicated measurement and extraction procedure, and result in a large number of parameters to describe fully the operation of the transistor. An example of this approach is the *BSIM* MOSFET model [23], with the 'level 4' model containing around four hundred extractable parameters to describe the detailed physics of short-gate MOSFET operation. Even the BJT/HBT physically-based compact models contain many parameters – the *AgilentHBT* model uses about one hundred extractable parameters.

There are several potential problems with a model that comprises a large number of model parameters. The parameter extraction process generally takes a long time, and can be open to interpretation by the modeler. It may also be difficult to know which parameters are the most important and that should be determined accurately for a given device application.

Before we leave this discussion of compact or equivalent circuit models, it is worth pointing out that it is common to try and place physical meanings and origins on the circuit elements in the model. The presumed physical origins of the components of the MESFET equivalent circuit model of Fig. 2.2 are illustrated in Fig. 2.5. The extrinsic or parasitic elements have also been included. While this can be useful in relating circuit performance to device design, it should be remembered that these circuit elements are derived from two-port small-signal measurements, and not from a physically-based

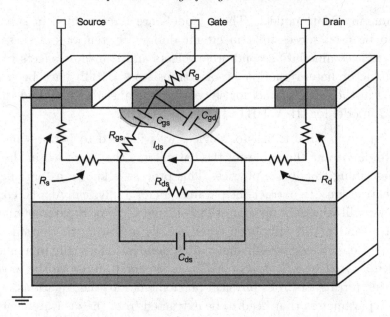

Fig. 2.5. Physical origins of the components of the equivalent circuit model of a MESFET shown in Fig. 2.2 based on [24]. © 1976 IEEE. Reprinted with permission.

model. This can lead to some confusion between small-signal and large-signal model parameters. For instance, the elements R_{gs} and R_{gd} (often called R_i and R_j, respectively) are often described as providing a 'charging path' with resulting characteristic time constants for the capacitances C_{gs} and C_{gd}, whereas they are simply small-signal component values determined from a measurement. Further, there is the often irresistible temptation to ascribe poor circuit performance to specific physical properties, leading to 'tweaking' of individual equivalent circuit component values to demonstrate this behaviour without paying any attention to how these element values depend on each other in the complete model.

2.3.3 Behavioural Models

As noted in the introduction, and illustrated in Fig. 2.1, we tend to reserve the label *behavioural model* for those modeling approaches that use a mathematical function or mapping to relate the measured output to input data. Behavioural models are generally used at higher levels of abstraction in the overall circuit or system design [25]. We have also noted that we often use curve-fitting of the voltage dependences for some of the elements in the

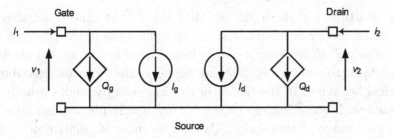

Fig. 2.6. Simple two-port representation that we adopt for the large-signal FET model.

equivalent circuit models, rather than try and estimate the true physical relationship. Although this appears to be a behavioural approach, we prefer to categorize these models as compact models.

2.3.4 Our Modeling Approach

We shall focus on creating compact models of RF power transistors. These models will be designed to preserve the dynamics of the transistor, while being simple to develop and extract. We adopt a two-port structure to describe the intrinsic part of the transistor under large-signal conditions, as shown in Fig. 2.6. The model consists of controlled current and charge sources that are each dependent on the instantaneous voltages V_{gs} and V_{ds} applied to the transistor (intrinsic) terminals. These current and charge source functions are determined from pulsed bias-dependent S-parameter measurements, as described in Chapter 3. The data is transformed, preserving the device dynamics, and is then fitted in the two-dimensional V_{gs}–V_{ds} space using appropriate mathematical functions. The model structure will preserve small-signal to large-signal consistency. The details of the construction of this large-signal, nonlinear model are presented in Chapter 6.

2.4 Memory Effects

The expression *memory effects* has become increasingly common currency in the discussion of the nonlinear behaviour of RF power amplifiers. Indeed, it has become something of a catch-all phrase to describe any of the distortions arising from the nonlinearities inherent in the active device – the RF power transistor. In the context of nonlinear systems the term 'memory' was proposed by Chua [26], to describe the influence on the output of a system at a time t of the input signal(s) not only at time t, but also spanning a finite history of the input signal, to some time in the past, $t-\tau$. This is known as a *fading memory*, as the influence of the input signals deep in the past fades to zero. Essentially, by memory effects we are describing the dynamical behaviour of the device or system, such dynamics usually being associated with either charge storage or hysteretic phenomena that occur over a wide range of timescales. By this we mean capacitive or inductive behaviour with characteristic times that are generally either of the same timescale as the signal frequency – *short-term memory* – or at much slower rates – *long-term memory*. From a modeling point of view, it is important that we are able to identify the causes of memory effects in the RF power transistor, and then describe them in our model of the device.

When we talk of memory effects in RF power amplifiers, we can identify four major sources; these are shown in Fig. 2.7. We can see that the causes of memory fall into two main categories: those inherent in the transistor itself, and those associated with the external circuitry necessary for the transistor to function. Initially we may be tempted to ignore the external causes, those due to the bias and perhaps any matching circuitry inside the package as being the responsibility of the circuit designer. But to do so would require us to ignore how our transistor model works in the circuit, which is a major validation of the model itself. In particular, we would miss the interaction between the long-term memory effects and the short-term (RF) response, which can only occur through a correct description of the nonlinear behaviour in our transistor model. We shall now outline the major sources of memory effects in the nonlinear device or system, and describe how we might accommodate them in the model of the RF power transistor.

2.4.1 Short-Term Memory Effects

The high frequency dynamics of the transistor are determined by the reactances associated with the transistor. In the usual description of a transistor model, these reactances comprise the capacitances and inductances associated with the extrinsic or parasitic elements of the transistor model, and also the charge storage within the transistor's active region, the intrinsic part of the model. The extrinsic components are usually considered to be linear elements, independent of bias, described simply by dv/dt and di/dt expressions for the capacitors and inductors, respectively. The intrinsic charge storage behaviour is usually nonlinear, dependent upon the instantaneous voltage or current, and so the dynamic behaviour will change with bias or

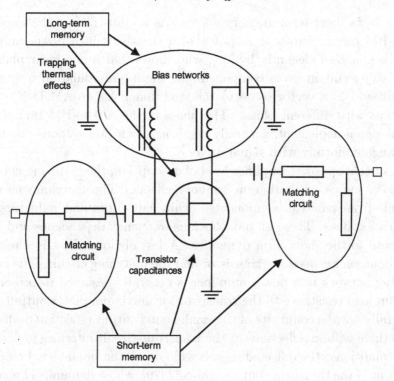

Fig. 2.7. Origins of short-term and long-term memory effects in a transistor circuit (adapted from [27]).

drive signal. When linearized, these nonlinear charge storage components can be described by capacitances in the equivalent circuit representation. For small-signal characterization, the short-term memory effects are simply the frequency response of the transistor. We note that the frequency response is bias-dependent, which requires that the capacitances describing the linearized charge storage behaviour in the transistor are also bias dependent: this is a bias-dependent linear transistor model, described in more detail in Chapter 6.

Under large-signal conditions, the voltage- or current-dependence of the charge storage functions becomes important. The changing dynamical behaviour with signal drive is manifest in measured quantities such as AM-to-AM (gain compression) and AM-to-PM (phase transfer characteristic) of the transistor, and hence the amplifier or system as a whole. The phase transfer characteristic is usually defined as the change in phase of the S_{21} of the transistor from its small-signal value as the signal drive is increased. The AM-to-PM effects are essentially the nonlinear behaviour that is often referred to as short-term memory effects. As an illustration, the impact of AM-to-PM on the two-tone response of a transistor in a power amplifier circuit is analyzed elegantly by [28], who shows that a nonlinear phase response will result in extra components at the intermodulation frequencies that will add in a vector sense to the traditional AM-to-AM IM3 components, but with different phase. The phase of the AM-to-PM IM3 components is signal-dependent, generally resulting in an IM3 response that does not change uniformly with signal drive.

At a system or power amplifier level of description, the matching networks are also a source of short-term memory effects. The matching networks are built from reactive components – chip capacitors and inductors – or transmission lines. These all have obvious frequency dependence and hence contribute to the short-term dynamics. A less obvious effect is a result of the interaction between the transistor and the matching circuit. The output matching network in a power amplifier is generally designed to present the optimum load resistance to the transistor for maximum power output. This is generally not the conjugate of the small-signal output reflection coefficient, and so there will be reflections at the transistor-circuit interface under large signal conditions: the reflected signals will not see the small-signal reflection coefficient of the transistor, but the *hot-S22* [29], whose dynamical behaviour is different from the small-signal case: short-term memory effects. The transistor model must be able to accommodate these small-signal and large-signal behaviours.

2.4.2 Long-Term Memory Effects

The long-term memory effects are due to dynamics that take place on a timescale that is much longer than the period of the RF signal. In RF power amplifiers, 'long-term' is generally considered to be on the order of the timescale of the signal envelope, or even longer. Within the transistor, there are considered to be two main causes of long-term memory: thermal effects, and charge trapping. Additionally, we have a circuit-dominated effect, related to the bandwidth of the DC bias network.

2.4.2.1 Thermal Effects

Under conditions of constant drive, the transistor channel heats up uniformly in cross-section. When driven using a modulated information signal, we may find that at some instant the signal is high amplitude, and hence high energy: the transistor channel heats up a little in response to this signal. A short time later, the signal has returned to a low value, but the channel has not cooled down instantaneously: it is still at a slightly higher temperature. Because of this local change in temperature, some of the transistor's parameters will be slightly different; for instance, the gain may be slightly reduced from the equilibrium-temperature value. This later signal will see the reduced gain, and therefore the output from the transistor will be slightly reduced from the expected, equilibrium-gain value. The transistor parameters are exhibiting a memory of the previous signal. The time constants associated with thermal transients are generally of the order of milliseconds to microseconds. This timescale is significantly longer than the RF period, and is closer to the timescale of the information signal in the envelope: RF information channel bandwidths are in the range 200 kHz to 20 MHz and beyond. These long-term memory effects can be seen in the AM-to-AM characteristics of RF power amplifiers, as a 'spread' around the mean gain compression curve [30], as shown in Fig. 2.8. The thermal memory effects can be included in the transistor model through a dynamic coupling of the electrical and thermal signals.

2.4.2.2 Charge Trapping

Imperfections and defects in the semiconductor occur in the channel itself, at interfaces between the semiconductor and oxide, at the surface of the semiconductor, and at the channel–buffer interface. These imperfections often manifest themselves as available states that can capture and release electrons or holes. The trapping and release are governed by local potentials and temperature. The action of trapping or releasing an electron is effectively changing the charge density in the channel of the transistor. The rate of trapping and release is on a timescale of kilohertz through megahertz, depending on the nature of the trapping center. The trapping mechanism is, therefore, one that can change the signal current in response to local voltage changes, on a long timescale: a long-term memory effect.

Fortunately, LDMOS transistors do not suffer from such trap-related phenomena [31], and so memory effects due to trapping are negligible. On the other hand, gallium arsenide and gallium nitride FETs display several trap-related phenomena. One example is the well known dispersion of the small-signal transconductance and output conductance in GaAs FETs: the values of these parameters fall significantly from DC to RF, with the transition occurring between about 10 kHz and 10 MHz. At frequencies above this, the traps are unable to respond to the voltage signal. Other phenomena seen in GaN power transistors include knee collapse and walkout under RF drive, these effects recovering after the drive is removed [32].

2.4.2.3 DC Bias Network

In the strict sense, this may be considered to be a circuit design issue: the DC bias network provides a low impedance path for the DC bias connections that is simultaneously a high impedance to the RF signal. The bandwidth of the low impedance path is known as the video bandwidth, and it is limited by the reactive components that define it. In other words, this path has inductance and capacitance that control the frequency response of the video bandwidth from DC to a few tens of MHz. Any signal components in this frequency range will experience *memory effects*.

The appearance of signal components in this frequency range is a result of the even-order nonlinearities in the transistor, which should be captured by the model. The finite impedance of the DC bias connections means that these low-frequency components will be impressed upon the transistor, causing small changes in the bias conditions on a timescale appropriate to the signal envelope. The small bias deviations will result in changes in the transistor's RF behaviour – its gain, for example – which will affect the output RF signal. These changes will occur at a slow rate compared with the RF signal frequency, and again fall under the rubric *long-term memory effects*. Essentially, the even-order components in the video bandwidth are being remixed with the RF signal through this gain modulation effect: they re-appear at the IM3 frequencies, with amplitude and phase that reflect the frequency response of the video bandwidth. This can lead to asymmetry in the IM3 responses, which is often taken to be a characteristic of long-term memory effects [30].

2.4.3 Measuring Memory Effects

Short-term memory effects are relatively straightforward to characterize. Typically, one would measure the RF frequency response of the transistor, and the AM-to-AM and AM-to-PM characteristics. The frequency response is largely determined by the package and matching networks, and the small-signal capacitances in the transistor model. We have outlined how the package reactances such as the bondwire arrays, MOS capacitors, and the like can be determined using segmentation techniques and EM simulation, and the device capacitances can be determined from the S-parameter characterization required for the model extraction. Short-term memory effects should, therefore, be reasonably straightforward to capture in our model.

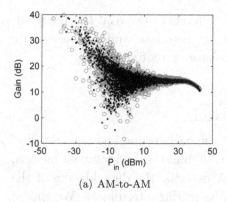

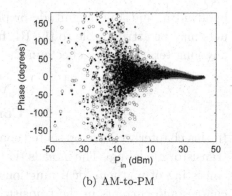

(a) AM-to-AM   (b) AM-to-PM

Fig. 2.8. Long-term memory effect on the compression characteristic of a power amplifier; a memoryless device or circuit should have a response that is a single line. The memory effects here are indicated by the spread of the compression characteristic: the actual point response depends on the signal value at some previous instant of time. This compression characteristic is obtained using a modulated signal [33]. © 2006 IEEE. Reprinted with permission.

Long-term memory effects are often characterized by asymmetry in the intermodulation products. In other words, the upper and lower IM3 products have different magnitudes and phases from each other. A typical measurement to observe these long term memory effects is to perform two-tone measurements at high RF power, and sweep the tone spacing. If long-term memory effects are present, the IM3 responses will vary with the tone spacing, which translates to baseband frequency response. The identification of the origins of the long-term memory are more difficult. Thermal effects can be observed by making pulsed measurements, and varying the pulse width and duty cycle, to change the IM response. The transistor model can account for these thermal effects with a suitable electro-thermal model, the construction of which is described in Chapter 6.

A number of techniques have been proposed for measuring the video bandwidth of the circuit in which the transistor is embedded, and identifying the limiting components and effects in this bandwidth [34].

It is important to stress that some care should be exercised in carrying out and interpreting swept-tone-separation two-tone measurements or simulations of RF power transistors. Generally, RF power transistors, particularly those designed for wireless infrastructure applications, have narrow RF bandwidths of the order of 60 MHz. Once the two-tone spacing exceeds about 20 MHz, the separation of the two IM3 products will exceed the RF bandwidth, and any magnitude or phase differences between the IM products may be attributable to the RF frequency response, and not necessarily any long-term memory effects in the transistor or its fixture.

2.5 Conclusions

In this chapter we have presented some of the background and framework of transistor modeling. The models that we shall build will be compact models, consisting of mathematical functions that describe the dependences of the charges and currents in the transistor on its intrinsic terminals. We embed that compact intrinsic model in the manifolds, bondwire arrays and package to obtain a model of the packaged transistor that can be used in circuit simulation. We can accommodate effects such as long-term and short-term memory by accurate modeling of the electro-thermal and charge trapping behaviour.

In the following chapters we will describe the measurements that are required for the characterization of the transistor, and how the electrical and thermal models of the transistor are defined and determined. We shall then present some practical methods for implementing the compact model in the simulator, and finally we shall outline verification and validation techniques to demonstrate the performance and veracity of the transistor model.

References

[1] M. Reiser, "Large-scale numerical simulation in semiconductor device modeling," *Comput. Methods in App. Mech. Eng.*, 1, no. 1, pp. 17–38, 1972.

[2] C. M. Snowden and R. R. Pantoja, "GaAs MESFET physical models for process oriented design," *IEEE Trans. Microwave Theory Tech.*, 40, no. 7, pp. 1401–09, July 1992.

[3] E. M. Buturla, P. E. Cottrell, B. M. Grossman, and K. A. Salsburg, "Finite-element analysis of semiconductor devices: the FIELDAY program," *IBM J. Res. Dev.*, 25, no. 4, pp. 218–231, July 1981.

[4] J. Machek and S. Selberherr, "A novel finite-element approach to device modelling," *IEEE Trans. Electron Devices*, 30, no. 9, pp. 1083–1092, Sept. 1983.

[5] http://www.synopsys.com/products/tcad/tcad.html

[6] http://www.silvaco.com/

[7] B. Troyanovsky, Z. Yu, L. So, and R. W. Dutton, "Relaxation-based harmonic balance technique for semiconductor device simulation," in *Digest of Technical Papers, 1995 IEEE/ACM Int. Conf. on Computer-Aided Design*, San Jose, CA, Nov. 1995, pp. 700–703.

[8] G. Golub and C. van Loan, *Matrix Computations*, 3rd edn. London, UK: Johns Hopkins University Press, 1996.

[9] G. H. Loechelt and P. A. Blakey, "A computational load-pull system for evaluating RF and microwave power amplifier technologies," in *IEEE MTT-S Int. Microwave Symp. Dig.*, Boston, MA, June 2000, pp. 465–468.

[10] D. Denis, C. M. Snowden, and I. C. Hunter, "Coupled electrothermal, electromagnetic, and physical modeling of microwave power FETs," *IEEE Trans. Microwave Theory Tech.*, 54, no. 6, pp. 2465–70, June 2006.

[11] G. Dambrine, A. Cappy, F. Heliodore, and E. Playez, "A new method for determining the FET small-signal equivalent circuit," *IEEE Trans. Microwave Theory Tech.*, 36, no. 7, pp. 1151–59, July 1988.

[12] J. Wood and D. E. Root, "Bias-dependent linear scalable millimeter-wave FET model," *IEEE Trans. Microwave Theory Tech.*, 48, no. 12, pp. 2352–2360, Dec. 2000.

[13] W. R. Curtice, J. A. Plá, D. Bridges, T. Liang, and E. E. Shumate, "A new dynamic electro-thermal nonlinear model for silicon RF LDMOS FETs," in *IEEE MTT-S Int. Microwave Symp. Dig.*, Anaheim, CA, June 1999, pp. 419–422.

[14] W. R. Curtice and M. Ettenberg, "A nonlinear GaAs FET model for use in the design of output circuits for power amplifiers," *IEEE Trans. Microwave Theory Tech.*, 33, no. 12, pp. 1383–94, Dec. 1985.

[15] H. Statz, P. Newman, I. W. Smith, R. A. Pucel, and H. A. Haus, "GaAs FET device and circuit simulation in SPICE," *IEEE Trans. Electron Devices*, 34, no. 2, pp. 160–168, Feb. 1987.

[16] A. E. Parker and D. J. Skellern, "A realistic large-signal MESFET model for SPICE," *IEEE Trans. Microwave Theory Tech.*, 45, no. 9, pp. 1563–71, Sept. 1997.

[17] D. E. Root, S. Fan, and J. Meyer, "Technology independent large signal non quasi-static FET models by direct construction from automatically characterized device data," in *Proc. 21st European Microwave Conf.*, Stuttgart, Germany, Sept. 1991.

[18] W. Shockley, "A unipolar field-effect transistor," *Proc. IRE*, 40, pp. 1365–76, 1952.

[19] R. A. Pucel, H. A. Haus, and H. Statz, "Signal and noise properties of Gallium Arsenide microwave field effect transistors," in *Advances in Electronics and Electron Physics*. New York, NY: Academic Press, 1975, 38, pp. 195–265.

[20] T. Ytterdahl, Y. Cheng, and T. A. Fjeldly, *Device Modeling for Analog and RF CMOS Circuit Design*. Chichester, UK: John Wiley & Sons, 2003.

[21] C. C. McAndrew, J. A. Seitchik, D. F. Bowers, *et al.*, "VBIC95, the vertical bipolar inter-company model," *IEEE J. Solid State Circuits*, 31, no. 10, pp. 1476–83, Oct. 1996.

[22] M. Iwamoto, D. E. Root, J. B. Scott, A. Cognata, P. M. Asbeck, B. Hughes,

and D. C. D'Avanzo, "Large-signal HBT model with improved collector transit time formulation for GaAs and InP technologies," in *IEEE MTT-S Int. Microwave Symp. Dig.*, Philadelphia, PA, June 2003, pp. 635–638.

[23] W. Liu, *MOSFET Models for SPICE Simulation, including BSIM3v3 and BSIM4*. New York, NY: Wiley-IEEE Press, 2001.

[24] C. A. Leichti, "Microwave Field Effect Transistors–1976," *IEEE Trans. Microwave Theory Tech.*, 24, no. 6, pp. 279–300, June 1976.

[25] J. Wood and D. E. Root, *Fundamentals of Nonlinear Behavioral Modeling for RF and Microwave Design*. Norwood, MA: Artech House, 2005.

[26] L. O. Chua, C. A. Desoer, and E. S. Kuh, *Linear and Nonlinear Circuits*. New York, NY: McGraw-Hill, 1987.

[27] E. Ngoya and A. Soury, "Envelope domain methods for behavioral modeling," ch. 3, in *Fundamentals of Nonlinear Behavioral Modeling for RF and Microwave Design*, J. Wood and D. E. Root, Eds. Norwood, MA: Artech House, 2005.

[28] S. C. Cripps, *RF Power Amplifiers for Wireless Communications*, 2nd edn. Norwood, MA: Artech House, 2006.

[29] J. Verspecht, "Everything you've always wanted to know about Hot-S22 (but were afraid to ask," in *IEEE MTT-S Int. Microwave Workshop 'Introducing New Concepts in Nonlinear Network Design'*, Seattle, WA, June 2002.

[30] N. B. de Carvalho and J. C. Pedro, "Large- and small-signal IMD behavior of microwave power amplifiers," *IEEE Trans. Microwave Theory Tech.*, 47, no. 12, pp. 2364–74, Dec. 1999.

[31] J.-M. Collantes, "Modelisation des transistors MOSFETs pour les applications RF de puissance," Ph.D. dissertation, University of Limoges, 1996.

[32] B. M. Green, V. Tilak, V. S. Kaper, J. A. Smart, J. R. Shealy, and L. F. Eastman, "Microwave power limits of AlGaN/GaN HEMTs under pulsed bias conditions," *IEEE Trans. Microwave Theory Tech.*, 52, no. 2, pp. 618–623, Feb. 2003.

[33] J. Wood, M. LeFevre, D. Runton, J. C. Nanan, B. H. Noori, and P. H. Aaen, "Envelope-domain time series (ET) behavioral model of a Doherty RF power amplifier for system design," *IEEE Trans. Microwave Theory Tech.*, 54, no. 8, pp. 3163–72, Aug. 2006.

[34] B. Noori and S. Rumery, "The Low Frequency Probe: a new technique for measuring the modulation bandwidth of power amplifiers," in *IEEE Topical Symposium on Power Amplifiers for Wireless Communications*, Long Beach, CA, Jan. 2007.

3

Electrical Measurement Techniques

3.1 Introduction

Obtaining precise measurement data at microwave frequencies is a demanding task. Complex equipment and elaborate calibration procedures are needed, and a significant fraction of the time needed to generate a model is spent on the collection of measurement data. Measurement quality and accuracy are paramount, as they are the basis for generation and validation of models.

Historically, significant advances in transistor modeling have coincided with the development of new measurement techniques. The more obvious examples include: the introduction of the vector network analyzer to measure small-signal scattering parameters [1,2]; mechanical and electronic load-pull systems for mapping the small-signal (for example, noise parameters) and the large-signal (power, linearity, and so forth) performance parameters as a function of the impedance presented to the transistor [3]; pulsed DC [4–6] and S-parameter [7,8] systems used to overcome complex transistor dynamics and dispersive phenomena. More recently, the development of the large-signal vector network analyzer has enabled the characterization of transistors under realistic large-signal modulations [9–11].

In this chapter, we elaborate on general issues related to the measurement environment, including a description of the fixtures used during the model extraction and validation processes. In addition, we shall present a description of the different calibration schemes used during the measurement process followed by an explanation of the de-embedding process. Subsequently, we shall describe measurement techniques that are essential for the generation and validation of high-power transistor models. The techniques are presented in detail with specific insight into their usefulness in the characterization, extraction, and validation stages of model development.

71

3.2 Electrical Reference Planes

The electrical reference planes define physical locations in the circuit at which the measurements of voltage and current are known to be as accurate as our measurement instruments allow. This accuracy is achieved by *calibrating* the instruments against known standards that are placed at these reference points. By 'known standards', we mean components whose values can be determined either by virtue of their physical or mechanical construction, or whose values are compared regularly against nationally defined standard components – these latter are known as traceable standards. The purpose of calibration is to measure the known standards, and compare our actual measurements against these known standards to determine the systematic errors in our measurement instruments; then, knowing these errors, we can correct our measurements of unknown components to obtain an accurate estimation of their value at the specified reference planes.

At DC, the measurement instrumentation often has a built-in calibration; this is performed at the front-panel connections so that the instrument is 'accurate' at its own connections. This built-in self-test often consists of the measurement of a known internal voltage reference, which can be made stable and temperature insensitive using IC design techniques. We now have a known measurement reference at the instrument plane. While this is a good starting point, to perform practical measurements on a transistor we need to connect the device to the DC bias and the rest of the measurement circuit: this requires bias networks, cables, connectors, and so forth, which will all contribute some resistive loss that must be accounted for in order to make accurate measurements at the transistor terminals. We define the DC reference plane at the transistor itself, and calibrate our measurement system by placing a short circuit at this point: the measured voltage and current at the instrument plane determine the loss in the intervening cables and connectors. Subsequently, when the current–voltage characteristics of the transistor are to be determined, by measuring the current supplied and knowing this resistive loss, the voltage drop along the cabling can be determined, and hence the terminal voltage at the transistor is found.

For RF measurements, whereas the principle is the same, the actual procedure is a little more involved. The calibration of the vector network analyzer (VNA) is a prerequisite to obtaining accurate S-parameter measurements. A discussion of the calibration of the VNA could be the subject of a book in itself, and so we will only outline the basics here, focusing more on the practical details and implementation of the calibration techniques.

3.2.1 SOLT Calibration

A basic VNA measurement setup is shown in Fig. 3.1. The VNA takes the signals that are input to its R, A, and B ports, and takes the complex ratio of these signals to determine the S-parameters. This arrangement allows us to measure the input reflection coefficient, $S_{11} = A/R$, and the forward transmission coefficient, S_{21}, B/R of the device-under-test (DUT). The reverse S-parameters require the DUT to be disconnected and rotated so that its ports are reversed to that of the forward measurement. Such a setup is typical of three-channel (detector) VNAs such as the venerable Hewlett Packard (now Agilent Technologies) 8753. More modern VNAs swap the position of the source internally by means of switches, but this simple structure illustrates one of the fundamental means of calibrating the VNA. Conceptually, the circuit in Fig. 3.1 can be analyzed using signal flowgraph techniques [12] to accommodate the mismatches at component connections, the coupler directivity and losses, and so forth. The signal flowgraph representation is usually simplified to yield the error model shown in Fig. 3.2, in which we can identify the following errors associated with reflection, transmission, and crosstalk:

- (i) $e_{00} = e_D$ = Directivity error
- (ii) $e_{10}e_{01} = e_R$ = Reflection or Tracking error – these two signals, e_{10} and e_{01}, cannot be separated in the analysis.
- (iii) $e_{11} = e_S$ = Source match error
- (iv) $e_{22} = e_L$ = Load match error
- (v) $e_{23} = e_T$ = Transmission error
- (vi) $e_{30} = e_X$ = Crosstalk error

This is the forward path of the well-known 12-term error model for the VNA. A calibration in which known standards are substituted for the DUT is used to determine these error terms. This must be performed at every frequency of interest. Once the error vectors are known, the measured S-parameters can be corrected to yield the true S-parameters of the DUT.

Let us consider just the reflection terms: a one-port measurement. The errors are collected into a fictitious error adapter that modifies the reflection coefficient of the DUT, and the modified value is measured by a perfect reflectometer. There are three error terms, so by removing the DUT and replacing with three known standards we obtain three simultaneous equations that can be solved for e_D, e_R, and e_S, the error adaptor network. Typically, the three standards are a $\underline{S}$hort-circuit, an $\underline{O}$pen-circuit, and a matched (Z_0) $\underline{L}$oad, yielding the *Short-Open-Load* or *SOL* one-port calibration. This procedure is illustrated in Fig. 3.3. The physical location in the measurement

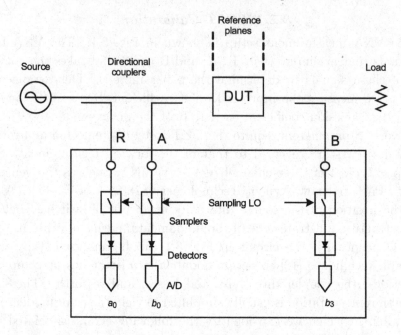

Fig. 3.1. A simple transmission-reflection vector network analysis setup. This is a '3-sampler' setup, as found in the Agilent 8753.

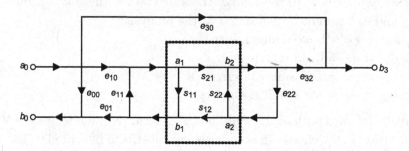

Fig. 3.2. Signal flowgraph representation of the transmission-reflection test setup shown in Fig. 3.1. The error signals are labeled, unlabeled flows have unity magnitude. From [13], used with permission of the author.

system at which the connections to the DUT and the standards are made defines the *reference plane*. For a two-port DUT we will have input and output reference planes, which are the locations of the input and output connections of the DUT.

The transmission error terms can be calculated by placing a known standard in place of the DUT. A simple known standard is a length of transmission line, which in some circumstances can be of zero length, in other

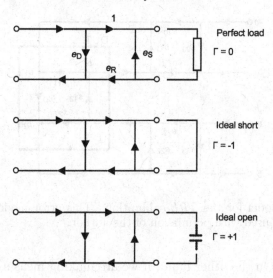

Fig. 3.3. Illustrating the *Short-Open-Load* calibration procedure.

words, the two reference plane ports where we would place the DUT are connected directly together. This is a Through connection. Knowing the reflection error terms, and the S-parameters of the Through, the load match and transmission error terms can be calculated. Finally, the crosstalk error term is measured by removing any physical connection between the reference planes: this can be achieved effectively by placing matched loads on both ports. We now have the six error terms.

For a full two-port characterization using a four-channel VNA, we place the Short, Open and Load at the reference planes at each port, and then connect the two reference planes with a Through: seven connections in all for twelve error terms. This is the *Short-Open-Load-Through* or *SOLT* calibration method.

One of the drawbacks of the SOLT calibration method is that the electrical characteristics of each of the standards must be known to a high degree of precision. If we are working in a connectorized environment, using, for example APC-7 seven millimeter precision connectors, then a calibration kit that includes short, open, load, and through components that are traceable standards is available from manufacturers of these precision components, such as Agilent Technologies or Maury Microwave.† We can then trust the standards to have the electrical characteristics that are defined in the

---

† Our mention of these manufacturers does not imply any recommendation.

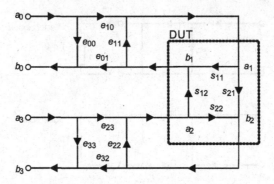

Fig. 3.4. Error model for the *TRL* calibration. This error model has only eight terms. From [13], used with permission of the author.

calibration kit. On the other hand, if we are making measurements of transistors directly on wafer using electrical probes, then we must use either an impedance standard substrate (ISS), which contains the various calibration components as microstripline or coplanar waveguide elements in thin- or thick-film form on an insulating dielectric substrate, or build our own 'standards' onto the semiconductor wafer, next to the DUT. Such standards are clearly not traceable, and so the accuracy of our measurements can be compromised. We will discuss 'on-wafer' standards and wafer-probing RF measurements in more detail later, but first we will outline some more robust calibration principles, which can yield very accurate measurements but relax the constraints on the characteristics of the standards.

3.2.2 *TRL Calibration*

It is possible to reduce the 12-term error model to an 8-term error model, given certain assumptions, and provided that we use a four-channel VNA. The schematic of this error model is shown in Fig. 3.4: there are now only two 4-term error adaptors between the DUT and the ideal reflectometers, and the signal source is switched between a_0 and a_3 to generate the forward and reverse S-parameters. One of the key assumptions is that the switch is perfect and does not change the port match term as the switch changes state from forward to reverse. In fact, this is not a significant limitation as it is possible to ratio out any imperfections mathematically. It is also assumed that the crosstalk term can be ignored, or found later in a separate measurement.

The key to the reduction in the number of error terms, and hence measurements, without sacrificing accuracy, comes from treating the system as a two-port measurement instrument for all the calibration standards. A signal flowgraph treatment of the system then shows that the DUT is always embedded in the two error adaptors, and the measured (T_M) and true (T) DUT parameters are related through:

$$T_M = \frac{1}{e_{10}e_{32}} \begin{bmatrix} -\Delta_A & e_{00} \\ -e_{11} & 1 \end{bmatrix} T \begin{bmatrix} -\Delta_B & e_{22} \\ -e_{33} & 1 \end{bmatrix} \qquad (3.1)$$

where

$$\Delta_A = e_{00}e_{11} - e_{10}e_{01} \qquad (3.2)$$

and

$$\Delta_B = e_{22}e_{33} - e_{32}e_{23} \qquad (3.3)$$

and we use cascade T-parameters instead of S-parameters for this analysis. This formulation was first presented by Engen and Hoer [14] in their development of a calibration method for a six-port network analyzer, and has been developed further for VNAs by Eul and Schiek [15]. From eq. 3.1, we can identify seven error terms in the cascade of matrices. There are three error terms associated with port 1 (Δ_A, e_{00}, e_{11}), three associated with port 2 (Δ_B, e_{22}, e_{33}), and a transmission term ($e_{10}e_{32}$). We calibrate this system, in other words, find the values of these error terms, by measuring just three two-port standards. These three measurements will give us twelve observations from which we need to find only seven terms. This gives us some freedom in the choice of two-port calibration standards that we can use.

Following Eul and Schiek [15], we note that only one of the three connections or standards needs to be known completely, that is, all four of its S-parameters, accounting for four of the seven unknowns. This standard can be realized simply by using a zero-length Through component: connecting the reference planes together directly. The second standard is a transmission line. This standard requires us to know the electrical length and the characteristic impedance, and provides us with two of the remaining unknowns in the calibration. The electrical length is related to the physical length and can therefore be estimated accurately. The characteristic impedance of the line is usually chosen to be 50 Ω, to provide a match. In this calibration, the characteristic impedance of the line is the characteristic impedance of the calibration. The final standard is a reflect standard: either an Open or Short can be used, all that we need to know for the calibration is whether the reflection coefficient of the standard is close to +1 or −1. Ideally, identical reflects should be provided at both ports, and in some calibration routines this is assumed, although strictly speaking it is not necessary. The reflect standard is used to determine the last unknown. This calibration technique is known as *Through-Reflect-Line* or *TRL*.

The TRL calibration is widely used as it requires fewer connections than SOLT, less absolute knowledge of the standards, and in fact has been shown to be more accurate, showing lower residual errors [16]. A limitation of this method is that the frequency range over which the calibration is valid is determined by the length of the transmission line standard. Ideally, the line standard should have an electrical length of 90° for best accuracy, but is usable for lengths that are not close to a half-wavelength, that is 0° or 180°, corresponding to one cycle of the Smith chart. In practice, we usually set the limits on the line length to be 20° and 160°, which then determines the frequency range of the calibration. To obtain a wider frequency range, we must use several transmission lines of different lengths, which means that we have to make additional connections to accomplish the calibration. A widely-used software for managing the calibration process and calculating the error adaptors for TRL calibration is *MultiCal*, available from the National Institute of Standards and Technology (NIST) [16].

Commercially available and traceable calibration kits for TRL comprise Open, Short, Through, and several Line standards to cover the specified frequency range. These standards are the same ones that we would use in SOLT calibration, except that we do not need to know the specifications of all of the standards in so much detail. The calibration kits are available in several types of precision coaxial connector, for example APC-7, APC-3.5 and 2.4-mm, depending on the test frequencies of interest. The calibration components are generally made to more exacting tolerances than the regular connectors, and care must be exercised in their use: 'connector care' procedures should be strictly adhered to [17]. We use these calibration kits for calibrating the VNA when we want to make measurements on connectorized components or on transistors that have been mounted in a test fixture, as described in Section 3.3.

3.2.3 On-Wafer Measurements

If we want to make S-parameter measurements on a device that is still on the semiconductor wafer, then we have several choices regarding the calibration techniques that we can use. First, we can choose between using an ISS that is usually supplied by the probe and calibration software provider [18] (see Fig. 3.5), or designing and building a custom calibration kit onto the semiconductor wafer. In either case, the standards are not traceable, but can be constructed with enough precision to permit TRL calibration techniques to

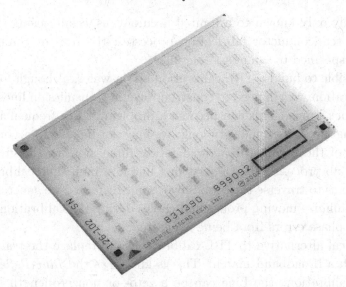

Fig. 3.5. Impedance standard substrate, ISS. Reproduced with the permission of Cascade Microtech Inc.

be used successfully. The use of ISS calibration kits is widespread: they are simple to use, reliable, and do not require access to the semiconductor fab design rules, as is necessary for an on-wafer design. One drawback of the ISS is that the substrate material has very different dielectric properties from the semiconductor wafer material. Usually the ISS is alumina, a low-loss medium that is well suited to microwave circuit fabrication. The transistors that we are interested in characterizing are typically built on silicon or gallium arsenide (GaAs) substrates. Gallium arsenide substrates are semi-insulating and have relatively low loss, and are able to support near-ideal transmission line propagation. Even so, the difference in dielectric properties between GaAs and the alumina of the ISS can lead to inaccuracies in the calibration [19]. The silicon substrates are, in contrast, relatively lossy even at RF frequencies. This loss is not accounted for in the ISS and as a result the error adaptors may not capture the loss properly. Generally, the ISS calibration is made at the probe tips, minimizing any substrate effects on the calibration accuracy. We then have to move the reference planes to the device ports by using *de-embedding* techniques, as outlined in Section 3.3.5. Building the calibration kit on-wafer generally leads to more satisfactory measurements, as the dielectric and loss properties of the substrate are built into the calibration standards, resulting in self-consistent calibration and measurement. The calibration kits require knowledge of the impedance parameters of the lumped element standards: Open, Short, Match. These are generally only known to a limited accuracy, especially using standards built from semiconductor fabrication processes; the resistor component is often only specified to a modest tolerance.

It is possible to build a TRL calibration kit on-wafer, although for a wide-band calibration we will need to construct several transmission lines to cover the frequency range. This can become a problem at low frequencies where the line lengths can become very long: we may not be able to accommodate the length of the line in the allowed reticle dimension for the given semi-conductor fab process. Also, a very long line in a multi-line calibration kit will require us to traverse the probes over some distance: we generally try to avoid or minimize moving probes and cables during the calibration process, to prevent phase errors from being generated.

A practical alternative to TRL calibration is to replace the transmission line(s) with a broadband match. This is known as the *Line-Reflect-Match* or *LRM* calibration: the Line can be a zero- or nonzero-length Through. In the VNA the broadband match can be modeled in the TRL calibration procedure as a very long line: in the Agilent 8510, for example, the match is mimicked by a line of ≈ 1 second delay. This is equivalent to a line of several thousand kilometers in length. The LRM calibration kit has the advantage of being easier to construct than the TRL kit, and it requires less space on the wafer or ISS. Further, the whole calibration kit can be designed so that the probes do not need to be moved at all in terms of the in-out direction, or so that this movement is very small, minimizing cable movement and potential for phase errors.

An on-wafer calibration kit for GaAs FET characterization is shown in Fig. 3.6 [20]. In this design, care has been taken to transform the coplanar electromagnetic (EM) fields that exist at the ground-signal-ground (GSG) probe tips into the microstrip EM field distribution that is seen by the transistor DUT in circuit applications. This is done by designing a 50 Ω coplanar waveguide (CPW) transmission line structure for the probes to land upon, and then transforming this to a 50 Ω microstrip transmission line: the transformer is logarithmic, and via-holes through to the back side ground metallization are placed on the CPW grounds. The reference planes for this calibration kit are at the end of the microstrip lines – this is where we place the FET structures for test. This reference position can be en-sured by using a zero-length Through component in the calibration kit: this component is well defined electrically, but has the small drawback that the probes need to be moved inwards by the length of the DUT, causing a small cable movement. A nonzero-length Through component can be used; the reference plane is then defined to be exactly in the center of the line, and

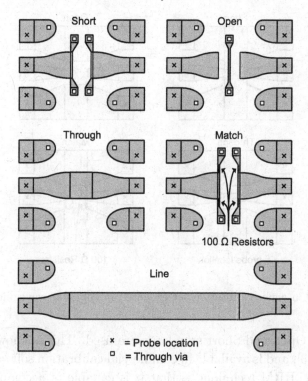

Fig. 3.6. On-wafer calibration set executed in microstrip.

can be rolled back to the end of the microstrip launch. This action requires more knowledge of the electrical properties of the line component – electrical length, loss, characteristic impedance, and so forth. Inaccuracies in this physical data are probably similar in effect to any small phase errors that may be introduced by moving the probes and cables through a few hundred microns to touch the zero-length Through. The remaining standards used in the LRM calibration are a 50 Ω load (Match standard), and an Open reflect: this latter is preferred over the Short reflect in this microstrip calibration kit, because any misplacement of the through via-holes during processing of the back side of the wafer could lead to asymmetric Reflect standards, resulting in errors in the calibration.

On silicon, the lossy dielectric substrate has a frequency-dependent loss, and the lumped-element resistor may be a complex impedance: calibrations based on a known resistive match such as SOLT or LRM, may be unreliable [21]. The probe pads can also introduce a significant lossy capacitance that must be de-embedded from the measurement for an accurate calibration [22].

An extension of the LRM method uses an additional Reflect component,

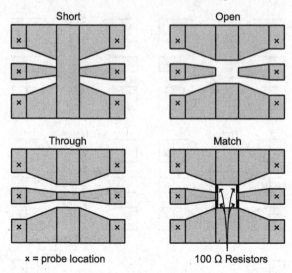

Fig. 3.7. On-wafer calibration set executed in coplanar waveguide CPW.

so that both Open and Short standards are used. This is known as LRRM calibration [23], and is available in commercial calibration software [18]. The advantage of LRRM technique is that it is capable of accommodating the reactive losses that are inherent in the Match component of the calibration kit, giving superior results to regular LRM, especially at higher frequencies.

It is also possible to design an on-wafer CPW calibration kit, as shown in Fig. 3.7. Here the Reflect standard is a Short as it is easy to realize in this stripline medium. The Match component is also easy to realize using 100 Ω resistors between the signal and ground lines. Since all of the calibration kit components are realized using photolithographic techniques in the fabrication process, their dimensions can be carefully controlled. This is suitable for GaAs and GaN transistors, which are built on a semi-insulating substrate material; LDMOS transistors have a source connection through the substrate, and so this CPW approach is not appropriate.

These on-wafer approaches to calibration can also be used in principle for in-fixture measurements, by constructing fixture calibration standards for TRL/LRM. This is generally satisfactory if the DUT is biased using external bias tees, which are usually placed outside the VNA couplers and therefore do not affect the RF calibration. If the gate and drain biases are included on fixture, these bias networks must be accommodated in the calibration standards: this can make the standards more difficult to define, so TRL/LRM approaches are recommended.

3.2.4 Verification

Once we have measured all of the standards and computed the error coefficients, we have a calibrated VNA; we now need to verify that the calibration is accurate. Usually the verification begins by measuring some or all of the standards used in the calibration measurements. Of course, we expect to see the VNA make accurate measurements of these standards, by which we mean that the measurement should return the standard definition in the calibration kit, and not the ideal value. For example, if the Short standard is defined as having an inductance of 20 pH, the measurement of the Short after calibration should produce data corresponding to a 20 pH inductor. This is a verification of the correctness of the implementation of the calibration algorithm. While this is a necessary aspect of the verification, it is not sufficient. We complete the verification by measuring other independent standard components, not used during the calibration, whose properties we already know, and are traceable or can be calculated from their dimensions.

In a connectorized environment, we will often use Line standards of lengths other than those used in the calibration measurement, and an *Offset Short* (or *Offset Open*) component: constructed from one of the Line standards and the Short or Open standard. These verification components enable us to get an assessment of the accuracy of the measurements around the center of the Smith Chart – using the Line – and close to the edge of the Smith Chart – using the Offset Short; this latter component also gives us a measure of the phase accuracy of the calibration.

Similar verification components can be constructed on-wafer or in-fixture. An interesting approach to the design of on-wafer or ISS verification standards is to treat them as two-port standards. An *Offset Load* standard can be constructed by placing 50 Ω resistors between signal and ground conductors of a CPW line: this produces a 25 Ω resistance in parallel with the 50 Ω VNA ports, yielding a 17 Ω resistance seen in port reflection and transmission measurements. This gives a value of 6 dB for each of S_{11}, S_{12}, S_{21}, and S_{22}. This standard is easy to construct and is a straightforward measure of the magnitude accuracy. The on-wafer Offset Short can also be made using similar artwork, with metal instead of resistor, to verify the phase accuracy of the calibration, knowing the round-trip delay from the length of the offset. In a microstrip environment, the Offset Load and Short can introduce some inductance, associated with the through-vias to the back side ground metal. This will produce a frequency-dependent Load 'verification' standard, although at microwave frequencies up to about 30 GHz this is not serious. An Offset Open is a better choice for the phase verification standard.

For measurement of high power transistors, where the calibrated system impedance may be much smaller than 50 Ω, the offset load can be made smaller, to verify the calibration accuracy at higher reflection coefficient loads.

3.2.5 E-Cal

Electronic calibration, or 'E-Cal', is a calibration procedure that can be performed with some modern VNAs. The network analyzer controls a switch-box that connects the various calibration standards to the VNA ports. Only one physical connection to the VNA's ports is made, so connect-disconnect errors are minimized. The switch-box contains low-loss switches and high-quality reflection (one-port) and transmission standards. Usually, three or four reflection standards are provided, covering Open, Short, and Load standards, and two Line standards of different lengths. *E-Cal* is now available for frequencies up to 67 GHz, and the reduction in connection-disconnection events makes it attractive for the calibration of multi-port VNAs [24].

3.2.6 Non-50 Ω Calibration

Virtually all of the published VNA calibration techniques have been concerned with measurement in a 50 Ω system. Some calibration methods, such as TRL, can be used successfully in non-50 Ω environments, as the Line standard (or Match in LRM) defines the characteristic impedance of the measurement. At microwave and millimeter-wave frequencies, power transistors are generally sufficiently small that a 50 Ω environment makes practical sense, but the power transistors used for RF wireless infrastructure PAs generally exhibit very high reflection coefficients, and therefore measurement in a low characteristic impedance environment is necessary to reduce measurement errors due to reflections and mismatches. Non-50 Ω calibration requires the use of transformers or pre-match tuners to convert the low device-plane impedance to the regular 50 Ω environment of the test equipment. This in itself does not pose any problems for the calibration process, though care is required in the definition of the reference planes and impedances, and the error coefficients at these planes must be converted to the reference rather than the system (50 Ω) impedance if the measured data is to be interpreted correctly. The use of fixtures for accurate measurements in a non-50 Ω environment is discussed in detail later.

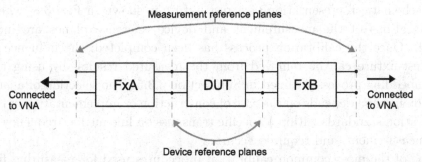

Fig. 3.8. A schematic representation of a device within a test-fixture: FxA and FxB represent the left and right halves of the test-fixture located on either side of the device-under-test.

3.3 Measurement Environment

The development and validation of several different types of fixtures are required to perform the diverse set of measurements used for model extraction. The needs vary from fixtures that require broad frequency characteristics for S-parameter measurements that are used during the model extraction process, to fixtures that require a substantial amount of impedance transformation from 50 Ω, and are capable of dissipating high levels of DC and RF power during loadpull measurements used for model validation. In this section, we cover the important features of the fixture design for a number of measurement interfaces, from the perspectives of model extraction and validation.

3.3.1 Introduction to Fixturing Concepts

At microwave frequencies, S-parameters are regularly employed to describe the characteristics of a circuit. The majority of circuits in planar media are, unfortunately, impossible to measure directly, since the coaxial connectors of the vector network analyzer cannot be connected directly to the circuit. Instead, the measurements are made at a plane physically distant from the device, and the circuit is said to be embedded in a network or test-fixture [25]. The purpose of the test-fixture is to modify the electromagnetic fields propagating within the coaxial cables to the field configuration within the microstrip line, with low insertion loss [26]. In practice, the finite losses and the discontinuity introduced by the test-fixture will also modify the measurements of the device-under-test, and these effects must be removed.

A schematic representation of a test-fixture is shown in Fig. 3.8, where the locations of the measurement and device reference planes are indicated. Once the calibration process has been completed, the influence of the test-fixture can be removed from the measured results, by using the de-embedding process outlined in Sub-section 3.3.5. The selection of a calibration method depends on the ease of construction or implementation of the calibration standards within a specific transmission line media, frequency of the measurement, and required accuracy.

One of the most common calibration procedures used for measuring fixtured devices involves the use of a two-tier calibration scheme. Generally, the first tier of the calibration is performed with a commercially available calibration kit, at the ends of the coaxial cables. After the first-tier calibration has been completed and verified, the second-tier calibration is carried out. The purpose of the second tier of calibration is to shift the calibrated reference plane from the coaxial cable ends to another location that is closer to the device-under-test; this reference plane is typically in a microstrip environment.

During the second-tier calibration, the S-parameters of the different calibration standards are measured and collected, and then the S-parameters of the input and output fixture halves can be computed easily. The end result is a set of portable fixture S-parameters. The key advantage in having a set of portable fixture S-parameters is that subsequent measurements can be made using only the first-tier calibration and appropriate de-embedding of the test fixture S-parameters to arrive at the S-parameters of the device-under-test.

3.3.2 Probe-Based Fixturing

Measurements performed directly on-wafer using coplanar probes are widely used to characterize active and passive circuits. Many of the measurement techniques discussed in upcoming sections employ probe-stations and coplanar probes. The stations are well suited for software automation allowing many measurements to be taken quickly. By probing the wafer directly, one can avoid issues relating to manufacturing or packaging individual dies for testing. Details of the calibration methodology used with coplanar probes for on-wafer devices were covered in detail earlier.

On-wafer measurements are widely used during the model extraction process, for example, the measurement of frequency- and bias-dependent, continuous or pulsed S-parameters, capacitance measurements with C–V meters, and continuous or pulsed DC measurements. The combination of high accuracy, excellent repeatability, very broadband frequency response, and environmental control (of the probe-station) result in on-wafer probing being an essential measurement technique. If a large number of on-wafer measurements are required, for example, when generating a large database of models and data for a statistically-based model, on-wafer probing is the ideal technique, since it can be fully automated with a computer-controlled probe-station.

The basic components of a probe-station include a large round metal plate, the chuck, which is connected electrically to ground. Wafers are positioned on top of the chuck for mechanical support. The temperature of the chuck can be adjusted using a temperature control system. Coplanar measurement probes are mounted on micro-manipulators that are used to control precisely the position and height of the probes, thereby enabling extremely repeatable measurements. A high-magnification optical microscope is mounted on top of the station; this allows us to view the circuits to be measured, and to ensure that the probes are in good contact with the circuit. The entire assembly is usually placed on top of a vibration-dampening table to reduce any external vibrations that may damage the probes or the wafer-under-test. Cables attached to the probes can be connected to a VNA, loadpull tuner or other measurement test-set. Additional features of a probe-station to facilitate measurements include vacuum lines for holding the device-under-test in place, an enclosed environmental chamber enabling a flow of clean dry air to avoid frost accumulation for extended low temperature measurements, and the availability of an inert gas to surround the coplanar probes to avoid potential metal oxidation. A photograph of a probe-station is provided in Fig. 3.9(b).

The coplanar probes consist of a coaxial connector (for example a 3.5 mm or SMA) mounted onto a low-loss coaxial cable contained within the plastic housing: the probe-body. The probe-body is then bolted to the micro-manipulators. Internal to the body, a low-loss coaxial cable connects the coaxial connector to the probe tips as illustrated in Fig. 3.9(a). Another set of tips is connected to the exterior of the coaxial cable protruding through the body. These tips are used to provide ground continuity between the measurement instrumentation and the circuit-under-test. Designations of the style of probe follow the number and arrangement of tips connected to a probe. For example, a coplanar probe with two ground tips on either side of the signal tip is called a ground-signal-ground (GSG) probe. In addition to the type of probe, it is important to specify the pitch of the probe, which is defined as the distance between the probe tips. The pitch determines the size of the on-wafer pad that can be probed. Great care must be taken

(a) Examples of GSG probes (courtesy of Cascade Microtech Inc. and GGB Industries Inc.)

(b) Probe-station (courtesy of Cascade Microtech Inc.)

Fig. 3.9. Photographs of the equipment used to perform probe-based measurements.

during the use of the probes because they are extremely easy to damage during the measurement. For detailed handling and operating instructions of GSG probes the reader is referred to [27, 28].

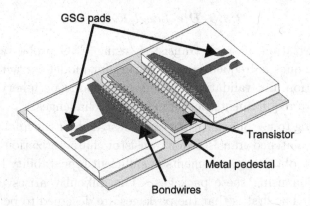

Fig. 3.10. A package adapter for use with GSG probes. In this illustration a transistor is bonded to the metal-carrier and bondwires are attached exactly as implemented within the overall design.

In addition to measurements that are performed directly on-wafer, there are also devices that are designed to be wire-bonded into a circuit. Consequently, they do not have the access pads for probing, or required calibration standards. For these devices it is desirable to perform the measurements with the same bondwire configuration as implemented within the circuit, so that the bondwires become part of the circuit being characterized. A package adapter designed for this purpose is illustrated in Fig. 3.10. The adapter consists of GSG pads connected to microstrip transmission line on an alumina substrate. The adapter substrates are bonded to a metal carrier, with the device to be characterized mounted in between the carriers. Bondwires are then used to connect to the device. A calibration substrate is then used to establish reference planes at the ends of the microstrip transmission lines on the adapter substrates.

The characterization of transistors using GSG probes and on-wafer probing techniques is arguably the best method for transistor characterization due to the excellent measurement repeatability, the wide commercial availability of probes, and the accuracy of on-wafer calibration. However, high-power transistors often require a packaged approach since the physical width of the die to be measured can be 25–50 times the probe pitch. The effect on transistor performance from current spreading, low-impedances and high-power levels complicates the measurement. Additionally, the output power for a high-periphery transistor can exceed the maximum current limits and damage a probe. Therefore, a package-based measurement approach in addition to probe-based measurement is often required.

3.3.3 Die-Level Fixtures

The characterization of small transistors using GSG probes and on-wafer probing techniques is frequently used during for model extraction. For the model extraction and validation of moderately large periphery transistors, the test-fixture requirements pose a new set of challenges.

It is tempting to use one of the packages designed for the final packaging of the internally matched transistor for transistor characterization. In practice, it is difficult to obtain measurements of sufficient repeatability for transistor characterization using these packages. The difficulty arises primarily for three reasons. The first is that the packages are designed to be used in low-cost, high-volume power amplifier manufacturing lines, where the package will be soldered onto a circuit board. Because of this, there is a variability in the package dimensions due to manufacturing tolerances, which means that the package placement within the test-fixture is not consistent. And since the test-fixture must be re-used, soldering each packaged device into the test-fixture is not practical. The second difficulty is that the package leads are very wide, significantly wider than the 50 Ω transmission line of the test-fixture. The third difficulty arises from the thickness of the package flange. In many instances, the thickness of the flange requires a recessed area in the heatsink for the leads of the package to connect to the test-fixture. This abrupt change in current flow creates an electrical discontinuity in the return current path which affects the circuit performance [29].

A test-fixture commonly used to measure large die-level devices is illustrated in Fig. 3.11 [30]. Central to the method is a specially designed 50 Ω microstrip carrier assembly that allows measurement reference planes to be rigorously established. The metal carrier serves as the ground plane and as a mechanical support. The uniform ground-plane avoids any of the aforementioned issues related to discontinuities in the ground-plane [31]. Bonded to this conductor is an alumina substrate that contains the 50 Ω microstrip transmission line. Bond-pads at the ends of the microstrip allow bondwires to connect the transistor to the fixture. To accommodate different sizes of transistors, carrier assemblies of various dimensions are constructed by changing the length of the metal carrier and width of the bond-pads on the alumina substrate.

The assemblies illustrated in Fig. 3.12 are designed such that they can be inserted into a test-fixture for measurement. The metal carrier is bolted to a mechanical support and coaxial-to-microstrip launchers connect the cables of vector network analyzer to the alumina microstrip lines as illustrated in Fig. 3.11. Calibration of the test fixture is accomplished through a two-tiered

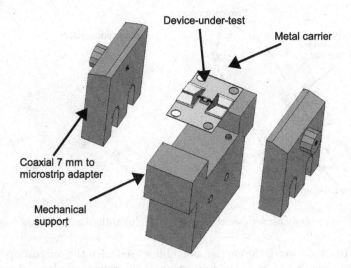

Fig. 3.11. An exploded view illustrating the assembly of the microstrip test-fixture. The assembly is complete when the carrier assembly and the adapters are bolted to the mechanical support [32]. © 2005 IEEE. Reprinted with permission.

TRL calibration [33]. The measurement reference planes are established at the ends of the narrow microstrip, just prior to the bond-pad, as indicated in Fig. 3.12.

The mechanical support shown in Fig. 3.11 is designed to accommodate a thermocouple mounted directly beneath the metal carrier, to establish a thermal reference plane. The temperature can be controlled through convection cooling or through more elaborate liquid cooling techniques, where a liquid is passed through the metal support. With a feedback loop between the thermocouple and the refrigeration unit, the temperature can be kept constant throughout a measurement.

In addition to transistor characterization, this fixture is very well suited for the measurement of passive components (for example bondwires, MOS capacitors). Furthermore, since many packages employ a plastic encapsulant as a means of protecting the circuit, a molding system placed around the carrier assembly allows the performance variation with the plastic to be examined. An illustration of a carrier assembly with an encapsulating material is provided in Fig. 3.12(b). For transistors of up to a few millimeters of gate periphery, these fixtures work very well. A different fixturing approach is required for measuring full-sized, high-power packaged transistors; this is outlined below.

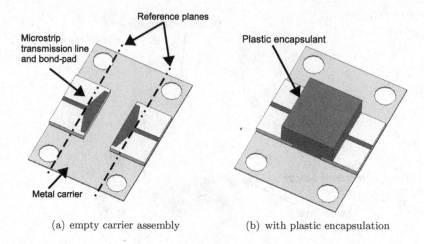

(a) empty carrier assembly   (b) with plastic encapsulation

Fig. 3.12. Illustrations of the carrier assemblies used with the microstrip test-fixture illustrated in Fig. 3.11. Note the indication of the reference plane of each package after calibration. A molding system (not shown) allows plastic to be injected and when cured forms a cube located between the reference planes as shown in (b) [32]. © 2005 IEEE. Reprinted with permission.

3.3.4 Package-Level Fixtures

We have presented two different kinds of fixtures that are used during the transistor model extraction and validation. These two fixtures are geared towards the measurement of smaller transistors or building blocks, rather than the full size die found in commercially-available high power packaged transistors, such as in [34]. Here we describe the design and use of a fixture that can accommodate a packaged transistor; that is, a complete transistor product. A good package-level fixture design must have the necessary characteristics to accommodate all the measurements that are needed to validate the nonlinear model of the packaged transistor.

Typically, three types of measurement are required to validate the packaged transistor model; these are described in Chapter 9. The first and most fundamental validation exercise is the comparison of measured and simulated small-signal S-parameters as a function of frequency and bias. The second category consists of large-signal measurements, typically loadpull measurements or fixed-impedance power bench measurements. In this category of measurements, the fixture design must accommodate high levels of dissipated and delivered output powers, as well as providing very good low-frequency decoupling, to prevent oscillations and to optimize the performance of the transistor by reducing low frequency memory effects. The third set of validation measurements are those in which the thermal environment must be carefully monitored. In this last category, the fixture design must include thermocouples or the ability to force a set of thermal conditions.

It is difficult to obtain accurate measured S-parameters of packaged power transistors because their input and output impedances are very low. Even after pre-matching the die within the package, the input and output impedances are often around 1 to 2 Ω. It is well known that when the impedance of a circuit is significantly different from that of the system impedance, under high reflection coefficient conditions, the measurement accuracy of the VNA is degraded. Additionally, the loadpull systems used for handling high power levels of base-station transistors are primarily based on mechanical tuners. Such high-power tuners have a limited impedance range, and obtaining accurate measurements at the edge of the Smith chart under high-power conditions is a challenging exercise [35].

One solution for improving the measurement accuracy for both small-signal S-parameter measurements and large-signal loadpull measurements is to reduce the mismatch between the system impedance reference plane and the packaged transistor by incorporating a pre-matching circuit into the test-fixture, to transform the system impedance to a lower value that is closer to the packaged transistor input and output impedances. For example, a 10-to-1 transformation ratio will produce a 5 Ω system impedance at the package-to-microstrip interface, although this comes at the expense of a more complicated fixture design and calibration procedure. The benefit of placing a pre-matching fixture in between the mechanical tuners and the packaged device is the expansion of the impedance tuning range. Incorporating a lower characteristic impedance at the package reference plane is not limited to passive loadpull systems [36]: for active loadpull systems, the incorporation of a pre-matching fixture means a significant reduction in the amount of available power needed to characterize the transistor [37].

Often, a single-section quarter-wave transformer is used to obtain the frequency transformation. One disadvantage of this otherwise simple approach is that the transformer is of narrow bandwidth. Further, because the transformer is effectively transparent at the even multiples of the fundamental frequency ($2f_0$, $4f_0$, and so forth), there is a large mismatch around these frequencies, resulting in the measurement errors outlined above, and preventing accurate control of the impedances for harmonic tuning, or de-embedding of the harmonic impedances.

One of the many different designs of package-level fixtures is illustrated in Fig. 3.13. In this particular design, the distributed planar matching networks are constructed within the test-fixture. The entire test-fixture is constructed of a single printed circuit board (PCB) soldered to a metal carrier, which has

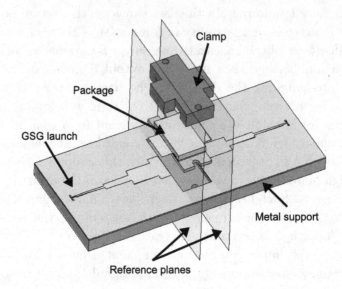

Fig. 3.13. An exploded view illustrating the assembly of the test-fixture used to obtain S-parameter measurements of large packaged transistors [38]. © 2006 IEEE. Reprinted with permission.

a recessed area to accommodate the package flange as indicated. The depth of the recessed area is such that the leads of the package rest on top of the test-fixture when the package flange is securely fastened to the metal-carrier. A plastic block or clamp is used to exert pressure on the transistor package, to establish a conductive connection between the package flange and the metal carrier while simultaneously forcing contact between the package leads and the transmission lines of the test-fixture. Measurement repeatability is enhanced by using GSG probes and by manufacturing the entire calibration set of standards from a single PCB. An alternative to this fixture design uses a pair of microstrip-to-coaxial transitions instead of the GSG probes. This alternative approach is used for bench or loadpull measurements in which the mechanical tuners and couplers are components equipped with coaxial connectors. This fixture design is suitable for the measurement of small-signal S-parameters over a broad frequency range (provided by the bandwidth of the transformers) and large-signal measurements with a coaxial connector or probe-based loadpull system.

The fixture illustrated in Fig. 3.13 uses Chebyshev transformers designed for a specific bandwidth and transformation ratio. The test-fixture is designed to accommodate the transistor package geometry and to minimize any electrical discontinuities between the packaged device and the test fixture.

This discontinuity is minimized by selecting the width of the last section of the transformer to match the lead width of the packaged transistor. Thus, the transformation ratio is dictated by the width of the package lead and the substrate characteristics.

The calibration of a fixture containing a transformer uses non-50 Ω transmission line standards. The Through standard consists of the two sides of the test-fixture placed back-to-back, and the Line standard consists of two transformers separated by a length of transmission line. The Reflect standard is composed of transformers that are terminated in either open or short circuits. A two-tiered calibration approach is applied, where the characteristic impedance of the inner shell is based on the characteristic impedance of the lines used during the TRL calibration [14, 36]. Although standard TRL calibration techniques are used, a key step in the calibration is determining the characteristic impedance of the line standard as this is the impedance to which the S-parameters of the packaged devices are referenced after de-embedding. One method for obtaining the characteristic impedance of the calibration transmission lines is through the determination of the propagation constant from measured data and a measurement of the low-frequency capacitance [39, 40], and another technique employs time-domain reflectometry to good effect [41, 42].

We have highlighted several important aspects of fixture design for the electrical measurements often encountered during the model extraction and validation of packaged transistors. For completeness we point out some important design aspects that need to be considered if the fixture is to be used in thermal measurements or thermal monitoring conditions.

During the validation of an electro-thermal model the temperature of the heatsink directly beneath the package flange must be known, and able to be held constant over all of the impedance states presented to the transistor by the tuners. Typically, the cooling of high-power transistors is performed using fixtures that rely on fans circulating air around the fixture. For higher-power devices, convection cooling is not sufficient for temperature regulation and liquid cooling integrated into the fixture is often used. A photograph of a liquid cooled test-fixture is provided in Fig. 3.14, where water flows from a chilling unit through a serpentine trench that is milled into the heatsink, as illustrated in Fig. 3.15(a). This ensures immediate cooling or heating of the device. The temperature at the bottom of the package flange is monitored by a thermocouple inserted from the back side of the heatsink. The thermocouple provides temperature feedback to the refrigeration unit to maintain the desired temperature at the package flange. A close-up view of the pre-matching network and biasing circuit is shown in Fig. 3.15(b).

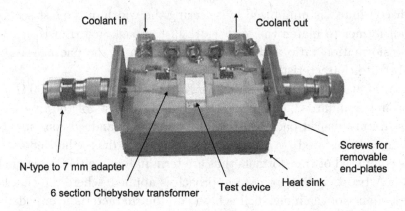

Coolant in | ↑ Coolant out

Screws for
removable
end-plates

N-type to 7 mm adapter

6 section Chebyshev transformer   Test device   Heat sink

Fig. 3.14. A photograph of a liquid cooled loadpull test-fixture [42]. © 2000 IEEE. Reprinted with permission.

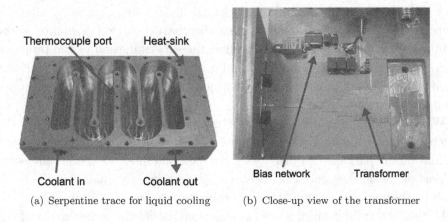

Thermocouple port   Heat-sink

Coolant in   Coolant out

Bias network   Transformer

(a) Serpentine trace for liquid cooling   (b) Close-up view of the transformer

Fig. 3.15. Photographs detailing the matching network and biasing circuit of the test-fixture shown in Fig. 3.14. In (a), the bottom of the fixture is shown with the cover removed to show the serpentine channel for liquid cooling and the location of the thermocouple port [42]. © 2000 IEEE. Reprinted with permission.

3.3.5 De-embedding and Segmentation Techniques

We have taken some care in the establishment of the reference planes so that we are able to make accurate measurements, or simulations using a circuit or electromagnetic field solver, of a component defined at these planes. We have seen that a typical high-power RF transistor contains many components. For example, the RF power transistor shown in Fig. 1.1 comprises the package itself, and then for each of the three LDMOS dies, there are two bondwire arrays and a MOS capacitor that form the input matching network, the transistor die, and another two bondwire arrays and a MOS capacitor for the output matching network. The structure as a whole is too complex for us to simulate all at once, and this complexity places too great a computational burden on the simulator or modeling tool to make this an efficient exercise. Instead, we can manage the computational expense by systematically *segmenting* the transistor into its constituent parts, and modeling each component individually, using the most appropriate tool for the job: using a full-field three-dimensional EM simulator for the bondwire arrays, a planar EM solver for the capacitors, and a large-signal compact model for the transistor. The complete product model is then created by cascading these individual model elements together to obtain the overall system response.

The key to the success of this approach is that first, the individual components must somehow be separated from each other; and second, the component when analyzed by itself must behave identically to when it is embedded within the complete device. A critical issue is that the planes at which the circuit is divided must be carefully selected such that the field configurations on either side of the plane are matched up, otherwise, an artificial discontinuity will be added to the analysis results. Thus, it is important that the reference planes, within the measurement system or the simulator, are carefully defined.

The outcomes from the EM simulations of the passive components will generally be in the form of S-parameter files or equivalent-circuit models. After we perform the measurements of the whole transistor, for the purpose of extracting the large-signal model, these component blocks will need to be *de-embedded* from the measured data, to arrive at the transistor die reference planes. The de-embedding can be performed mathematically, as explained in the example below, or carried out in the circuit simulator using de-embedding S-parameter blocks.

A schematic representation of a device located within a test-fixture is illustrated in Fig. 3.8; essentially, a number of two-port devices are cascaded together. The two-port labeled FxA represents the left-half of the test-fixture, FxB represents the right-half, and the DUT represents the transistor whose S-parameters are to be measured.

Mathematically, this experimental setup can be viewed as a set of cascaded $ABCD$ matrices, represented by eq. 3.4. To de-embed the test-fixture from the measured results, the $ABCD$ matrix of the measurements, M, is pre- and post-multiplied by the inverse $ABCD$ matrices of FxA and FxB as indicated in eq. 3.5

$$[\mathbf{M}] = [\mathbf{FxA}][\mathbf{DUT}][\mathbf{FxB}] \qquad (3.4)$$

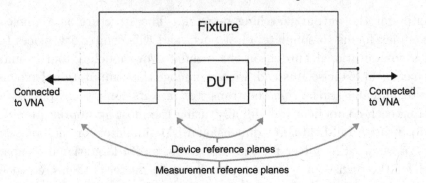

Fig. 3.16. A schematic representation of a device within a test-fixture when the ports of the test-fixture are coupled together.

$$[\mathbf{DUT}] = [\mathbf{FxA}]^{-1}[\mathbf{M}][\mathbf{FxB}]^{-1} \qquad (3.5)$$

A further practical issue is that any inter-element coupling, such as the mutual inductance between arrays of bondwires, must also be characterized for the transistor to be properly de-embedded. The de-embedding formulation presented above assumes that the two halves of the test-fixture are not electromagnetically coupled together. If the test-fixture has been designed such that the two halves are coupled, the de-embedding algorithm is significantly more complex. This situation is rarely encountered when measuring a device within a microstrip test-fixture, as the large distances between the ports of the test-fixture prevent significant coupling from occurring. Often, a semiconductor die must be placed within a package for measurement, but it is the S-parameters of the die only that are wanted. Coupling occurs between the bondwires used to connect the device to the package.

Once the S-parameters of the package and wires are known, either from simulation or measurement, they can be de-embedded from the S-parameters of the entire device. Alternately, if the S-parameters of the package and the die are known, they can be embedded to yield the S-parameters of the entire device. The equations for embedding and de-embedding a circuit, when significant inter-port coupling is presented, are derived below. To be consistent with the terminology used earlier, the circuit exhibiting inter-port coupling is referred to as the fixture. A schematic representation of a circuit contained within a test-fixture exhibiting significant inter-port coupling is illustrated in Fig. 3.16. This derivation is an expanded version of the one presented by Dobrowolski [43].

To embed the S-parameters of the fixture, **T**, around those of the circuit, **S**, and compute the S-parameters of the entire network, $\mathbf{S_{net}}$, the fixture S-parameters must first be partitioned into sub-matrices. The sub-matrices are grouped according to the location of the ports, and they are labeled as internal or external. The matrix representing the entire system can be written as

$$\begin{bmatrix} b_e \\ b_i \end{bmatrix} = \begin{bmatrix} T_{ee} & T_{ei} \\ T_{ie} & T_{ii} \end{bmatrix} \begin{bmatrix} a_e \\ a_i \end{bmatrix} \quad (3.6)$$

where a_e, b_e and a_i, b_i are vectors representing the outgoing and incoming waves for the network T. The internal network, labeled as the DUT in Fig. 3.16, also describes a system of equations written in matrix form as

$$b_s = Sa_s \quad (3.7)$$

Since the two networks, S and T, are connected together, the following constraints are introduced:

$$a_i = b_s \quad (3.8)$$

$$b_i = a_s \quad (3.9)$$

The waves leaving T are the waves entering S. The S-parameters of the entire network can be computed from eqs. 3.6, 3.7, 3.8, and 3.9 by first eliminating b_i. This results in

$$a_i = (S^{-1} - T_{ii})^{-1} T_{ie} a_e \quad (3.10)$$

and then by eliminating a_i in eq. 3.6 we get

$$b_e = [T_{ee} + T_{ei}(S^{-1} - T_{ii}^{-1})T_{ie}]a_e \quad (3.11)$$

Recognizing that the ratio $b_e/a_e = S_{net}$, the solution for the S-parameters of the entire network results:

$$S_{net} = T_{ee} + T_{ei}(S^{-1} - T_{ii}^{-1})T_{ie} \quad (3.12)$$

The package effects can be de-embedded from the entire network by re-arranging eq. 3.12 to solve for S, which yields

$$S = [T_{ii} + [T_{ei}^{-1}(S_{net} - T_{ee})T_{ie}^{-1}]^{-1}]^{-1} \quad (3.13)$$

Even after this exercise, we may need to perform further segmentation and de-embedding at the transistor die level. A typical high-power RF transistor die has a large gate periphery, and the signals to the gate and drain electrodes of the transistor are fed using large manifold structures as shown in Fig. 1.1. Again, these structures may be simulated using a planar EM simulator to obtain the S-parameters of each structure, and the manifold can then be de-embedded so that we can focus on modeling the transistor itself.

3.4 Measurements for Model Extraction

In this section, we introduce several measurement techniques used to collect the necessary data for the extraction of transistor models. These measurements can be categorized into electrical and thermal measurements. In the electrical measurement category, we begin with a description of the most basic form of transistor characterization: DC measurement of the transistor current–voltage relationships under both continuous and pulsed excitations. To complement the DC characteristics of the transistor, the frequency response is typically captured through the measurement of small-signal S-parameters, also under continuous and pulsed RF excitation. The S-parameters are generally measured as functions of the applied DC bias. Techniques for performing thermal measurements are explained in detail in Chapter 5.

3.4.1 DC Measurements

The most basic form of transistor characterization is the measurement of the voltage–current relationships. For Si LDMOS, GaAs MESFET, and GaAs PHEMTs this means the measurement of the drain terminal voltage–current relationship as a function of the applied gate voltage. For GaAs FETs the gate terminal voltage–current relationship must also be characterized. An accurate measurement of the I_d–V_{ds} and I_g–V_{gs} curves is extremely important since the main nonlinearities of a FET are contained in its voltage–current relationships, and therefore the measurements are an important foundation for the large-signal model development of the transistor.

Continuous DC measurements are very practical and are widely used to characterize small periphery transistors (that is, gate peripheries resulting in drain currents of around 1 A or less) owing to the small amount of current and hence power required to capture the voltage and current characteristics. Continuous DC measurements usually require simple measurement setups, provide fast measurement times and can typically be made with a very low noise measurement floor. The measurement of large periphery transistors (drain current values in excess of 1 A) is quite difficult under constant DC excitation because of the large power required and significant amount of self-heating that occurs, which poses a risk of possible destruction or damage to the device-under-test.

For self-heating effects to be described properly by the nonlinear transistor model, we need to account for the dynamic temperature dependences of the model parameters that are caused by the electrical signal. During the extraction procedure, isothermal model parameters can be extracted at

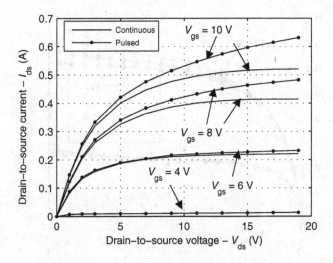

Fig. 3.17. An example of I_{ds}–V_{ds} curves for an LDMOS transistor. The differences between constant and pulsed DC measurement are visible.

various temperatures, provided that the electrical and thermal dependences of the model parameters are de-coupled.

One method to determine isothermal model parameters is to measure the devices under pulsed conditions. Evidence of self-heating during the characterization under continuous DC excitation is evident by comparing the I_d–V_{ds} curves under continuous and pulsed DC excitations, as plotted in Fig. 3.17. Pulsing the DC voltage with narrow pulse widths and a low duty cycle, while measuring the voltage–current characteristics of the transistor, overcomes the heating that is encountered when measuring large periphery transistors [44]. Although it is possible to de-embed the thermal effects of the non-isothermal data gathered during the measurement of a transistor under continuous DC conditions, the measurement of the device under pulsed conditions is straightforward, and avoids potential errors that can be introduced during the de-embedding process, such as errors made in the determination of the thermal resistance.

There are several good reasons for performing pulsed DC characterization of transistors. For silicon LDMOS transistors, the first advantage is that the thermal and electrical behaviours of the transistor can be decoupled, enabling an accurate measurement of the current–voltage characteristics. A second advantage is that the safe measurement range of the transistor can be extended significantly compared with steady DC measurements. This allows us to access regions of the I_d–V_{ds} characteristics corresponding to

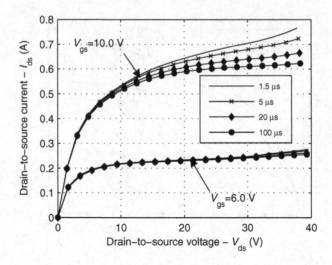

Fig. 3.18. An example of constant and pulsed DC drain current for an LDMOS transistor. I_d is plotted against V_{ds} for pulse widths of 1.5, 5, 20, and 100 μs at V_{gs} of 6.0 and 10.0 V

high dissipated powers, typical of power amplifier operation, that would destroy or permanently damage the transistor if applied continuously.

The characterization of GaAs and GaN FETs presents an additional challenge, since these transistors may suffer from low frequency dispersion of the *I–V* characteristics; that is, a change to the output conductance or transconductance as a function of frequency. The III–V FETs have 'slow states': these are thought to be attributable to surface or interface traps, or unwanted impurities in the substrate or buffer layer, which can only capture or release charge at a relatively slow rate compared with the RF operating frequency of the device, but can respond to the slow rate of change of bias under DC characterization. Consequently, the device *I–V* characteristics are very different at DC and RF. Indeed, when coupled with the thermal effects, the transistor's low frequency dynamics can be very complex. The idea of using pulsed DC measurements to capture the dominant drain current nonlinearity in GaAs MESFETs and HEMTs was originally proposed by Platzker *et al.* [6], and has been pursued more recently by Parker and Rathmell [45]. Accurate measurement of the effective high-frequency drain and gate currents can be carried out using pulsed DC measurements, and used to provide an accurate model of the main nonlinearity of the transistor. The low-frequency dynamics are also of interest for power amplifier applications, since they will affect the video bandwidth of the amplifier. Gallium arsenide

FETs and HEMTs have been shown to exhibit very different low-frequency behaviour when characterized using pulsed DC measurements from different quiescent bias points [46], and this needs to be accounted for in a successful model.

The first step in performing a pulsed DC characterization is to determine the minimum pulse width, sampling location and duty cycle that will guarantee an isothermal environment. The sensitivity of the drain current to pulse width for an LDMOS transistor is plotted in Fig. 3.18, where pulsed DC drain current is plotted against drain-to-source voltage for 1.5, 5, 20, and 100 µs pulse widths and gate-to-source voltages of 6.0 V and 10.0 V. Very little change occurs for pulse widths less than 1.5 µs. For each pulse width the pulse repetition period was modified to maintain a constant duty cycle of 0.1% [44].

Not surprisingly, significant differences also occur in constant versus pulsed DC values of G_{ds} and G_m. A clear comparison of the constant and pulsed DC output conductance G_{ds} against V_{ds} for V_{gs} varying from 4.0 V to 10.0 V is shown in Fig. 3.19. The constant and pulsed DC transconductance G_m against V_{gs} for V_{ds} of 5.0 V and 20.0 V is shown in Fig. 3.20. Notice how the constant and pulsed DC values of I_d, G_{ds}, and G_m diverge from each other as the power dissipation (that is, the product of V_{ds} and I_d) increases [44]. By examining the differences of the output conductance and transconductance between constant DC and pulsed DC, one cannot immediately conclude that all the observed differences are induced by thermal effects, since low frequency dispersion or trapping could also be affecting the measured results. The amount of low frequency dispersion effects in silicon LDMOS devices are significantly smaller compared to GaAs devices [47,48]. Therefore the vast majority of the differences observed in Fig. 3.20 can be attributed to thermal effects.

3.4.2 Continuous DC Measurement System

The most rudimentary instrument at the disposal of the modeling engineer to perform DC measurements of a transistor is the DC curve tracer. The early curve tracer models did not have the necessary functionality to record the I–V data, so its usefulness as a data collection instrument was somewhat limited. Its main advantage is its versatility and ease of use to measure and characterize transistors.

Nowadays, the DC parametric analyzer is the most common instrument used for measuring the DC characteristics of transistors. Most modern DC parametric analyzers perform a periodical auto-calibration automatically to

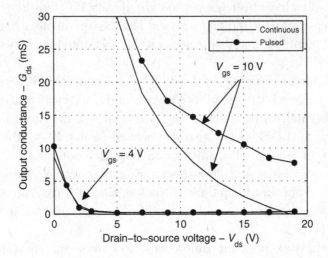

Fig. 3.19. An example of constant and pulsed DC output conductance of an LDMOS transistor. G_{ds} versus V_{ds} is plotted for V_{gs} of 4.0 V and 10.0 V.

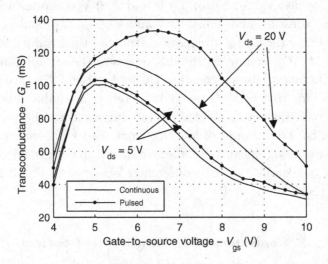

Fig. 3.20. An example of constant and pulsed DC transconductance of an LDMOS transistor. G_m versus V_{gs} is plotted for V_{ds} voltages of 5.0 V and 20.0 V.

maintain measurement accuracy. This is accomplished by the DC parametric analyzer disconnecting its outputs from the DUT, measuring possible offset voltages and currents, and then comparing these with a very stable voltage or current reference, and making the corrections as appropriate. This type of auto-calibration does not require the interaction of the user. The

DC parametric analyzer's current and voltage capabilities are well suited for the characterization of small transistors (owing to its current handling capabilities) and to situations which require the accurate measurement of very low current levels, in the order of nano- to femto-amperes.

The DC power required to characterize large gate periphery transistors is often beyond the capability of the DC parametric analyzer. Under those circumstances a DC measurement system can be assembled with the appropriate power supplies, and voltage and current meters. For very high current values it is sometimes necessary to replace the current meter by a very low resistance element, which has been very accurately characterized so that the current can be computed by measuring the voltage drop across it. In addition, oscillation suppression components, which are frequently needed to overcome stability issues during the transistor characterization, must be included in the gate and drain bias lines or bias networks of the DC measurement system.

High-precision DC measurement systems utilize Kelvin measurements to avoid inaccuracies associated with the series contact resistances of the measurement leads or electrodes. This measurement procedure, also known as the four-wire method, consists of two separate lines, *viz*, a *force* line and another connected in parallel, the *sense* line. The force line carries the current from the meter, while the sense lines allow the meter to measure the voltage across the DUT. With this type of configuration, any voltage drop due to the resistance of the force lines and their contact resistances is ignored by the meter. To improve the accuracy of the Kelvin measurement method, the point where the force and sense lines are tied together must be as close as possible to the DUT.

3.4.3 Pulsed DC Measurement System

Typical pulsed DC measurement systems are capable of generating short voltage pulse widths (around 200 ns to 1 µs minimum) with low duty cycles and high voltages and currents. There are several commercially available systems [49–51], and a number of systems have been developed but not commercialized by corporations and academia [4–8].

To measure high voltage LDMOS devices, the main purpose of the DC pulsed system is to overcome self-heating effects. The minimum pulse width should be in the 1 µs range, with the maximum input (gate) pulsed voltages of around 20 V and maximum output (drain) pulsed voltages in excess of 100 V, while being able to supply at least 10 A of current at duty cycles of around 0.1 to 1 percent.

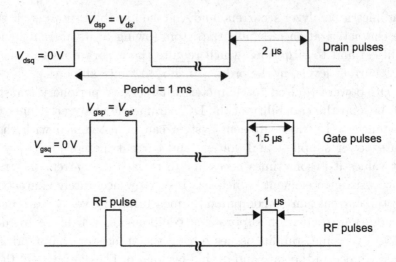

Fig. 3.21. Examples of the pulse timing used during pulsed-DC and pulsed-RF measurements

If the application is the characterization of microwave GaAs FET power devices, the maximum voltage requirements are somewhat less, perhaps around 5 V for the input and 20 V for the output, with maximum current capability of 1 to 2 A. To make sure that all low frequency dispersion phenomena are properly addressed, the minimum pulse width should be significantly smaller, around 200 ns.

A diagram of a typical train of pulses used during the characterization of LDMOS devices is shown in Fig. 3.21. The gate voltage pulses are seen to be contained within the drain voltage pulses, allowing the transistor to be biased following a zero bias current path; that is, the drain voltage is applied first while the gate voltage is still 0 V, producing no drain-to-source current, and thus reducing the possibility of device damage or oscillation issues. The diagram also shows the quiescent bias point from which the pulses are being pulsed; this is 0 V for both the gate and drain. A schematic diagram of the equipment used for a pulsed DC measurement system is shown in Fig. 3.22.

During the characterization of LDMOS devices, where the amount of low frequency dispersion of the output conductance and the transconductance is negligible, the choice of the initial DC quiescent point from which to pulse the input and output voltage is, to a certain extent, arbitrary. Since the purpose of pulsed DC measurements for the characterization of LDMOS transistors is to remove self-heating, a convenient zero-power dissipation point can be selected. For GaAs FETs that exhibit dispersion characteristics, the value of the initial DC bias point has a significant effect on the

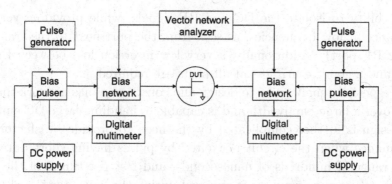

Fig. 3.22. A schematic of the typical equipment used for pulsed DC measurements.

measured (pulsed) drain-to-source and gate-to-source currents, and special care must be taken in its selection [46].

By controlling the surrounding temperature of the device during the pulsed measurements, and the fact that a negligible amount of self-heating is produced, the junction temperature can be controlled, allowing the transistor characteristics to be measured under specific thermal conditions. When performing on-wafer measurements, the temperature of the wafer chuck can be used to set the isothermal environment of the transistor during the pulsed DC measurement process. Similar temperature-setting arrangements can be made when using fixture devices, either by water-cooled fixtures, temperature-controlled chucks, or systems that can force hot or cold air on top of the device.

3.4.4 Small-Signal S-Parameter Measurements

The measurement of small-signal S-parameters is perhaps the most common and useful way to provide a linear description of a circuit and its frequency response. S-parameter measurements are fundamental to the development of linear and nonlinear models and they are often divided into two broad categories by the type of component measured: active or passive. The basic distinction is that active components require the injection of a DC component in addition to the RF signal that is used to determine the frequency-dependent S-parameters.

The combination of a DC and an RF signal is accomplished with the integration of a bias circuit into the RF circuit, yielding a new three-terminal circuit, called a 'bias tee.' This component has one port for the RF input, another for the DC input, and the third port for the DC plus RF output, which is connected to the transistor. The quality of the bias tee is governed by its ability to isolate the DC and RF signals, while providing very low impedance at low frequencies, as seen from the port next to the transistor (DC & RF port). Additionally, a very low insertion loss between the RF ports, and a good return loss at all ports are required.

It is especially challenging to design and construct a bias network that operates over a large bandwidth and is capable of handling large DC currents. The design is further complicated by the need to provide satisfactory RF performance when the circuit is excited by pulses having short durations, in the range of hundreds of nanoseconds, and fast rise-times, of the order of tenths of a nanosecond, as is required during the measurement of pulsed DC and pulsed S-parameter measurements.

As outlined in Section 3.4.1, a major problem that must be overcome during the characterization of high-power devices is the self-heating of the transistor. This is also the case when measuring bias-dependent S-parameters at high-frequencies, to extract isothermal model parameters. Often, S-parameter measurements are performed on-wafer, presenting an additional challenge to obtaining isothermal measurements, since the heatsink is not optimal, in contrast with fixtured and packaged measurements. The temperature of the device-under-test is controlled using a cooling station or temperature-controlled chuck. A very effective and often-used method to overcome the effects of self-heating is to excite the transistor with DC and RF pulses that are substantially smaller than the thermal time constants of the transistor. If the temperature rise is reduced to a negligible amount, then we have achieved an isothermal measurement environment. This allows us to maintain thermal consistency between the pulsed DC and S-parameter measurements.

Performing pulsed S-parameter measurements instead of continuous S-parameter measurements is not without its disadvantages. A major drawback is a drastic reduction of the measurement dynamic range. If wide-band detection is used as the basic architecture of the pulsed S-parameter test-set, the dynamic range is fairly independent of the pulse width used during the measurement, but its value is well below the value that is obtained with continuous S-parameter test sets (around 65 dB compared with around 100 dB). More modern pulsed VNA architectures employ a narrow-band detection scheme, where the dynamic range is proportional to the signal energy received. This means that the dynamic range is reduced as the pulse width is narrowed. The reduction in dynamic range is because the narrow-band filter is attenuating everything but the fundamental tone of the pulsed signal. As the duty cycle decreases, less energy is present in the fundamental tone and hence the reduction in dynamic range [52].

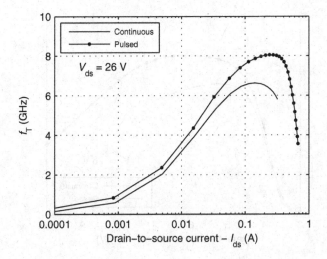

Fig. 3.23. Constant and pulsed DC and RF f_T vs. I_d for V_{ds} of 26.0 V.

Differences between pulsed and continuous S-parameters are evident when analyzing the measured short-circuit current gain unity frequency, or the *transition frequency*, f_T, of the transistor as a function of I_d. Figures 3.23 and 3.24 show f_T versus V_{gs} and I_{ds} of an LDMOS transistor. To calculate the value of f_T, the S-parameters were measured at the desired bias, between 0.5 GHz and 8 GHz, and then converted to H-parameters, where the value of f_T was interpolated between the adjacent frequency points in which the magnitude of H_{21} became less than unity. The drain current was measured at the same time under pulsed and continuous conditions, allowing us to plot f_T versus I_d under continuous and pulsed conditions. Clear differences are present in the peak value of f_T and its roll-off characteristics. Those differences arise from the strong thermal dependence of the transistor's transconductance.

3.5 Measurements for Validation

Several of the most common measurements used during the model validation process are highlighted in this section. Large-signal loadpull measurements are by far the most common method used to validate high power RF transistor models. Their popularity arises from their versatility: loadpull measurements can be performed under many different kinds of signal excitations, power levels, and frequencies in both on-wafer and in-fixture environments. Loadpull systems have been available commercially to the RF and microwave

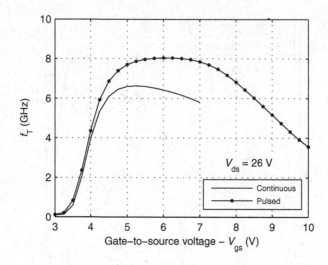

Fig. 3.24. Constant and pulsed DC and RF f_T vs. V_{gs} for V_{ds} of 26.0 V.

engineer for many years. The section will conclude with the description of an instrument that is gaining acceptance in today's marketplace as a great complement to a frequency-domain and impedance-plane-only model valida-tion strategy: the large-signal network analyzer. With this instrument the frontier of time-domain measurements is starting to be explored and a new way to characterize transistors under large-signal conditions is providing additional insights into the accuracy and range of validity of models.

3.5.1 Loadpull Measurements

A key aspect of any transistor modeling and validation activity is the deter-mination of the region of validity of the model, under realistic bias, power level and signal excitation conditions. Within the context of large-signal model validation, the loadpull measurement technique is extremely popular in the RF and microwave engineering community. Its flexibility in char-acterizing transistors of different sizes, at different frequencies and under arbitrary stimuli makes it today's preferred tool for transistor model val-idation. Loadpull test-benches allow control over the specific impedance terminations presented to the device, at the fundamental and the harmonic frequencies, while measuring the performance figures of merit of the tran-sistor. The measurements are plotted as contours of equal performance as a function of the input and output impedance termination super-imposed

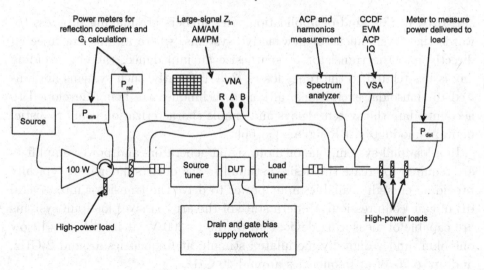

Fig. 3.25. A schematic diagram of a typical loadpull system [53]. © 2006 IEEE. Reprinted with permission.

on a Smith chart. During the design, plotting the various transistor characteristics simultaneously allows a designer to make compromises to meet the required design specifications.

An illustration of a typical passive loadpull measurement system is provided in Fig. 3.25. The impedances presented to the transistor are varied by the mechanical tuners, which contain a slab transmission line loaded by one or more sliding stubs. The position of the stubs and their proximity to the center conductor is controlled by precision stepper motors. Via computer control, a wide range of reflection coefficients can be presented to the transistor. However, the range of impedances that can be realized by the loadpull system is limited by the losses of the system from the mechanical tuner reference plane to the active transistor reference plane. These losses are introduced by connectors, pre-matching fixtures, transistor packaging, and so forth. For very high-power devices, the optimal impedances are often around 1 Ω, and the losses limit the range of impedances that can be presented by the loadpull system. This is a major disadvantage of passive loadpull system with mechanical tuners, compared with active loadpull systems, which can present impedances at the very edge of the Smith chart, and even beyond [54]. Although the impedance-tuning restrictions due to losses limit the tuning range of the passive tuner, the peak power handling capability of these tuners is critical for the measurement of high-power transistors and is the main reason why electrical tuners are not more frequently employed in this kind of application. Researchers are making progress towards the development of new active systems, where signals are injected directly into the transistor to synthesize an impedance, thereby avoiding any issues related to the tuner losses [37]. Presently, such systems are limited to sinusoidal excitations, although techniques are being developed to accommodate the system delays and enable characterization using realistic, digitally-modulated RF signals [55, 56].

In a loadpull system, one or more signal generators and power amplifiers are required to create the desired stimuli to the device-under-test, typically providing enough available input power to drive the transistor into several dB of gain compression. Modern state-of-the-art (passive) loadpull systems are capable of measuring devices up to 100 to 200 W at impedances below one ohm, under digitally modulated stimuli, at frequencies around 2 GHz, and up to 20 W at frequencies around 40 GHz.

Prior to performing loadpull measurements, a system calibration must be performed in order to know the exact impedances and power quantities presented to the device-under-test at the frequencies of interest. Each tuner needs to be connected to a vector network analyzer to measure its two-port S-parameters for the specified set of tuner positions and frequency points. In addition, the S-parameters of the remaining passive components of the system, such as the input and output couplers, isolators and attenuators, need to be measured accurately. Once the tuner and the remaining S-parameters of the pre-matching fixture halves are determined and properly loaded into the system software, the loadpull system can then be assembled for a power calibration. The power calibration of the RF source is conducted to measure the available power at the input of the source tuner over a programmed power range. A zero-length through line is usually connected in place of the device-under-test and the fixture to perform the power calibration.

The final step prior to commencing the transistor characterization in a loadpull bench is the verification of the loadpull system. The complete system can be verified by measuring its transducer gain under sourcepull and loadpull conditions of a Through or Delay (Line) standard. Since S-parameters of both the input and output tuners, load strings and fixture halves, and the Through standard, are all known from calibration, the transducer gain of the system can be computed by cascading the appropriate sets of S-parameters. By comparing the calculated transducer gain to the measured transducer gain, the errors in the system can be determined. From experience, we have determined that an acceptable value of the difference in measured and calculated transducer gain, or ΔG_t, needs to be less than 0.2 dB in the areas of interest [35, 57]. If the value of ΔG_t is small, it can

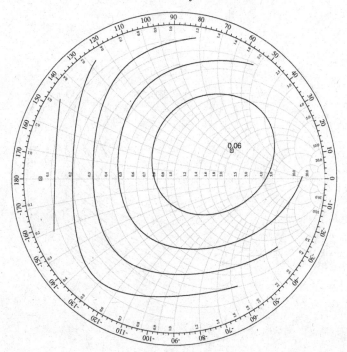

Fig. 3.26. Typical ΔG_t contours. The minimum ΔG_t is 0.06 dB and the contours increase in 0.05 dB steps outward [42]. © 2000 IEEE. Reprinted with permission.

be deduced that the calibration and the error boxes for the pre-matching fixture are accurate. Fig. 3.26 shows loadpull contours of ΔG_t while setting the input tuner to an impedance of 5 Ω at the calibrated reference plane. Similar results were achieved when measuring sourcepull contours of G_t and a load tuner setting of 5 Ω at the calibrated reference plane.

In the previous section, we highlighted the importance of having fixtures that provide a significant impedance transformation, to enable the user of a loadpull system to be able to reach the low impedance values of the packaged transistors with a reasonable amount of system accuracy. Another important component of the loadpull fixture, in addition to its pre-matching charac-teristics, is its low frequency bias decoupling. To minimize low-frequency memory effects (more details can be found in Chapter 2), the bias network must provide a very low impedance to the active device at frequencies several times the bandwidth of the signal being used to test the device.

An often-overlooked detail while collecting loadpull data for large-signal model validation is the information about the impedances presented by the measurement system to the device-under-test at frequencies other than the fundamental frequency. It is well known that the impedances presented

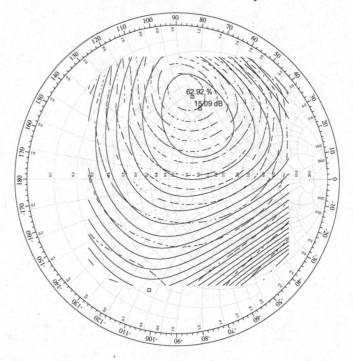

62.92 %
15.09 dB

Fig. 3.27. An example of loadpull contour plots superimposed on a Smith chart.

to the device at the harmonic frequencies and at low frequencies have a significant effect on the transistor's RF performance [58]. Therefore, the simulation environment must replicate the measurement conditions in which the transistor was characterized at all the relevant frequencies [59].

Traditionally, scanning through the loadpull data was a fairly simple exercise. Today's commercially available software that is used to control loadpull systems and to gather and display the measured data provides greater flexibility and capability. For example, the measurement and display of performance contours at a constant output power can be displayed easily. With the advent of digitally modulated signals in second- and third-generation cellular communication systems, loadpull systems have been adapted to measure and analyze new kinds of transistor characteristics, such as adjacent channel power [60]. More recently, with the trend for more peak power from high power RF transistors, loadpull systems are being modified to measure and analyze the ratio of the peak to average power (PAR) [53]. In a similar fashion, contours of constant output PAR can be measured and displayed, resulting in a very compact snapshot of the transistor performance. The results of an example loadpull measurement are plotted in Figs. 3.27 and

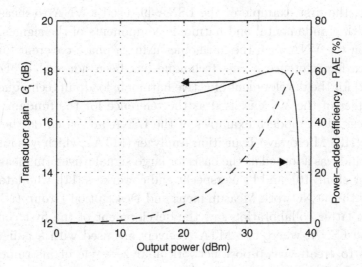

Fig. 3.28. An example of transistor performance for a single set of impedances.

3.28, where the gain and efficiency of a transistor are plotted as functions of the output power.

3.5.2 Large-Signal Network Analyzer

The large-signal network analyzer (LSNA) can be thought of as a vector-corrected and time-aligned microwave oscilloscope. It works in a similar way to a VNA, by measuring incident and reflected voltage waves applied to the device-under-test. The VNA operates at small signal levels, in the linear regime, and determines the ratios of the incident and reflected waves in a complex ratiometer: these ratios are the scattering parameters or S-parameters. In the case of the LSNA, we are generally measuring the DUT under large-signal conditions, and we must allow for the possibility of harmonics of the original wave being generated in the reflected and transmitted waves – indeed, if there are any sourcepull effects we may have harmonics of the incident wave as well. We now need to measure the absolute magnitudes and relative phases of all of the harmonic components, to make any sense of the large-signal transfer and reflection characteristics of the DUT. The LSNA captures the information in the frequency domain, like a VNA, but (usually) the visualization of the measurement is done in the time domain, like an oscilloscope: the LSNA is carrying out the Fourier analysis of the scattered waves, at microwave frequencies.

In fact, the first example of the LSNA [9] used a VNA to measure successively the fundamental and harmonic components of the signals, by synchronizing the VNA with the harmonics using a phase-coherent fundamental source for reference. This technique has been adapted by Ferrero *et al.* [54,61] for the development of active harmonic loadpull techniques, where one sampler of the VNA is used as the reference for the fundamental and harmonics. Most modern examples of the LSNA are based on the Hewlett Packard (HP) Microwave Transition Analyzer (MTA), which was introduced in 1992, and was adopted as the basis for large-signal waveform measurement by Tasker *et al.* [10], and by Verspecht and coworkers [11], who later collaborated with the Network Measurement and Description Group (NMDG) of HP. This latter collaboration saw the development of the first commercial prototype LSNA, where two MTA receivers are used with a pair of reflectometers to create a two-port network analyzer-style of instrument, with 40 GHz of RF bandwidth. This instrument measures directly the magnitude and phase of the incident and reflected waves at all harmonic or intermodulation frequencies, with the phase being coherently referenced so that reconstruction of the large-signal time domain waveform can be carried out. This signal-processing and timing hardware is an essential piece of the instrument.

The MTA-based LSNA requires a calibration to be able to measure the absolute magnitude and phase of the four a and b waves at the two ports. The calibration comprises three parts: the first part is a regular VNA calibration, to establish the measurement reference planes and determine the linear error coefficients of the system. The second part comprises a power calibration. This is usually done at a remote port, especially for on-wafer measurements: a one-port SOL calibration is carried out first, to enable the measurement at this port to be re-referenced or de-embedded to the device measurement reference planes. Then a (calibrated) power meter is attached and the signal source swept over power and harmonic frequencies, to establish the actual power detected at the samplers. Finally, for the third part, a phase calibration is carried out by attaching a calibrated phase reference unit at this same port, and the absolute phase of the samplers can be determined. The phase reference unit is a step recovery diode (SRD) driven by a high-amplitude square wave, generating very sharp pulses at a specific repetition rate, at harmonics of the drive frequency, which is 600–1200 MHz in the commercial LSNA [62]. This effectively gives a lower limit to the RF bandwidth of this instrument. The phase reference unit is a traceable calibration: the unit can be recalibrated at NIST using a nose-to-nose calibration method with a sampling oscilloscope to measure the pulses.

For many of the wireless applications, this phase calibration can be omitted: the samplers used in the LSNA (and HP8510) have a phase response that is flat to within about one degree, to 20 GHz, and so the phase correction is negligible for RF signals of around 2 GHz fundamental [63]. This allows a little more freedom in the low-frequency range of the instrument, and a simpler calibration procedure.

The 'microwave oscilloscope' description of the LSNA is a reality with modern digital sampling oscilloscopes (DSO), which claim 13–14 GHz RF bandwidth. By connecting a four-channel DSO behind a pair of reflectometers, after suitable calibration we have an instrument that can display the time-domain current and voltage waveforms of the reflected and transmitted microwave signals of the DUT: this is essentially LSNA functionality. In fact one of the original papers on high-frequency time-domain waveform measurement used a coupler and DSO to measure the waveforms and the voltage and current at the gate and drain of an HF transistor [64]. The drawback with the DSO approach is the limited dynamic range, attributable to the limited range of the analogue-to-digital converters (ADC) typically found in these instruments: only 8-bit, compared with the 14-bit ADCs in the MTA.

The research and development in nonlinear microwave measurements using the LSNA instrument has resulted in new developments of the instrument itself, including loadpull, pulsed capability, and extended RF and IF bandwidth, resulting in second-generation LSNA, as well as the characterization of nonlinear components and the development of new device and system models. These models for microwave transistors are of both compact and behavioural flavours.

The LSNA has been used for transistor modeling, in terms of collection of data for modeling, and model verification. Gaddi *et al.* [65] used LSNA measurements to create a nonlinear electro-thermal model as described in [47], and showed that the model created from LSNA measurements compared well with the traditionally derived model, though both models showed some shortcomings in the charge model, a characteristic of this model configuration (see Chapter 6 for a fuller discussion of this topic). Curras-Francos and co-workers [66, 67] used the LSNA with a complex variable load to generate measurements for extracting a table-based HEMT model based on nonlinear current and charge generators. The model performance in predicting output power, harmonic performance and small-signal S-parameters was excellent, similar to the non-quasi-static model generated using more traditional bias-dependent S-parameters. Schreurs *et al.* [68] used an LSNA with two RF sources, one each for gate and drain signals, to generate a loadpull effect that enabled a rapid coverage of the output characteristic space of the FET, and used artificial neural networks to model the FET terminal characteristics, with promising results. This modeling technique has since been applied to the generation of RF 'black-box' behavioural models of RFIC amplifiers [69].

The LSNA has also been used for large-signal model validation [70]. For example, the transistor load-lines under active drive conditions can be mapped out from the measured terminal currents and voltage, and compared with model predictions [67]. The use of the LSNA for model validation is described more fully in Chapter 9.

References

[1] G. F. Engen and R. W. Beatty, "Microwave reflectometer techniques," *IEEE Trans. Microwave Theory Tech.*, 7, no. 3, pp. 351–355, July 1959.

[2] G. F. Engen, "Advances in microwave measurement science," *Proc. IEEE*, 66, no. 4, pp. 374–384, Apr. 1978.

[3] J. M. Cusack, S. M. Perlow, and B. S. Perlman, "Automatic load contour mapping for microwave power transistors," *IEEE Trans. Microwave Theory Tech.*, 22, no. 12, pp. 1146–1152, Dec. 1974.

[4] T. M. Barton, C. M. Snowden, J. R. Richardson, and P. H. Ladbrooke, "Narrow pulse measurement of drain characteristics of GaAs MESFET's," *Electron. Lett.*, 23, pp. 686–687, 1997.

[5] M. Paggi, P. H. Williams, and J. M. Borrego, "Nonlinear GaAs MESFET modeling using pulsed gate measurements," *IEEE Trans. Microwave Theory Tech.*, 36, no. 36, pp. 1593–1597, Dec. 1988.

[6] A. Platzker, A. Palevsky, S. Nash, W. Struble, and Y. Tajima, "Characterization of GaAs devices by a versatile pulsed IV measurement system," in *IEEE MTT-S Int. Microwave Symp. Dig.*, Dallas, TX, May 1990, pp. 1137–1140.

[7] J. F. Vidalou, J. F. Grossier, M. Chaumas, M. Camiade, P. Roux, and J. Obregon, "Accurate nonlinear transistor modeling using pulsed S-parameters measurements under pulsed bias conditions," in *IEEE MTT-S Int. Microwave Symp. Dig.*, Boston, MA, June 1991, pp. 95–98.

[8] J. Scott, J. G. Rathmell, A. Parker, and M. Sayed, "Pulsed device measurements and applications," *IEEE Trans. Microwave Theory Tech.*, 44, no. 12, pp. 2718–2723, Dec. 1996.

[9] U. Lott, "Measurement of magnitude and phase of harmonics generated in nonlinear microwave two-ports," *IEEE Trans. Microwave Theory Tech.*, 37, no. 10, pp. 1506–1511, Oct. 1989.

[10] M. Demmler, P. J. Tasker, and M. Schlechtweg, "On-wafer large-signal power, S-parameter and waveform measurement system," in *Tech. Dig. of the 3rd International Workshop on Integrated Nonlinear Microwave and Millimeterwave Circuits (INMMIC)*, Duisburg, Germany, Oct. 1994, pp. 153–158.

[11] T. V. den Broeck and J. Verspecht, "Calibrated vectorial nonlinear network analyzers," in *IEEE MTT-S Int. Microwave Symp. Dig.*, San Diego, USA, June 1994, pp. 1069–72.

[12] G. H. Bryant, *Principles of Microwave Measurements*. Stevenage, UK: Peter Peregrinus, Ltd, 1993.

[13] D. Rytting, "Network analyzer error models and calibration methods," in *NIST Short Course on Microwave Measurements and Instrumentation*, Boulder, CO, Dec. 2003.

[14] G. F. Engen and C. A. Hoer, "Thru-reflect-line: an improved technique for calibrating the dual six-port automatic network analyzer," *IEEE Trans. Microwave Theory Tech.*, 25, pp. 987–993, Dec. 1979.

[15] H.-J. Eul and B. Schiek, "A generalized theory and new calibration procedures for network analyzer self-calibration," *IEEE Trans. Microwave Theory Tech.*, 39, no. 4, pp. 724–731, Apr. 1991.

[16] R. B. Marks, "A multiline method of network analyzer calibration," *IEEE Trans. Microwave Theory Tech.*, 39, no. 7, pp. 1205–1215, July 1991.

[17] http://na.tm.agilent.com/pna/connectorcare/Connector_Care.htm/

[18] http://www.cascademicrotech.com/

[19] D. Williams and R. B. Marks, "Compensation for substrate permittivity in probe-tip calibration," in *44th ARFTG Conference Digest*, Boulder, CO, Dec. 1994, pp. 20–30.

[20] J. Wood and D. E. Root, "Bias-dependent linear scalable millimeter-wave FET model," *IEEE Trans. Microwave Theory Tech.*, 48, no. 12, pp. 2352–2360, Dec. 2000.

[21] D. F. Williams and R. B. Marks, "On-wafer characteristic impedance measurement on lossy substrates," *IEEE Microwave Guided Wave Lett.*, 3, no. 6, pp. 175–176, June 1994.

[22] E. P. Vandamme, D. Schreurs, and C. van Dinther, "Improved three-step de-embedding method to accurately account for the influence of pad parasitics in silicon on-wafer RF test structures," *IEEE Trans. Electron Devices*, 48, no. 4, pp. 737–742, Apr. 2001.

[23] A. Davidson, K. Jones, and E. Strid, "LRM and LRRM calibrations with automatic determination of load inductance," in *36th ARFTG Conference Digest*, Monterey, CA, Dec. 1990, pp. 57–63.

[24] K. Wong and R. S. Grewal, "Microwave electronic calibration: transferring standards lab accuracy to the production floor," *Microwave J.*, 37, no. 9, pp. 94–105, Sept. 1994.

[25] R. F. Bauer and P. Penfield, "De-embedding and unterminating," *IEEE Trans. Microwave Theory Tech.*, 22, no. 3, pp. 282–288, Mar. 1974.

[26] R. L. Eisenhart, "A better microstrip connector," in *IEEE MTT-S Int. Microwave Symp. Dig.*, Dallas, TX, June 1982, pp. 318–320.

[27] "A guide to better vector network analyzer calibrations for probe-tip

measurements," Cascade Microtech Inc., http://www.home.agilent.com/upload/cmc_upload/All/TECHBRIEF4.pdf

[28] S. Wartenberg, "RF coplanar basics," *Microwave J.*, Mar. 2003.

[29] P. Aaen, J. Plá, and C. A. Balanis, "Increased feedback due to package mounting," in *13th Topical Meeting on Electrical Performance of Electronic Packaging (EPEP)*, Portland, OR, Dec. 2004.

[30] G. D. Vendelin, A. M. Pavio, and U. L. Rhode, *Microwave Circuit Design using Linear and Nonlinear Techniques*, 2nd edn. New York, NY: Wiley-Interscience, 1992, ch. 5, pp. 313–334.

[31] P. H. Aaen, "Simulation and modeling of matching networks within RF/microwave power transistors," Ph.D. dissertation, Arizona State University, Tempe, AZ, 2005.

[32] P. H. Aaen, J. A. Plá, and C. A. Balanis, "On the development of CAD techniques suitable for the design of high-power RF transistors," *IEEE Trans. Microwave Theory Tech.*, 53, no. 10, pp. 3067–3074, Oct. 2005.

[33] T. Liang, J. A. Plá, P. H. Aaen, and M. Mahalingam, "Equivalent-circuit modeling and verification of metal-ceramic packages for RF and microwave power transistors," *IEEE Trans. Microwave Theory Tech.*, 47, no. 6, pp. 709–712, June 1999.

[34] Freescale Semiconductor Inc. http://www.freescale.com/rf

[35] B. Noori, "Accurate load-pull measurements," in *IEEE Radio and Wireless Symp., Workshop on High-power RF devices and amplifiers*, San Diego, CA, Jan. 2006.

[36] J. F. Sevic, "A sub 1-ohm load-pull quarter-wave pre-matching network based on a two-tier TRL calibration," in *52nd ARFTG Conference Digest*, Rohnert park, CA, Dec. 1998, pp. 73–81.

[37] Z. Aboush, J. Lees, J. Benedikt, and P. Tasker, "Active harmonic load-pull system for characterizing highly mismatched high power transistors," in *IEEE MTT-S Int. Microwave Symp. Dig.*, Long Beach, CA, June 2005, pp. 1311–1214.

[38] P. H. Aaen, J. A. Plá, and C. A. Balanis, "Modeling techniques suitable for CAD based design of internal matching networks of high-power RF/microwave transistors," *IEEE Trans. Microwave Theory Tech.*, 54, no. 7, pp. 3052–3059, July 2006.

[39] R. B. Marks and D. F. Williams, "Characteristic impedance determination using propagation constant measurement," *IEEE Microwave Guided Wave Lett.*, 1, no. 6, pp. 141–142, June 1991.

[40] J. Plá, W. Struble, and F. Colomb, "On-wafer calibration techniques for measurement of microwave circuit and devices on thin substrates," in *IEEE MTT-S Int. Microwave Symp. Dig.*, 3, May 1995, pp. 1045–1048.

[41] D. Descher, "Measuring parasitic capacitance and inductance using TDR," *Hewlett-Packard J.*, no. 11, pp. 1–19, Aug. 1996.

[42] P. Aaen, J. Plá, D. Bridges, and E. Shumate, "A wideband method for the rigorous low-impedance loadpull measurement of high-power transistors suitable for large-signal model validation," in *56th ARFTG Conference Digest*, Boulder, CO, Dec. 2000.

[43] J. A. Dobrowolski and W. Ostrowski, *Computer-Aided Analysis, Modeling and Design of Microwave Networks: the Wave Approach*. Norwood MA: Artech House, 1996.

[44] J. Plá, "Characterization and modeling of high power RF semiconductor devices under constant and pulsed excitations," in *Proc. Fifth Annual Wireless Symposium*, San Jose, CA, Feb. 1997, pp. 467–472.

[45] A. E. Parker and J. G. Rathmell, "Bias and frequency dependence of FET characteristics," *IEEE Trans. Microwave Theory Tech.*, 51, no. 2, pp. 588–592, Feb. 2003.

[46] A. E. Parker and D. E. Root, "Pulse measurements quantify dispersion in PHEMTs," in *URSI Int. Symp. Dig. Signals, Systems and Electronics*, Pisa, Italy, Oct. 1998, pp. 444–449.

[47] W. R. Curtice, J. A. Plá, D. Bridges, T. Liang, and E. E. Shumate, "A new dynamic electro-thermal nonlinear model for silicon RF LDMOS FETs," in *IEEE MTT-S Int. Microwave Symp. Dig.*, Anaheim, CA, June 1999, pp. 419–422.

[48] S. Akhtar, P. Roblin, S. Lee, X. Ding, S. Yu, J. Kasick, and J. Strahler, "RF electro-thermal modeling of LDMOSFETs for power-amplifier design," *IEEE Trans. Microwave Theory Tech.*, 50, no. 6, pp. 1561–1570, June 2002.

[49] M. Dunn and B. Schaefer, "Link measurements to nonlinear bipolar device modeling," *Microwaves & RF*, pp. 114–130, Feb. 1996.

[50] Accent Inc. http://accentopto.com/Accent/productDetail.asp?product_id=21

[51] "AU6700 series pulsed IV RF modeling system," Auriga Measurement Systems LLC., http://www.auriga-ms.com/docs/AU6700Mss4.pdf

[52] "Agilent PNA microwave network analyzers, application note: 1408-12," Agilent Technologies Inc., http://cp.literature.agilent.com/litweb/pdf/5989-4839EN.pdf

[53] B. Noori, P. Hart, J. Wood, P. H. Aaen, M. Guyonnet, M. Lefevre, J. A. Plá, and J. Jones, "Load-pull measurements using modulated signals," in *Proc. 36th European Microwave Conf.*, Manchester, UK, Sept. 2006.

[54] B. Hughes, A. Ferrero, and A. Cognata, "Accurate on-wafer power and harmonic measurements of mm-wave amplifiers and devices," in *IEEE MTT-S Int. Microwave Symp. Dig.*, Albequerque, NM, June 1992, pp. 1019–22.

[55] T. Williams, J. Benedikt, and P. J. Tasker, "Experimental evaluation of an active envelope load pull architecture for high speed device characterization," in *IEEE MTT-S Int. Microwave Symp. Dig.*, Long Beach, CA, June 2005, pp. 1509–1512.

[56] A. Ferrero, Progettazione ed Alta Frequenza (PAF) - E-mail: Pafmicro@uol.it., private communication.

[57] J. Paviol, E. Kueckels, R. Varanasi, and L. Dunleavy, "PA load pull error limits using delta Gt contours," in *IEEE Topical Workshop on Power Amplifiers for Wireless Communications*, San Diego, CA, Dec. 2003, pp. 73–81.

[58] S. C. Cripps, *RF Power Amplifiers for Wireless Communications*, 2nd edn. Norwood MA: Artech House, 2006.

[59] J. F. Sevic, C. McGuire, G. R. Simpson, and J. A. Plá, "Data-based load pull simulation for large-signal transistor model validation," *Microwave J.*, no. 3, Mar. 1997.

[60] J. F. Sevic, M. B. Steer, and A. M. Pavio, "Large-signal automated load-pull of adjacent-channel power for digital wireless communication systems," in *IEEE MTT-S Int. Microwave Symp. Dig.*, 2, San Francisco, CA, June 1996, pp. 763–766.

[61] A. Ferrero, F. Sanpietro, U. Pisani, and C. Beccari, "Novel hardware and software solutions for a complete linear and nonlinear microwave device characterization," *IEEE Trans. Instrum. Meas.*, 43, no. 2, pp. 299–305, Apr. 1994.

[62] Maury Microwave, Inc. http://www.maurymw.com/

[63] P. J. Tasker and J. Verspecht, private communication.

[64] M. Sipila, K. Lethinen, and V. Porra, "High-frequency periodic time-domain waveform measurement system," *IEEE Trans. Microwave Theory Tech.*, 36, no. 10, pp. 1397–1405, oct 1988.

[65] R. Gaddi, J. A. Plá, J. Benedikt, and P. J. Tasker, "LDMOS electro-thermal model validation from large-signal time-domain measurements," in *IEEE MTT-S Int. Microwave Symp. Dig.*, Phoenix, AZ, May 2001, pp. 399–402.

[66] M. Fernandez-Barciela, P. J. Tasker, Y. Campos-Roca, M. Demmler, H. Massler, E. Sanchez, M. C. Curras-Francos, and M. Schlechtweg, "A simplified broad-band large-signal nonquasi-static table-based FET model," *IEEE Trans. Microwave Theory Tech.*, 48, no. 3, pp. 395–405, Mar. 2000.

[67] M. C. Curras-Francos, "Table-based nonlinear HEMT model extracted from time-domain large-signal measurements," *IEEE Trans. Microwave Theory Tech.*, 53, no. 5, pp. 1593–1600, May 2005.

[68] D. Schreurs, J. Wood, N. Tufillaro, D. Usikov, L. Barford, and D. E. Root, "The construction and evaluation of behavioral models for microwave devices based on time-domain large-signal measurements," in *Int. Electron Devices Mtg. Tech. Dig. (CD-ROM)*, San Francisco, USA, Dec. 2000.

[69] J. Wood, D. E. Root, and N. B. Tufillaro, "A behavioral modeling approach to nonlinear model-order reduction for RF/microwave ICs and systems," *IEEE Trans. Microwave Theory Tech.*, 52, no. 9, pp. 2274–84, Sept. 2004.

[70] D. Schreurs and J. Verspecht, "Large-signal modeling and measuring go hand-in-hand: accurate alternatives to indirect S-parameter methods," *Int. J. RF and Microwave CAE*, 10, no. 1, pp. 6–18, Jan. 2000.

4

Passive Components: Simulation and Modeling

4.1 Introduction

In the design and modeling of a microwave packaged power transistor, linear models of the package and matching networks are combined with a nonlinear model of the transistor. The resulting performance is dictated by the impedances presented by the matching networks to the transistor at the fundamental, harmonic and low-frequency terminations. These matching networks are often composed of arrays of small-diameter bondwires, metal-oxide-semiconductor (MOS) capacitors, and packages. The passive components provide the necessary low-loss impedance transformation essential for the successful operation of the RF power amplifier.

While the discussion here focuses on the modeling of the matching networks of a packaged LDMOS transistor used in wireless infrastructure applications, the techniques described are general enough to be applied to transistors that are based on other device technologies, or that are used for higher frequency applications. In this chapter, we will describe modeling techniques for each component of the matching network – that is, the package, bondwires, and internal matching capacitors, and show how modern computer-aided design tools can be used to support the modeling and design of the matching network.

4.2 Packages

Typically, the packages that are used for wireless infrastructure RF power amplifier applications are constructed from high-conductivity metals and low-loss dielectrics. The package is one component of the low-loss matching network located between the transistor die and the microstrip matching circuitry on the printed circuit board of the amplifier.

123

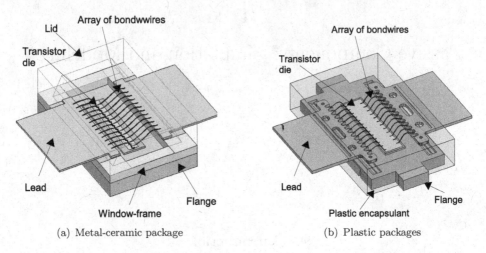

(a) Metal-ceramic package   (b) Plastic packages

Fig. 4.1. Illustrations of a power transistor using ceramic and plastic packages.

Stringent thermal-mechanical design practices are required to ensure that the package can dissipate the substantial heat-flux generated by the transistor [1]. Low thermal resistance packages are required (typically, around 0.5 W/°C), since all of the energy not converted into RF power is dissipated as heat. For example, a 200 W amplifier operating at 50% efficiency dissipates 100 W of power as heat through the package.

The packages, as illustrated in Fig. 4.1, are designed for the leads to rest on top of the microstrip transmission lines on the printed circuit board (PCB). The back side of the flange contacts the heatsink of the power amplifier forming a conductive electrical connection to the bottom conductor of the microstrip and a conductive thermal connection to the heatsink, which permits heat to flow away from the packaged transistor.

4.2.1 Analysis of an Empty Package

In general, the width of the package leads approximately match the total combined width of the dies mounted within the package. The largest packaged transistors have lead widths in excess of 500 mil. The combination of lead width, high ceramic permittivity and thin substrate of the window-frame forms a short transmission line having a low characteristic impedance, which is often approximated by an equivalent package capacitance.

The package capacitance can be extracted directly from the simulated or measured Y-parameters. Assuming a Π-network of capacitors, the shunt

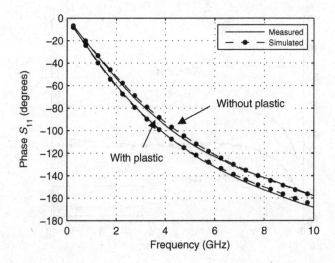

Fig. 4.2. The phase of the input reflection coefficient of the empty microstrip package. The encapsulant has a relative dielectric constant of $\epsilon_r = 3.5$ and a loss tangent of 0.005. The relative dielectric constant of the window-frame is 9.73.

capacitance, C_{pkg}, is expressed as,

$$C_{\text{pkg}} = \frac{\text{Im}(Y_{11} + Y_{12})}{\omega} \tag{4.1}$$

$$C_{\text{m}} = \frac{-\text{Im}(Y_{12})}{\omega} \tag{4.2}$$

where C_{pkg} is the extracted bond-pad capacitance to ground, C_{m} is the extracted capacitance between the input and output bond-pads, Y_{ij} are the elements of the two-port admittance matrix, and ω is the radial frequency.

As an example, we performed S-parameter measurement and simulation of the empty package illustrated in Fig. 3.12, in the test-fixture shown in Fig. 3.11, with and without a plastic encapsulant. Finite-element simulations were conducted by replicating the package geometry within a finite-element-based simulator. The phase of the reflection coefficient and the magnitude of the transmission coefficient are plotted in Figs. 4.2 and 4.3. It is worth noting that the effects of the plastic encapsulation on the package are an increased bond-pad capacitance and reduced isolation between package leads.

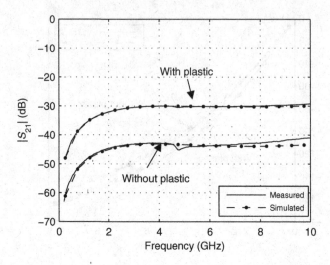

Fig. 4.3. Plots of the measured vs. simulated transmission coefficient of the empty microstrip package. Results for both encapsulated and non-encapsulated packages are provided.

4.3 Bondwires

The primary role of a bondwire is to provide a conductive electrical connection between a semiconductor die and the external environment. In many devices, such as microprocessors, digital signal processors, and so forth, the purpose of the bondwire is two-fold: to provide a means for the transmission of signals, and to distribute the DC power. The bondwire is typically considered to be a parasitic element, which degrades circuit performance. Estimates of its parasitic effects are crucial during the design process to avoid serious signal integrity issues.

In RF power transistors, the bondwire also functions as a low-loss element of an impedance-matching network. Since the shape of the wire, number of wires, spacing between wires, and wire radius are all easily controlled by automatic wire-bonding machines, it is the element that can be most easily adjusted during the design. Minor adjustments to the bondwire geometry can result in the matching network altering the transistor performance. Their characterization is paramount in achieving a successful design, since without accurately modeling these effects *a priori*, the design process is reduced to empirical tuning through re-manufacture.

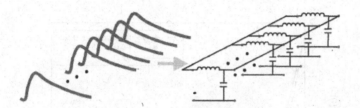

Fig. 4.4. Equivalent circuit of an array of bondwires. The coupling between the bondwires has been neglected to simplify the drawing.

4.3.1 Electrical Examination of the Bondwire Array

Often the equivalent circuit of the bondwire is assumed to be a series inductance, with a loss mechanism implemented through the addition of a series resistance. The small diameter wire is typically placed high above the ground plane, which results in a large inductance but a small capacitance. Using image theory, the per-unit-length inductance, L, and capacitance, C, can be computed from the transmission line parameters of a two-wire line [2]:

$$C = \frac{2\pi\epsilon}{\cosh^{-1}\left(\frac{H}{r}\right)} \tag{4.3}$$

$$L = \frac{\mu}{2\pi} \cosh^{-1}\left(\frac{H}{r}\right) \tag{4.4}$$

These values are exact for a conducting cylinder of infinite length, radius r, located at height H above an infinite ground plane, where μ is the permeability and ϵ is the permittivity of the medium surrounding the wire. For values of $H \gg 2a$, L dominates and C is typically deemed to be negligible: the high characteristic impedance approximation [3]. For example, a cylinder having a radius of 25 µm, and placed 500 µm above the ground plane, the per-unit-length inductance is 0.74 nH/mm and capacitance is 15 fF/mm.

Single bondwires are rarely employed in power transistors. Instead, large numbers of bondwires are placed in parallel to form an array. Often, the first step in developing an equivalent circuit for an array is to assume that each wire behaves independently, and to connect many inductors (wires) in parallel. This simple equivalent circuit cannot always model the behaviour of the array correctly. If the array of wires is analyzed as a set of coupled transmission lines, the importance of the capacitive coupling to the ground plane becomes apparent.

The equivalent circuit illustrating this arrangement is shown in Fig. 4.4, where each bondwire is replaced by the equivalent circuit of an electrically

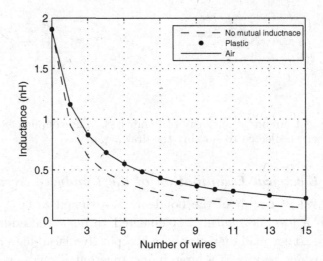

Fig. 4.5. Equivalent inductance of wire arrays for different numbers of conductors, with and without the presence of the plastic encapsulant. The effect of the mutual inductance is also shown.

short transmission line. While this pedagogical example neglects the mutual inductances and capacitances within the wire array, it illustrates that the relative importance of the capacitance and inductance will change as a function of the number of wires. The parallel connection results in the inductance decreasing and capacitance increasing as the number of wires in the array increases.

Electromagnetic simulations of a set of coupled conductors placed above a ground plane illustrate this behaviour. Since the coupled set of conductors supports a TEM-wave in a homogeneous medium, it can be considered as a single transmission line [4]. The equivalent inductance and capacitance, as a function of the number of conductors, extracted directly from the Z-parameters of the coupled transmission line using eqs. 4.11 and 4.13 (see Section 4.3.3), are plotted in Figs. 4.5 and 4.6.

The total inductance of the bondwire array is shown in Fig. 4.5. In the absence of mutual coupling between the wires, the total inductance curve will simply follow the pure mathematical value of a collection of inductors in parallel. The mutual inductance will increase as the wires are placed closer to one another and it will have the effect of increasing the total inductance of the bondwire array. The total capacitance of the array increases linearly with the number of wires, because the set of conductors supports a TEM wave, so at any point transverse to the direction of propagation, the wires

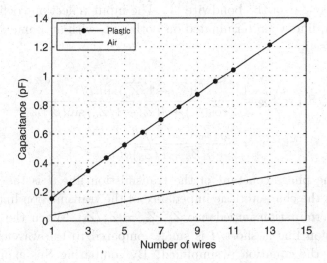

Fig. 4.6. Equivalent capacitance of wire arrays for different number of conductors, with and without the presence of the plastic encapsulant.

are at the same potential. With no voltage difference between the wires, there are no effects due to coupling capacitance. This indicates that the capacitance of the wire is directly to ground and as a result it scales linearly with the number of wires.

The fact that as the number of wires in the array increases, the inductance decreases and the capacitance increases linearly, is a consequence of the nature of the coupled set of transmission lines (in other words, the speed of an electromagnetic wave in a homogeneous medium is constant). It also implies that for arrays with large numbers of bondwires, the capacitance plays a more significant role, and cannot be neglected.

The effects of the plastic in over-molded packages must also be considered. These packages are designed such that the plastic completely encapsulates the wires. Since the capacitance is directly proportional to the relative permittivity of the medium, the total capacitance increases, while the inductance remains unaffected. The effect of the plastic on the equivalent inductance and capacitance of the set of coupled microstrips is also shown in Figs. 4.5 and 4.6, where the set of coupled microstrips were re-simulated within a dielectric having a relative dielectric constant of $\epsilon_r = 4$.

The analysis of the bondwire array is further complicated by the impedances presented at the ends of the set of wires. Through an analysis of the S-parameters of a transmission line terminated in equal impedances Z_t, Mouthaan has shown that the termination impedances can dictate the electrical behaviour of the bondwire [3]. The input reflection coefficient for a transmission line when terminated on both ends with an impedance of Z_t can be expressed as:

$$S_{11} = S_{22} = \frac{j(\overline{Z_o} - 1/\overline{Z_o})\sin(\beta l)}{2\cos(\beta l) + j(\overline{Z_o} + 1/\overline{Z_o})\sin(\beta l))} \qquad (4.5)$$
$$\approx \frac{j(\overline{Z_o} - 1/\overline{Z_o})\beta l}{2 + j(\overline{Z_o} + 1/\overline{Z_o})\beta l}$$

where β is the phase constant of the transmission line, l is the length of the line, $\overline{Z_o}$ is the characteristic impedance of the transmission line normalized by the termination impedance Z_t, $\overline{Z_o} = Z_o/Z_t$. When the length of the transmission line is short, l is small compared to the wavelength and $\sin(\beta l) \approx \beta l$: the equation is simplified. By comparing this expression to the S-parameters of a series inductor L_s and a shunt capacitor C_s, the qualitative behaviour of the line can be determined. The S-parameters for a series inductor L_s and a shunt capacitor C_s are given by [3]:

$$S_{11} = S_{22} = \frac{j\omega L_s/Z_t}{2 + j\omega L_s/Z_t} \qquad (4.6)$$

$$S_{11} = S_{22} = \frac{-j\omega C_s/Y_t}{2 + j\omega C_s/Y_t} \qquad (4.7)$$

Thus, if $\overline{Z_o} \gg 1$, the S-parameters of eq. 4.5 simplify to those of eq. 4.6, and the bondwire behaves like an inductor. Conversely, if the $\overline{Z_o} \ll 1$, the S-parameters of eq. 4.5 simplify to those of eq. 4.7, and the bondwire behaves like a capacitor. In other words, it was found that if the characteristic impedance of the bondwire array is much greater than the impedance of the termination, then the bondwire behaves inductively, whereas if the characteristic impedance of the bondwire is much smaller than the termination impedance, then it behaves capacitively.

4.3.2 Simulation of an Array of Bondwires

There exist numerous approximate formulas and methods in the literature [5–9] that attempt to compute the approximate behaviour of the self- and mutual inductance of a bondwire and arrays of bondwires through the application of the Biot–Savart law or Neumann's equation [10]. Both of these equations calculate the inductance of a conductive element formed in a closed loop. Since the bondwires form non-closed loops, the concept of partial inductance was introduced, which represents part of the total closed loop inductance. Rosa argues that in this sense the use of the Biot–Savart law and Neumann's formula is legitimate [10]. The majority of these formulas have been derived from Rosa's original work, where the wire was considered to be a straight wire positioned horizontally over a ground plane at an infinite distance [5]. Image theory can be used to account for a ground plane, and several correction terms have been added to account for short wire lengths and skin depths.

Similar formulas and approximations are used to compute the mutual inductance between two wires; however, these equations lack the ability to define precisely the three-dimensional nature of the bondwire, and concepts such as the average height of the bondwire arc above a ground plane must be introduced. Although these equations can be used to calculate bondwire parameters approximately, in many instances they do not provide the sufficient level of accuracy required for high power transistor design. Furthermore, these equations suffer from severe limitations when analyzing high frequency circuits accurately where the distributed nature of the bondwire, radiation and non-uniform current densities due to proximity effects become important or cannot be ignored. To compute the inductances for these types of geometry, we turn to computational electromagnetic techniques. Nevertheless, an understanding of the more common approaches to inductance calculation and their limitations can be useful.

One of the most common expressions for the inductance of an isolated conducting cylinder of length l and diameter d, is [10]

$$L = \frac{\mu_0 l}{2\pi}\left\{ ln\left[\frac{2l}{d} + \left(1 + \left(\frac{2l}{d}\right)^2\right)^{0.5}\right] + \frac{d}{2l} - \left[1 + \left(\frac{d}{2l}\right)^2\right]^{0.5} + \mu_r \delta \right\}$$

(4.8)

The mutual inductance between two wires of length l separated by center-to-center distance s is computed using,

$$M_g = \frac{\mu_0 l}{2\pi}\left\{ \ln\left[\frac{l}{s} + \left(1 + \left(\frac{l}{s}\right)^2\right)^{0.5}\right] + \frac{s}{l} - \left(1 + \left(\frac{s}{l}\right)^2\right)^{0.5} \right\}$$

(4.9)

For arrays of bondwires, the diagonal elements of the impedance matrix are computed using eq. 4.8, and the off-diagonal terms, or mutuals, are computed using eq. 4.9. However, these equations are of limited use since the calculations return an approximate inductance and the curvature of the bondwire cannot be incorporated.

To account for the curvature of the bondwire, a Neumann's inductance equation is often used to compute the mutual and self-inductances of bondwire [9]. Neumann's formulation computes the mutual inductance between closed filamentary loops C_1 and C_2,

$$L_{12} = \frac{\mu_o}{4\pi} \oint_{C_1} \oint_{C_2} \frac{d\mathbf{l_1} \cdot d\mathbf{l_2}}{r} \qquad (4.10)$$

where μ_0 is the permeability of free space and r is the distance between $d\mathbf{l_1}$ and $d\mathbf{l_2}$ differential sections of line. Similarly the self-inductance is computed using eq. 4.10, where C_1 is the surface of the wire and C_2 is the axis of the wire [11]. While this formulation enables the calculation of curved bondwires, it assumes the current density within the array is uniform and losses are implemented through skin depth calculations.

The partial element equivalent circuit (PEEC) technique has also been widely applied for the simulation of package leads and bondwires. The method discretizes the volume of metallic objects. The formulation is derived from magneto-quasi-static analysis, which originates from the assumption that the displacement current is negligible. For a bondwire, the method assumes that current within a long thin wire flows parallel to the wire surface. The skin effect is included by discretizing the cross-section of the wire into multiple current filaments with each one carrying a different current magnitude. Following a method-of-moments procedure, a system of equations is formed and solved for the inductance and series resistance of the bondwire. This method allows the calculation of the inductance of a curved bondwire as well as the mutual coupling between adjacent bondwires [12].

The main advantage that the PEEC method provides is that only the metallic volumes, that is, the wires, must be meshed into filaments. The remaining non-metallic volumes next to or enclosing the bondwires do not have to be meshed as is commonly done with finite-element (FEM) and finite-difference time-domain (FDTD) methods. Because of the reduced meshing requirements, the PEEC method has an advantage over these two methods especially for low frequency simulations with a small number of bondwires. Perhaps the most well known implementation of the PEEC method is *FastHenry*, a publicly-available computer program [12].

As the frequency increases, the displacement currents, radiation loss, non-uniform ground planes, capacitance to the ground plane, and surrounding dielectrics cannot be neglected, and full-wave simulation methods are required. Three of the most popular techniques that have been used to simulate bondwire and packaged circuits include the finite-difference time-domain method, the method-of-moments (MoM), and the finite-element method.

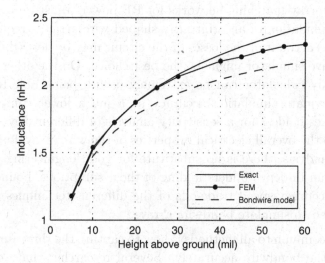

Fig. 4.7. Calculated inductance of a 2 mil diameter, 100 mil long wire vs. height above the ground plane.

Of the three, we have focused on the FEM because it is well suited for problems containing arbitrary materials, boundaries and three-dimensional geometries.

To gain insight into the relative accuracy of the aforementioned techniques, each method is used to simulate the inductance of a canonical geometry, which has a closed form solution that can be found directly from Maxwell's equations. One example is the infinitely long cylindrical conductor suspended above an infinite ground plane; the solution is given by eq. 4.4. For a wire of 2 mil diameter suspended 20 mil above a ground plane, the per-unit-length inductance is 18.7 pH/mil. Thus, the total inductance for a 100 mil bondwire is 1.87 nH. Values computed using eq. 4.8, Neumann's inductance equation, and FEM are all plotted in Fig. 4.7 as a function of the height above the ground plane.

In practice, bondwires are never perfectly flat, and so equations that do not account for the curvature of the bondwire are of limited use in the design of the matching network arrays in the RF power transistor. In addition, we need to accommodate the effects of any plastic encapsulant that surrounds the wire and bond-pad.

We have so far outlined the advantages and disadvantages associated with the various techniques that can be used to compute the electrical characteristics of (sets of) bondwire arrays. There is no perfect method that will address all of the technical challenges that we will encounter in the design of the bondwire array matching networks for RF power transistors. At times, an accurate simulation of an arbitrarily shaped wire array is required, and a very detailed account of the losses of the circuit may be desired. In such a case, a full-wave simulator might be the best choice. On the other hand, an inductance-only calculator can be extremely useful, being able to provide more-than-adequate simulation accuracy at a much lower computational cost; this makes it ideal for a sensitivity analysis of different key geometric dimensions to the overall electrical properties of the wire array, for example. In the final analysis, there is no substitute for good engineering judgment coupled with an in-depth understanding of the assumptions, boundary conditions and accuracy-speed trade-offs of the different techniques and tools commonly used to simulate bondwire arrays.

A problem common to all methods is how to capture the three-dimensional geometry of the bondwire accurately. Several researchers have developed techniques to overcome this burdensome task through the use of scanning electron microscope or digital microscope images [13, 14]. From a few digitized photographs, and several known dimensions and observation angles, the coordinate rotation matrices are computed and the bondwire geometry is extracted directly from the photographs. The size and complexity of a physically large packaged transistor, and the difficulty and time associated with obtaining photographs of sufficient quality, often limits the applicability of these techniques. Usually, though, the bondwire profiles are extracted in a manual process where a high-power optical microscope outfitted with a precision XYZ-table is used to focus directly on the bondwires at various positions along its length.

No matter which method is used, having a program to transform the measurements into a file format readable by the electromagnetic simulator is very desirable. Most CAD companies provide macro language capabilities or detailed specifications of their input file formats. With the measurement process and knowledge of the simulator input file format, it is possible to extract the geometry and enter it directly into the simulator.

An often-ignored issue during the modeling and simulation of the bondwire array is the nature of its connection to the surrounding environment. The bondwire array connects to the rest of the components in the packaged transistor; the metal manifold comprising the input lead, the top metal plate of the MOS capacitor, or the package window-frame, for instance. The most common way of describing the connection is to use a microstrip-like transmission line. The bondwire array-to-microstrip discontinuity will be detailed next.

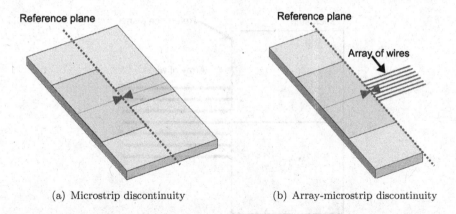

(a) Microstrip discontinuity   (b) Array-microstrip discontinuity

Fig. 4.8. Illustrations of the discontinuity between two microstrip transmission lines and between an array of wires and a microstrip line. The ground plane has been removed to simplify the drawing. Arrows indicate the sections of transmission line to be de-embedded to the reference plane.

4.3.3 The Bondwire-to-Microstrip Discontinuity

Different transmission lines have different field configurations, even if they support the same type of transmission mode (TEM, for example). Whenever two dissimilar transmission lines are connected together, changes to the field configurations at the interface are necessary to enforce continuity as the fields transition from the configuration of one transmission line to that of the other. This effect is observed even if the two transmission lines have the same characteristic impedances but different geometric dimensions. Changes in the electric field give rise to a capacitive effect, while changes to the magnetic field give rise to an inductive effect [15]. These extra capacitances and inductances can be viewed as representative of the electrical discontinuity produced by the disruption of the field configuration at the interface located between the two transmission lines. Such a discontinuity exists where the array of bondwires is attached to planar sections of the matching network, the package, MOS capacitor, or the die itself. The existence of a discontinuity between the package and the array of bondwires is analogous to the step discontinuity present when two microstrip transmission lines of unequal width are connected, as illustrated in Figs. 4.8(a) and 4.8(b). In the microstrip environment, the discontinuity can be characterized by simulating a circuit consisting of the two connected transmission lines. Each transmission line can then be de-embedded to the junction, and the resulting network parameters are those of the discontinuity. The complex three-dimensional geometry of the microstrip-to-bondwire interface

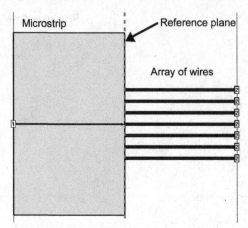

Fig. 4.9. An illustration of a model used to simulate the discontinuity between an array of wires and a microstrip line. The reference plane is established at the center of the model.

complicates this analysis: there is not a clear reference plane to divide the structure into microstrip and bondwire sections. The analysis of this discontinuity has only been included in full-wave simulations for small numbers of wires, and is only roughly approximated analytically [6, 16, 17].

In a similar manner the discontinuity for the package wire array can be analyzed. Shown in Fig. 4.9 is an illustration of a circuit that can be used to characterize the discontinuity within an electromagnetic simulator. The microstrip transmission line and the coupled set of wires are de-embedded to the interface between them. Transforming the remaining S-parameters to Z-parameters and then converting them to equivalent circuit parameters of an L-C-L T-network yields the following expressions

$$L_1 = \frac{\text{Im}(Z_{11} - Z_{12})}{\omega} \tag{4.11}$$

$$L_2 = \frac{\text{Im}(Z_{22} - Z_{21})}{\omega} \tag{4.12}$$

$$C = \frac{-1}{\omega \text{Im}(Z_{12})} \tag{4.13}$$

where Z_{ii} are the elements of the two-port impedance matrix.

The equivalent total discontinuity inductance, $L_{\text{total}} = L_1 + L_2$, is plotted in Fig. 4.10 for various package widths connected to an array of bondwires having a width of 20 mil, where the array width is defined as the center-to-center distance between the outermost bondwires. The discontinuity is

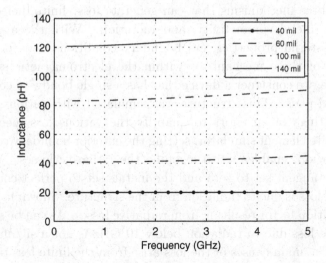

Fig. 4.10. Total equivalent inductance of the bondwire-to-microstrip discontinuity for varying package widths connected to a 20 mil wide bondwire array. Simulations were performed for an array of 2 mil diameter wires spaced 5 mil apart.

primarily inductive and the value of this inductance can be a large percentage of the total array inductance. For example, a 50 mil long, five-wire array, with 1 mil radius wires, 5 mil wire spacing, and located 20 mil above ground has an equivalent inductance of approximately 770 pH. Connecting this to a 140 mil wide package creates a discontinuity of 125 pH or about 1/6 the total array inductance, a significant percentage of the total array inductance, as shown in Fig. 4.10.

4.3.4 Examination of the Loss in the Bondwire Array

Our matching networks are designed with very high conductivity metals and low loss-tangent materials so as not to degrade the transistor performance. An effective simulation technique must be able to calculate these losses accurately. Since S-parameters are not sensitive to small losses, it is important to define a metric that is sensitive. One metric that can be employed is the conservation factor, CF,

$$CF = 1 - |S_{11}|^2 - |S_{21}|^2 \tag{4.14}$$

Another metric that can be used to examine the loss of a network is the maximum available gain (MAG), or loss for a passive network, which is calculated under the assumption that all losses due to input and output reflections are removed.

There are three mechanisms that can generate loss: finite dielectric loss tangent, finite metal conductivity, and radiation. With the appropriate selection of material properties and boundary conditions, it is possible to examine each of these mechanisms within the electromagnetic simulator. As an example, we construct a device that has a single bondwire connecting a pair of bond-pads. This structure was simulated with various material parameter settings in an effort to quantify the various loss mechanisms. Within the finite element simulator, setting the exterior boundary conditions to allow radiation, enables us to examine the radiative losses. With the dielectric loss tangent set to zero and the metals set to perfect conductors, any computed loss is due to radiation from the structure. Shown in Fig. 4.11 is the conservation factor resulting from radiative losses. As can be seen from this figure, the loss due to radiation below 10 GHz is not a significant loss mechanism. The main causes of the loss arise from the finite loss tangent of the ceramic and the metal conductivity.

Simulations with a finite metal conductivity and the dielectric loss tangent set to zero indicate a good agreement between the conservation factor computed from measured and simulated S-parameters. The losses are dominated by the finite conductivity of the metal. The ceramic loss tangent of $\tan \delta = 0.001$ (specified by the manufacturer) accounts for only a small fraction of the loss, as shown in Fig. 4.11. In fact, in this case, the losses due to the material loss tangent and radiation effects are small and can be neglected.

4.4 MOS Capacitor Modeling

While the bondwire array is the most common method of realizing a series or shunt inductive element in the internal matching network of the high-power transistor, a low-loss capacitive component is also required to complete the set of matching network components used to realize different sets of match topologies. The most common and versatile capacitor used in today's modern high power transistors is the metal-oxide-semiconductor (MOS) capacitor. These capacitors are designed as compact, high capacitance structures that exhibit low losses. They consist of a silicon dioxide layer deposited on top of a heavily-doped, highly conductive silicon substrate, which acts as one electrode of the capacitor. Even so, this substrate is still much more lossy than a deposited metal layer. A metal pattern is created on the top of the oxide to complete the construction of the capacitor. The back of the wafer is metallized to enable its attachment to a package flange and for the electrical connection to the bottom capacitor plate. An illustration of the

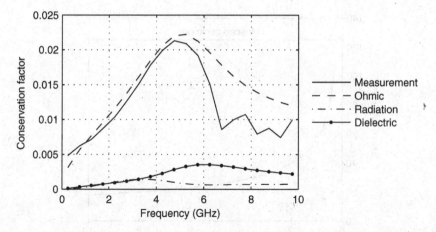

Fig. 4.11. Plots of the conservation factor for various material properties and boundary conditions. Ohmic losses only include loss due to finite metal conductivity ($\sigma = 4.1 \times 10^7$ S/m), radiative loss only includes loss due to radiation, and dielectric loss only includes losses due to a finite dielectric loss tangent (0.001).

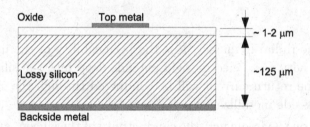

Fig. 4.12. An illustration of the cross-section of the MOS capacitor indicating typical dimensions. Bondwires are attached to the top metallization [18]. © 2005 IEEE. Reprinted with permission.

cross-section of a MOS capacitor is shown in Fig. 4.12. Typically, the capacitor is approximately the same width as the active die to accommodate the width of the bondwire array that connects to the transistor – hence its short and wide shape. Depending on design requirements, the capacitance can range from 10–500 pF, which can be controlled by varying the thickness of the oxide or the area of the top metallization.

The finite conductivity of the silicon results in a skin depth that is a large fraction of the substrate thickness. In Fig. 4.13, the skin depth is plotted as a function of frequency for a $\sigma = 80 \times 10^3$ S/m, using

$$\delta_s = \sqrt{\frac{2}{\omega \mu \sigma}}, \tag{4.15}$$

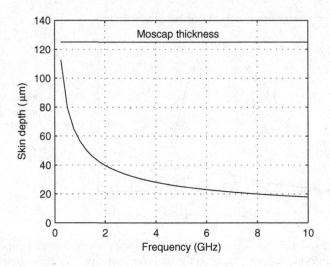

Fig. 4.13. The skin depth plotted against frequency for doped silicon with a conductivity of 8×10^4 S/m. The horizontal line represents the thickness of the material

where ω is the radial frequency, μ is the permeability of the material, and σ its conductivity. The electric field extends deep into the silicon and the majority of the return current will be conducted through the silicon rather than the back-side metallization.

Depending on the frequency, the conductivity of the silicon, and the thickness of the substrate, the electric field will decay to some fraction of its value within the substrate. The remaining electric field induces a current on the back-side metallization. The return current must flow through the silicon, a lossy return path, and the MOS capacitor introduces a frequency- and geometry-dependent loss. Shown in Fig. 4.14 is an illustration of a portion of a matching network containing the MOS capacitor and the current flow within it.

The MOS capacitor is part of a large class of structures called metal-insulator-semiconductor (MIS) structures. They have been well studied, but are notoriously difficult to model because of the complicated frequency-dependent behaviour introduced by the semiconducting substrate, as described above. It is well known that the semiconducting substrate can support three distinct modes of propagation. Whichever mode is dominant depends on the conductivity of the substrate, the frequency of operation, and the geometry of the transmission line. These factors significantly complicate the development of a simulation technique, and a model. A detailed

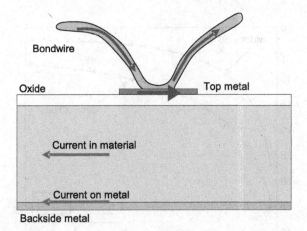

Fig. 4.14. An illustration of the cross-section of a MOS capacitor with attached bondwires. The current flow on the wires and within the silicon is indicated.

review can be found in [19–22]. In practice, we find that an approximate technique, described next, is more than sufficient.

4.4.1 MOS Capacitor Analysis

The layered structure of the MOS capacitor lends itself to simulation within a planar MoM-based simulator. The electrical behaviour of the capacitor can be approximated by a short transmission line [3]. The development of a transmission line model is simplified because of the high conductivity of the substrate, and the fact that the capacitors will be employed in matching networks operating below 10 GHz, where the dominant mode of propagation is the skin-effect mode, and the frequency-dependent behaviour of the transmission line is primarily dictated by skin depth penetration into the silicon.

A wafer was manufactured, which contained various GSG-probeable transmission lines and test capacitors: sufficient structures to permit the extraction and validation of the material parameters used in the MOS capacitors. The agreement between simulated and measured S-parameters is demonstrated through the simulation and measurement of several on-wafer transmission lines, independent of those used to extract the substrate parameters. The measured and simulated magnitudes of the input reflection coefficients for two of these transmission lines are shown in Fig. 4.15. These lines are 1000 µm long, have widths of 20 and 50 µm and are on top of a 1.38 µm thick oxide layer, which is in turn on top of a 125 µm silicon substrate.

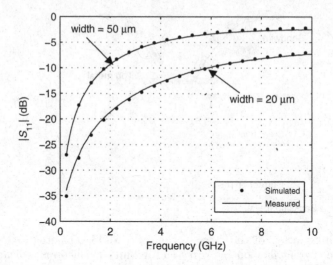

Fig. 4.15. A plot of the simulated and measured input reflection coefficients of two 1000 μm transmission lines. The two transmission lines have widths of 30 and 50 μm [18]. © 2005 IEEE. Reprinted with permission.

With the extraction of the material parameters and the validation in the simulator, simulating the MOS capacitor is then analogous to microstrip simulation, requiring only the appropriate artwork for the top metallization.

4.5 Example of the Use of Segmentation Techniques

As described in Section 3.3.5, the computational burden imposed by the electromagnetic simulation of the intricate matching network forces the development of segmentation procedures, whereby the network is partitioned into smaller sections for individual simulation. The partitioning methodology must be designed such that the sub-circuits, when cascaded back together, faithfully reproduce the response of the entire circuit. Owing to the three-dimensional nature of these circuits and the inter-element coupling resulting from the close proximity of the components to one another, the selection of these partitioning planes is not immediately obvious. The approach presented here separates the circuit at planes defined by the ends of the microstrip transmission lines connected to the bondwires. Initially, this approach appears to be problematic since the bondwires are attached past the end of the microstrip lines. This does not pose a problem as long as the de-embedded section of microstrip transmission line, past the ends of the wires, is included in the simulation of the components that are to be

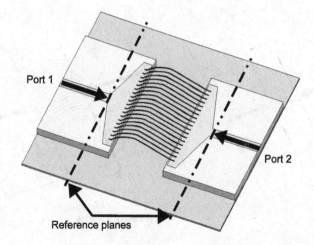

Fig. 4.16. A set of 19 bondwires connected between the leads of a package. The arrows indicate sections of transmission line that will be de-embedded from the simulated S-parameters. After de-embedding, the reference planes will be as indicated by the dashed lines [18]. © 2005 IEEE. Reprinted with permission.

cascaded on either side of the array of wires. This implies that when any element is excited within a simulator, it needs to be excited by a transmission line that has exactly the same characteristics as the one in the full matching network. As long as the circuits are excited identically, the electromagnetic fields on either side of the reference plane are the same and all of the electromagnetic simulation results can then be cascaded together, while faithfully reproducing the overall performance of the network.

4.5.1 Simulation of an Array of Bondwires

We demonstrate the segmentation approach by simulating an array of bondwires connected between bond-pads, as illustrated in Fig. 4.16. The microstrip transmission lines on either side of the bond-pads excite the array within the finite-element-based simulator. This circuit is selected since it is small enough that it can be simulated as a single entity. Differences between the simulated S-parameters of the entire structure and those computed from the segmentation approach indicate the accuracy of the segmentation method.

The ends of the bond-pads are selected as the segmentation planes, as illustrated in Fig. 4.17. To simulate the array of wires, the array is attached to two microstrip transmission lines which have the same width of the bonding

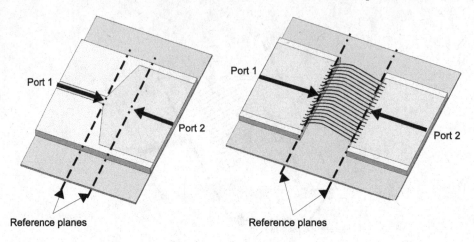

Fig. 4.17. An illustration of the segmentation of the package and bondwires that occurs at the reference planes indicated by the dashed lines. The arrows indicate sections of transmission line that will be de-embedded from the simulated S-parameters [18]. © 2005 IEEE. Reprinted with permission.

pad. These transmission lines are only used to excite the array of bondwires and are later de-embedded. As a result, the S-parameters contain the effects of the wires and the discontinuity between the microstrip and the array. This set of wires can then be cascaded with other planar circuit elements provided that they are excited with an identical transmission line.

Once the simulations of the three parts of the circuit are complete, the S-parameters of the entire circuit are computed by cascading the resulting two-port networks together. As can be seen in Fig. 4.18, the S-parameters from the segmented approach are in good agreement with those obtained from the simulation of the entire circuit.

4.5.2 Mutual Inductance

The arrays of bondwires within the package are often in close proximity to one another, and since they form small conductive loops that carry current, a significant amount of inductive coupling can occur. Once characterized, the coupling can be implemented within a circuit simulator through an appropriately-placed mutual inductor component.

The procedure used to obtain the mutual inductance builds on the techniques outlined in the previous sub-section to simulate individual arrays of bondwires. Once each of the arrays has been simulated, the mutual coupling

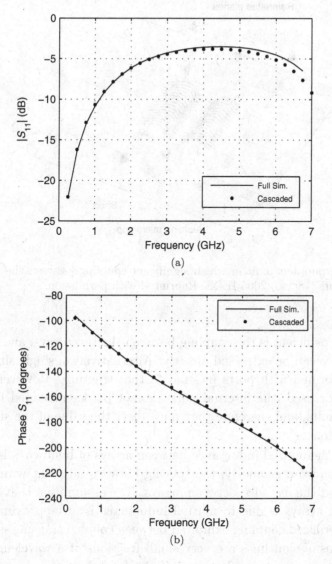

Fig. 4.18. A comparison of the simulated and measured input reflection coefficient for the device illustrated in Fig. 4.16. The device is simulated as an entire structure as well as using the segmentation procedure [18]. © 2005 IEEE. Reprinted with permission.

between the two arrays can be obtained by performing one additional simulation. In this simulation, the circuit element to which both bondwire arrays are connected is replaced by a short circuit. This circuit is simulated and then the the effects of the bondwire arrays are de-embedded. The resulting

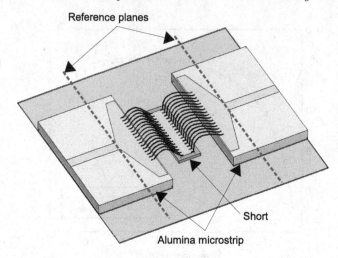

Fig. 4.19. An example circuit in which significant coupling between the arrays of bondwires occurs [18]. © 2005 IEEE. Reprinted with permission.

transmission coefficient is the coupling between the two arrays, and the mutual inductance can be extracted directly. Alternatively, a single simulation could be performed with ports internal to the structure. However, in the FEM simulator used, the internal ports cannot be de-embedded from the simulation results, and therefore they may affect the value of the simulated mutual inductance.

In general, the mutual inductance between arrays of bondwires is a function of the bondwire geometry, the spacing between adjacent wires within each array, and the distance between arrays. The assumption that the coupling between arrays is due to mutual inductance is an approximation to the more generalized coupling exhibited between coupled transmission lines. Since the transmission lines are very small fractions of a wavelength, this is a reasonable assumption. However, if the frequency of operation is increased, the bondwires become electrically large, and the coupling is no longer caused just by the mutual inductance. The coupling will also depend on the impedance in which the arrays are terminated [23].

As an example, the circuit illustrated in Fig. 4.19 contains two arrays of bondwires in close proximity to one another. After each set of arrays has been simulated individually, the mutual coupling between the two arrays is obtained by performing the additional simulation and de-embedding described above. The mutual inductance is plotted in Fig. 4.20, and we see that the inductance is a relatively smooth function of frequency.

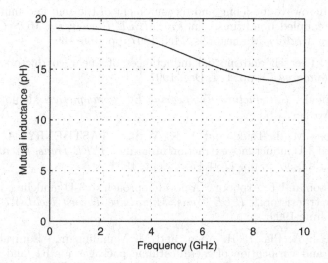

Fig. 4.20. The simulated mutual inductance between two arrays of bondwires.

References

[1] Z. Radivojevic, K. Andersson, L. Bogod, M. Mahalingam, J. Rantala, and J. Wright, "Novel materials for improved quality of RF-PA in base-station applications," *IEEE Trans. Compon., Packag., Technol.*, 28, no. 4, pp. 644–649, Dec. 2005.

[2] D. M. Pozar, *Microwave Engineering*, 2nd edn. New York, NY: John Wiley & Sons, 1998.

[3] K. Mouthaan, "Modeling of RF high power bipolar transistors," Ph.D. dissertation, Delft University, 2001.

[4] A. E. Ruehli and A. C. Cangellaris, "Progress in the methodologies for the electrical modeling of interconnects and electronic packages," *Proc. IEEE*, 89, no. 5, pp. 740–771, May 2001.

[5] S. March, "Simple equations characterize bond wires," *Microwaves & RF*, pp. 105–110, Nov. 1991.

[6] O. Gorbachov, "Evaluation of parasitic parameters for packaged microwave transistors," *Appl. Microwave & Wireless*, no. 4, pp. 78–88, Apr. 1999.

[7] N. Hassaine, Y. Shen, and P. Ntake, "Modeling and characterization of wire bonding and taped automatic bonding, analytical formulas and experimental validation," in *Proc. IEEE Radio and Wireless Conf. (RAWCON)*, Colorado Springs, CO, 1988, pp. 281–284.

[8] Z. He and L. Pileggi, "A simple algorithm for calculating frequency-dependent inductance bounds," in *Proc. IEEE Custom Integrated Circuits Conf.*, Rochester, NY, 1998, pp. 199–202.

[9] K. Mouthaan, R. Tinti, M. de Kok, H. C. de Graaff, J. L. Tauritz, and J. Slot-

boom, "Microwave modeling and measurement of the self- and mutual inductances of coupled bondwires," in *Proc. IEEE Bipolar/BiCMOS Circuit and Technology Meeting*, Minneapolis, MN, 1997, pp. 166–169.

[10] E. Rosa, "The self and mutual inductances of linear conductors," *Bulletin Bureau Standards*, 4, pp. 1–4, Dec. 1907.

[11] B. D. Popović, *Introductory Engineering Electromagnetics*. Menlo Park, CA: Addison-Wesley, 1971.

[12] M. Kamon, M. J. Tsuk, and J. K. White, "FASTHENRY: A multipole-accelerated 3-D inductance extraction program," *IEEE Trans. Microwave Theory Tech.*, 42, no. 9, pp. 1750–1758, Sept. 1994.

[13] T. Johansson and T. Arnborg, "A novel approach to 3-D modeling of packaged RF power transistors," *IEEE Trans. Microwave Theory Tech.*, 47, no. 6, pp. 760–768, June 1999.

[14] T. Liang, J. A. Plá, P. H. Aaen, and M. Mahalingam, "Equivalent-circuit modeling and verification of metal-ceramic packages for RF and microwave power transistors," *IEEE Trans. Microwave Theory Tech.*, 47, no. 6, pp. 709–712, June 1999.

[15] K. C. Gupta, R. Garg, I. Bahl, and P. Bhartia, *Microstrip Lines and Slotlines*, 2nd edn. Norwood, MA: Artech House, 1996.

[16] C. Schuster, G. Leonhardt, and W. Fichtner, "Electromagnetic simulation of bonding-wires and comparison with wide band measurements," *IEEE Trans. Adv. Packag.*, 23, pp. 69–79, Jan. 2000.

[17] F. Alimenti, P. Mezzanotte, L. Roselli, and R. Sorrentino, "Modeling and characterization of the bonding-wire interconnection," *IEEE Trans. Microwave Theory Tech.*, 49, pp. 142–150, Jan. 2001.

[18] P. H. Aaen, J. A. Plá, and C. A. Balanis, "On the development of CAD techniques suitable for the design of high-power RF transistors," *IEEE Trans. Microwave Theory Tech.*, 53, no. 10, pp. 3067–3074, Oct. 2005.

[19] H. Guckel, P. A. Brennan, and I. Palocz, "A parallel-plate waveguide approach to micro-miniaturized, planar transmission lines for integrated circuits," *IEEE Trans. Microwave Theory Tech.*, 15, no. 8, pp. 468–1727, Aug. 1967.

[20] E. Tuncer and D. P. Neikirk, "Highly accurate quasi-static modeling of microstrip line over lossy substrates," *IEEE Microwave Guided Wave Lett.*, 10, no. 2, pp. 409–411, Oct. 1992.

[21] H. Hasegawa, M. Furukawa, and H. Yanai, "Properties of microstrip line on Si-SiO$_2$ system," *IEEE Trans. Microwave Theory Tech.*, 19, no. 11, pp. 869–881, Nov. 1971.

[22] H. Hasegawa and S. Seki, "Analysis of interconnection delay on very high-speed LSI/VLSI chips using and MIS microstrip line model," *IEEE Trans. Microwave Theory Tech.*, 32, no. 12, pp. 1721–1727, Dec. 1984.

[23] E. M. T. Jones and J. T. Bolljahn, "Coupled-strip-transmission-line filters and directional couplers," *IEEE Trans. Microwave Theory Tech.*, 4, no. 2, pp. 75–81, Apr. 1956.

5

Thermal Characterization and Modeling

5.1 Introduction

In previous chapters, we have discussed the importance of capturing the transistor performance as a function of temperature and the changes that occur to the material parameters. While it is necessary to characterize and model this behaviour, transistors designed for high power densities also generate heat as a by-product of current flow within the junction area. This *self-heating* degrades the performance of the transistor and these effects must be included within our models.

To quantify the amount of self-heating we need only look at the definition of power-added efficiency η, $(P_{\text{out}} - P_{\text{in}})/P_{\text{dc}}$, where P_{in} and P_{out} are the input and output power at the operational frequency of the transistor, respectively, and P_{dc} is the supplied DC power. Power that is not converted from DC to RF or is transferred into the device, approximately $P_{\text{dc}}(1 - \eta)$, is dissipated as heat. The vast amount of generated heat flows away from the junction into the bulk semiconductor, down into the package flange, and then into the heatsink of the power amplifier. Maximum junction temperatures for transistors used within wireless base-stations are often specified to be around 150–200°C, depending on the device technology in use.

High-power transistors are operated close to the maximum limits of their electrical and thermal specifications, to extract the best possible device performance as economically as possible, in other words, to produce the maximum RF power per dollar of expenditure [1]. In these operating conditions the reliability of the transistor is at its worst: high temperatures accelerate the possible failure mechanisms in the device. Forecasting the operating lifetime of a high-power transistor is a major concern, since these components are typically the most expensive in the power amplifier, costly to replace and costly in terms of base-station down-time. The semiconductor industry has developed a rich tradition of thermal characterization and modeling with the goal of forecasting the device operating lifetime accurately.

Whereas obtaining the maximum temperature is often the focus of reliability studies, thermal models for use with compact transistor models require the ability to account dynamically for temperature change. As mentioned above, this temperature rise results from the power dissipated by the transistor, which depends upon several factors, including ambient temperature, drive-level, bias conditions, frequency, modulation scheme, and internal properties of the transistor such as the geometrical layout and the material parameters. In practice, these transistors are used in numerous applications and it is not possible to know *a priori* the operating parameters of the transistor, thus the compact thermal model, like the model for the drain current, must be able to account dynamically for changes in the aforementioned parameters.

To simulate the electro-thermal response of a transistor we need to solve simultaneously the coupled current-transport and heat-conduction equations. The solution process involves de-coupling the electrical and thermal dependencies to obtain isothermal model parameters, through the use of an electrical analogue circuit employing thermal resistors and capacitors, which represent the solution to the heat-conduction equation. The analogue circuit, or sub-circuit, is used to compute the instantaneous temperature rise of the device due to self-heating. As we shall see in the following sections, in its simplest form the sub-circuit consists of a current source, in parallel with a resistor and a capacitor [2]. In addition to the current source, which represents the instantaneous power dissipation of the device, the thermal circuit can also include a voltage source, which represents a point of constant or forced temperature, like the heatsink temperature. The instantaneous temperature of the transistor can then be computed by adding the calculated temperature rise and the heatsink temperature.

Other more specialized techniques exist where the coupling is implemented between electrical and thermal simulators directly. These programs use computationally intensive algorithms, which often take a long time to return results [3], making this approach impractical for circuit-level design work. Less intensive analytical formulations have been linked directly to circuit simulators. While fully analytical techniques can be difficult to formulate, once implemented they can be extremely fast [4].

Our preference is to use a thermal equivalent circuit approach, because it can be implemented easily within the electrical simulation environment using standard circuit techniques. Since nonlinear device models reside within circuit simulators, it is advantageous to perform electro-thermal simulations within the same simulation environment [5]. As long as the reference planes for the thermal circuit have been set appropriately, as will be discussed later in the chapter, and the circuit captures the thermal behaviour correctly, this approach allows a designer to simulate changes in temperature during the design of the transistor, in which it may be required to operate over a range of power amplifier classes, impedance states, or input signals. The capability of temperature prediction depends on the type of thermal model.

It is important to understand the differences between static and dynamic electro-thermal models. A static thermal model is, in essence, an electrical model capable of predicting the changes of electrical performance as a function of the operating temperature of the device. A static thermal model is often characterized at a fixed ambient temperature or range of temperatures with the effects of self-heating removed via pulsed measurement. The temperature of the model is set to a specific value and the model parameters are selected and fixed throughout the simulation. No temperature change due to the heat generated in the junction is taken into account. In contrast, models that exhibit self-heating take into account the ambient temperature and the instantaneous heat generated by the nonlinear drain current. While a static thermal model accounts for a specific set temperature, there is no way to account for the self-heating of the device.

A dynamic electro-thermal model is capable of predicting the instantaneous temperature variation experienced by the device while operating. This category of models require that the electrical model parameter changes versus temperature are known and incorporated into the model, and in addition a thermal model of the device must also be available. By combining the electrical and thermal models into a single framework, the inter-relation of the electrical and thermal dynamics can be explored and understood in a way that is impossible with a static thermal model.

The focus of this chapter is on the methods used to obtain and develop the thermal analogue circuit. This chapter begins with a review of the thermodynamic principles which govern heat flow and a discussion of the heat conduction mechanisms within semiconductor devices. A detailed discussion of several thermal measurement and simulation techniques used to capture the thermal characteristics of the device for both static and dynamic models is presented along with the introduction of thermal compact models and how to incorporate them into the nonlinear electrical model of the transistor to arrive at a nonlinear dynamic electro-thermal model.

5.2 Methods of Heat Transfer

Before delving into the details of heat transfer, a review of the basic definitions and thermodynamic principles is in order, as it is upon these principles
that we will build our understanding and ultimately our thermal modeling
strategy. Many introductory texts cover these concepts in great detail [6–8],
so we will only provide a brief review.

Temperature specifies the *hotness* or *coldness* of objects we encounter.
While this simple definition often suffices, we need to be more precise to
understand heat transfer. A more precise definition might be: the temperature of an object is related to the average translational kinetic energy
associated with the disordered microscopic motion of atoms and molecules.
It is a macroscopic property, like the pressure of a gas, which relates to an
average of microscopic quantities. Two objects at thermal equilibrium with
one another are said to be at the same temperature.

Heat is a measure of energy and is defined as energy in motion; it is a
form of energy associated with the motion and vibration of particles comprising matter. Nothing can be said to *have* or *store* heat. Heat-flux, often
represented by the symbol Q, is the rate of heat crossing a boundary and it
has units of W/m^2.

Whenever a heated object is placed in an environment with a temperature cooler than itself, heat transfer between the object and the environment
occurs. According to the second law of thermodynamics, in the absence of
applied work, heat is transferred between different objects having different
temperatures in the direction of the colder object. The flow of internal energy or heat from a region of higher temperature to one of lower temperature
occurs through the interaction of the adjacent particles (atoms, molecules,
ions, electrons, and so forth) in the intervening space.

The mechanism or mode of heat transfer depends on the type of interface
between the object and its environment. There are three fundamental modes
of heat transfer: conduction, convection, and radiation. Within power transistors, the high thermal conductivity of the materials and packages results
in conductive heat flow through the package and into the heatsink. Conduction is the dominant mode of heat transfer. While forced convection cooling
is crucial for thermal management of electronic equipment, we will only consider the thermal conduction into the heatsink. Thermal radiation does not
offer any significant cooling effects, but as will be seen in Section 5.3.1, it
is often used to measure the temperature on the die and surrounding areas
via thermal microscopy.

5.2.1 Heat Conduction

The heat generated by the transistor is predominantly removed through thermally conductive paths formed by the semiconductor die, the package and the heatsink of the power amplifier. A plot of the temperature contours within a packaged semiconductor die aids in the visualization of heat flow, as illustrated in Fig. 5.1(a). In Fig. 5.1(b) the simulated temperature contours of a heat source located on top of a silicon substrate are plotted. The substrate is attached to a thermally conductive flange. The heat flows perpendicularly to the isothermal contours.

Any heat flow through this type of thermal pathway is governed by the heat-conduction equation and the properties of the materials used to construct and package the transistor. Local changes in the thermal conductivity are visible as bends in the contour, so it is possible to see the interface between materials of different thermal conductivities.

The three-dimensional heat-conduction equation is derived from the first law of thermodynamics and conservation of energy, and it can be written in the following form [10]:

$$\nabla(\kappa \nabla T) + \dot{q} = \rho c \frac{dT}{dt} \tag{5.1}$$

where κ is the thermal conductivity in units of W/m· K, ρ is the material density (kg/m^3), c is the specific heat of the material (J/kg· s), and $\dot{q}$ is the rate of energy generation per unit volume of the heat source.

It is possible to include the temperature variation of the thermal conductivity using Kirchhoff's transformation to solve the heat-flow equation. The actual temperature rise can be found using the thermal conductivity at the initial temperature, $T = T_0$, and then computing the temperature rise through [11],

$$T = T_0 + \frac{1}{\kappa(T_0)} \int_{T_0}^{T} \kappa(T')dT' \tag{5.2}$$

To simplify the solution of this equation the thermal conductivity may be assumed to be homogeneous (uniform conductivity throughout the material) and constant with temperature. The degree to which this approximation is valid depends on the temperature ranges experienced by the various materials in the thermally conductive path. For small temperature variations, the use of a constant conductivity captures the majority of the effects while greatly simplifying the solution. As the range of temperatures experienced by the material increases, the more important the temperature dependence of the thermal conductivity becomes.

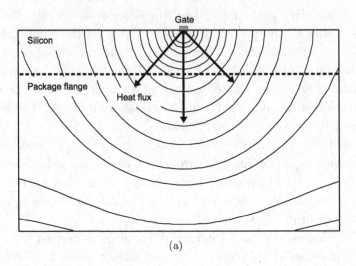

(a)

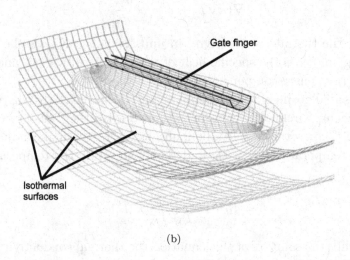

(b)

Fig. 5.1. Illustrations of the isothermal contours within an LDMOS transistor. In (a) isothermal contours are superimposed on a cross-section of the transistor mounted on top of a package flange. The simulated isothermal surfaces surrounding a single gate of the transistor are plotted in (b) [9]. © 2005 IEEE. Reprinted with permission.

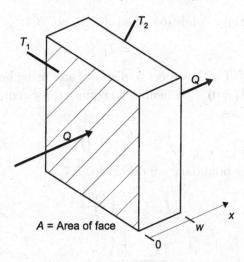

Fig. 5.2. Illustration of one-dimensional heat flow, with the temperature of the faces at $x = 0$ and $x = w$ given by T_1 and T_2, respectively.

If we make the assumption that κ is a constant, independent of position and temperature, then eq. 5.1 can be reduced to a simplified heat-conduction equation. Restricting the problem to heat flow in only one direction, along the x-axis, as illustrated in Fig. 5.2, eq. 5.1 becomes

$$\kappa \frac{\partial^2 T}{\partial x^2} + \dot{q} = \rho c \frac{\partial T}{\partial t} \tag{5.3}$$

In the steady state, the $\frac{\partial T}{\partial t}$ term of eq. 5.3 is equal to zero. After integration, the resulting equation for temperature T yields:

$$T = -\frac{\dot{q}x^2}{2\kappa} + A_1 x + A_2 \tag{5.4}$$

where A_1 and A_2 are constants of integration. This second-order differential equation can be solved given the following three common boundary conditions:

(i) at $x = 0$,

$$T = T_1 \tag{5.5}$$

(ii) at $x = 0$,

$$\frac{\partial T}{\partial x} = 0 \tag{5.6}$$

(iii) at $x = w$,

$$T = T_2 \tag{5.7}$$

Applying the boundary condition of eq. 5.5 to eq. 5.4:

$$A_2 = T_1 \qquad (5.8)$$

Differentiating eq. 5.4 with respect to distance and using boundary condition eq. 5.6 results in $A_1 = 0$, and hence the equation for temperature (eq. 5.4) in terms of x becomes:

$$T = -\frac{\dot{q}x^2}{2\kappa} + T_1 \qquad (5.9)$$

And finally, using boundary condition eq. 5.7,

$$T_1 - T_2 = \frac{\dot{q}w^2}{2\kappa} \qquad (5.10)$$

or

$$\frac{(T_1 - T_2)}{w^2/2\kappa} = \dot{q} \qquad (5.11)$$

The non-steady-state version of eq. 5.3 can be obtained by adding the $\frac{\partial T}{\partial t}$ term found in eq. 5.3 to eq. 5.11, to give

$$\frac{(T_1 - T_2)}{w^2/2\kappa} + \rho c \frac{\partial T}{\partial t} = \dot{q} \qquad (5.12)$$

Because eq. 5.12 is in terms of watts per unit volume, all terms on both sides of the equation have to be multiplied by the volume of the layer, V (area A times depth w) to obtain the equation in terms of total power dissipated through the layer, P:

$$P = \frac{(T_1 - T_2)2\kappa A}{w} + \rho c V \frac{\partial T}{\partial t} \qquad (5.13)$$

5.2.2 Thermal Circuit Development

Examining eq. 5.13 closely, we see that if the heat-flux Q is replaced with current I, the temperature T replaced with voltage V, and by substituting eqs. 5.14 and 5.15 into eq. 5.13, we can define two new quantities,

$$R_{th} = \frac{w}{2\kappa A} \qquad (5.14)$$

$$C_{th} = \rho c V \qquad (5.15)$$

and a very familiar equation appears, describing the current flowing through a capacitor shunted to ground and a series resistor:

$$I = \frac{(V_1 - V_2)}{R_{th}} + C_{th} \frac{dV}{dt} \qquad (5.16)$$

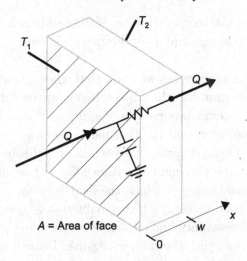

Fig. 5.3. Update of the one-dimensional, single layer diagram, with the thermal equivalent circuit.

The solution to the heat conduction problem presented in the previous sub-section is governed by the thermal resistance and capacitance as expressed in eqs. 5.14 and 5.15. It is worth noting that there is no thermal analogue for inductance; in contrast to electrical circuits where inductances can cause voltage overshoot, thermodynamics does not permit temperatures to overshoot their equilibrium values [8, 12].

This thermal–electrical analogy allows us to replace the heat conduction problem with the electrical circuit as illustrated in Fig. 5.3, and hence we can simulate the transfer of heat from one isothermal surface to that of another all within a circuit simulator.

The thermal analogues of resistance and capacitance need to be defined carefully [12, 13]. Thermal resistance is defined as the difference in temperature between two isothermal surfaces, or isotherms, divided by the total heat flow between them, and it has the units of degrees per watt (°C/W) [12]. This definition requires that all of the heat entering the plane defined by the first isothermal boundary must be equal to that passing through the second, there must not be net accumulation of mass or dispersal within the closed volume. The isothermal surfaces do not necessarily have to correspond to physical or geometrical surfaces, they are simply contours of equal temperature. Mathematically we can define thermal resistance as,

$$R_{\text{th}} = \frac{T_1 - T_2}{P} \tag{5.17}$$

where T_1 and T_2 are the temperatures at isothermal surfaces between which the heat-flux Q is passing, and P is the power, given by Q times the area of the surface.

From our practical experiences we know that when a heat source is applied it often takes some time for the object being warmed to reach a stable temperature. This occurs because the condition of equal heat-flux across both isotherms is not met, and the thermal capacitance (ρcV) between the isotherms is being charged, prior to thermal equilibrium. In addition to material properties, the thermal capacitance $C_{th} = \rho cV$ depends on size or volume of the material through which the heat is being conducted.

Under transient conditions, the heat conduction is governed by the combination of thermal resistance and capacitance, therefore we discuss thermal impedance rather than just thermal resistance. Instead of referring to the value of thermal capacitance directly, we use the thermal time constant, τ. The time constant refers to the time taken to reach $1 - e^{-1}$ or 63.2% of the equilibrium value, as defined in eq. 5.18. The temperature reaches about 95% of its equilibrium value in three time constants. For thermal circuits, time constants are often employed since they can provide significant insight into the heating or cooling curves. The thermal time constant or RC product is:

$$\tau = R_{th}C_{th} = \frac{w^2}{2\alpha} \tag{5.18}$$

where α is the thermal diffusivity

$$\alpha = \frac{\kappa}{\rho c}. \tag{5.19}$$

To further our understanding of the thermal behaviour of a packaged transistor, we shall examine a hypothetical equivalent circuit of a series of layered materials. Following the example from Sofia [13], a layered structure illustrated in Fig. 5.4(a) can be modeled with a cascade of thermal resistors and capacitors as illustrated in Fig. 5.4(b).

The thermal capacitances can only be detected during transient conditions, so we shall examine the response of the system in the time-domain resulting from a step excitation [13]. In Fig. 5.4(c), the instantaneous thermal impedance is plotted as a function of time. Immediately noticeable are three peaks within the time-domain response. Each peak occurs at approximately three times the time constant of each equivalent circuit.

The heat is generated in the die and flows through the die, into the die-attach region, and finally into the package flange. The thermal capacitance of the die is small, and so the die heats up rapidly, on the other hand, the

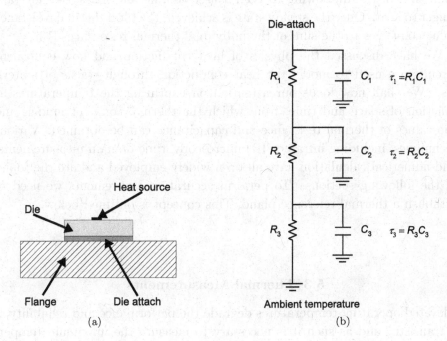

(a)   (b)

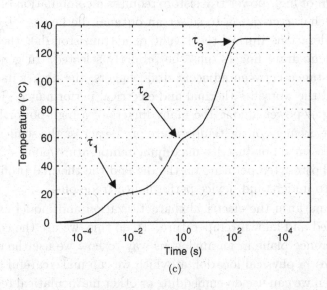

(c)

Fig. 5.4. The thermal analogue circuit of the packaged transistor shown in (a) is illustrated in (b). The time-domain response to a step excitation is plotted in (c). All figures have been derived from [13], © 1995 IEEE. Reprinted with permission.

values for the die-attach region and the flange of the package are much larger than that of the die, owing to their larger volume and mass, so they take longer to heat. Once the steady state is achieved, the total thermal resistance value converges to the sum of the individual thermal resistances [13].

We have discussed the physics of heat conduction and how equivalent circuits are used to model the heat conduction through stacks of materials. We shall now focus our attention on capturing the temperatures as function of space and time, from which the thermal compact models and the values of thermal resistance and capacitance can be obtained. Various techniques, including infrared (IR) microscopy, time-domain measurements, and numerical calculation have all been widely employed and are the topics of the following sections. To perform accurate measurements we need to establish a thermal reference plane. This concept is outlined below.

5.3 Thermal Measurements

Elevated operating temperatures degrade the performance and reliability of a transistor and as such it is necessary to measure the maximum temperature of the transistor die accurately. Electrical and thermal performances are often at odds with one another within the confined area available for the die. The design of high-power transistors requires a combination of thermal and electrical characterization to select an optimal die layout. To address thermal considerations during the layout of a transistor die, the spacing between adjacent drain fingers must be properly selected. In general, the smaller the distance between adjacent drain fingers, the higher its thermal resistance, and the worse its thermal and electrical performance. To ensure proper sharing of power among the many fingers of a high-power RF transistor, the measurement of the temperature profile across the semiconductor die is also necessary, not just its maximum temperature point. Measurements for the hottest temperature on the die and the thermal profile across it are critical to understand device performance trade-offs.

We need to perform the electrical characterization and model extraction at a well-defined and known temperature. To do this, we use the concept of a thermal reference plane in an analogous way to how we use the electrical reference planes: a physical location at which we can make careful measurements, and then we can use de-embedding or other mathematical techniques to infer the temperature at the point of interest, which in this case would be the channel of the transistor.

Pulsed DC and RF measurements are often made to characterize semi-conductor devices (see Chapter 3), in an attempt to minimize the temperature rise in the device. However, the definition of the pulse width and duty cycle inherently defines the thermal reference plane during the measurement. For example, by choosing an electrical measurement point inside the drain current pulse of around 10 microseconds after the switch-on time, we are effectively placing the thermal reference plane well inside the semi-conductor material substrate: the thermal energy has already traveled this far in the time available (see Fig. 5.5). In practice, we would normally use a much shorter pulse time (see Chapter 3). If we already know the thermal parameters – resistance and capacitance – of the semiconductor material, we can determine the junction temperature at the measurement time, and we can then determine the electrical model of the transistor at this isothermal condition.

A wide range of measurement techniques suitable for reliability analysis and compact thermal model development have been developed and they can be broadly categorized into two groups: steady-state and transient techniques. The steady-state technique captures the performance of the transistor after the operating temperature of the transistor has reached thermal equilibrium. An equivalent thermal resistance can be extracted from measurements of the dissipated power and the temperature change between two points, typically the case of the package and the transistor's junction. Steady-state techniques are used to capture the thermal profile across the die, and the maximum temperature. While these measurements are extremely useful, we need to be able to predict dynamic temperature changes and so the use of transient measurement techniques is required. To incorporate thermal dynamic effects properly in an electro-thermal compact model we need to obtain the thermal response of the transistor as a function of time. This is required in addition to the temperature gradients across the die, as well as the maximum temperature value and location on the die. These dynamic measurement techniques permit the extraction of the equivalent thermal impedance of the transistor. The thermal circuit can then be used to compute the temperature dynamically from any electrical signal and operating conditions of the transistor.

Each technique can also be classified as either a direct or inferred measurement based on whether the temperature is being measured directly or being inferred from an electrical measurement. The different physical phenomena and the advantages and disadvantages therein often dictate the applicability of each measurement technique [14]. Of primary concern are the minimum

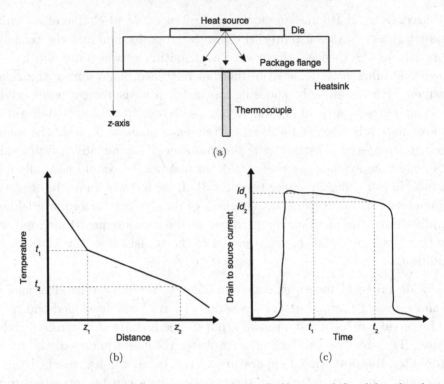

Fig. 5.5. The isotropic diffusion of heat in the half-space defined by the device, substrate, package and heatsink is illustrated in (a). The profile of temperature versus distance into the structure is provide in (b), at some instant t, showing how the different thermal resistances can affect the local temperature. The location of the measurement sample time in the drain current pulse shown in (c), corresponding to a depth d into the structure, and hence defining the thermal measurement reference plane in the device

temperature resolution, the smallest area or *spot-size* that can be measured, and the resolution in the time domain.

One inexpensive yet accurate method uses thermochromic liquid crystals which are deposited on top of the transistor [15–17]. Below a specific temperature these crystals are transparent; otherwise the crystal reflects a specific wavelength of light which depends on the temperature. The chemical make-up of the liquid crystals can be controlled to specify the range of temperatures over which it reacts. By viewing the reflected light through a microscope, very small spatial resolutions are possible. For this technique, the minimum detectable temperature resolution is approximately 1 °C and it is inexpensive to perform. The liquid crystal material can be difficult to apply evenly, especially for non-planar transistors. Using a similar methodology, although employing different physical principles, thermographic phosphors may also be used [14]. Since the materials are deposited on top of the transistor it is important to ensure that they do not affect its electrical performance.

Detection of the infrared (IR) energy from the heated surfaces of the transistor and package is commonly used as a non-invasive technique to obtain its temperature [14]. Infrared microscopy is similar to optical microscopy but special lenses are required to focus the longer infrared wavelengths. The IR radiation is gathered and focused on a liquid nitrogen cooled detector, where it is processed. With proper calibration, a colour image similar to the type generated by liquid crystals is obtained. Minimum spot sizes are approximately 5 μm^2 and measurement of the time-domain response is possible [18, 19].

Measurement with a spatial resolution of <3 μm and nanosecond time resolution is possible with very advanced techniques such as back-side laser interferometry [20, 21]. The measurement relies on the changes to the refractive index that occur due to temperature changes. Laser light is shone through the back side of the device, passes through the active area of the transistor, reflects from the top side, and returns. Changes in the refractive index cause a phase shift, which is proportional to the temperature rise along the path taken. Through standard interferometric techniques where this beam is combined with a reference beam, the phase shift can be measured and then translated to temperature [20, 22].

Temperatures can also be determined via electrical measurement. One of the most common techniques exploits the current–voltage relationship of a junction diode and its strong temperature dependence. The essence of using electrical measurements to detect temperature is to find a very temperature-sensitive parameter (TSP), either intrinsic to the device or in a device that is integrated next to the power transistor on the die. The Schottky diode within GaAs MESFETs and the source-to-drain diode of Si LDMOS FETs are often used for this purpose. The method requires a calibration where the diode voltage as a function of temperature is measured for a constant diode current. When the thermal steady state has been reached, the transistors are turned off and the constant diode current is injected. The diode voltage is measured as a function of time from which the values of thermal resistance and capacitance can be extracted [10].

The methods outlined above illustrate the richness of the techniques devoted to various types of thermal measurement and imaging. We shall now present in detail the theory, measurement equipment, and measured results for several techniques that we have used successfully in the development of compact thermal models for high-power transistors. We will first examine the theory behind IR thermography, discussing calibration and fixturing issues. Next, we shall expand on TSP measurement techniques. Both of these types of measurement have the ability to capture the necessary transient information required to develop the dynamic thermal circuit necessary for our compact model generation.

5.3.1 Infrared Microscopy

Infrared microscopy is often used to measure temperature directly on the transistor die while it is operating under realistic termination impedances and signal excitation. The IR microscope is often integrated within a test-bench that is used to measure the performance of the power amplifier. With this instrumentation, the temperature distribution across the die can be viewed as a function of power level, bias level, matching condition, frequency, and even by the selected modulation scheme: W–CDMA or IS–95, for instance.

The IR microscope has several lenses available, which allow the display of the overall temperature distribution of the transistor, as well as magnified views of relevant areas of the transistor. The most advanced equipment has spatial resolutions of less than 5 µm [19]. All the measurement data are digitized by the controlling computer and the temperature of any arbitrary scan line across a section of the die can be plotted. A photograph of an IR microscopy system is provided in Fig. 5.6.

The IR microscope converts the radiation detected by an indium antimonide sensor into temperature, using a calculated or fixed emissivity, and Planck's black-body radiation law [23]. The amount of radiation leaving a point on the surface, the *radiance*, is measured over a small spot size and is compared with the radiance from an ideal black-body radiator. The only issue with this approach is that the device-under-test is not a true black-body radiator and does not absorb all of the radiation impinging upon it. Thus a calibration comparing a black-body with the object to be measured must be performed to obtain a reference measure of the temperature. The calibration procedure uses an unpowered device, heated externally to two different temperatures [23]. From these two temperatures the emissivity of the object-under-test can be computed; emissivity is defined as the ratio of the radiant energy of an object to the radiant energy of a black-body object at the same temperature as the object.

Fig. 5.6. A photograph of an IR microscopy system. A test-fixture is mounted on the thermal chuck beneath the IR microscope. Software on the controlling computer is used to calibrate and capture, compute, and display the temperature [18]. © 2001 SPIE. Reprinted with permission.

Most modern IR microscopes provide pixel-by-pixel measurement with emissivity correction. However, for some semiconductors like silicon, which are translucent to the infrared portion of the electromagnetic spectrum, a coating that does not affect the RF performance must be applied to ensure a constant emissivity across the entire die [18, 24, 25].

An example of a thermal image of a transistor is shown in Fig. 5.7. In this example, the transistor is operating at DC and is dissipating 60 W. The temperature profile across the transistor die along scan line #1 is shown in Fig. 5.8, for the transistor under RF excitation at 2.0 GHz, and at DC: in each case the dissipated power is 60 W. From these figures, the areas of higher temperature can be seen clearly. These regions correspond to

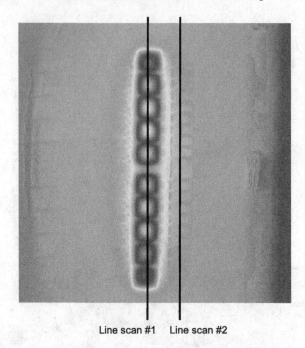

Line scan #1   Line scan #2

Fig. 5.7. A thermal image obtained using an IR microscope. The image is of an 80 mm silicon LDMOS transistor within a package under DC operation where 60 W are dissipated. Two scan lines are indicated where plots of the temperature as a function of position are generated. The first scan line crosses through the center of the die, while the second crosses the set of wires connecting the drain of the transistor to the package.

high local power dissipation, and hence higher current density. A properly designed high-power RF transistor will exhibit a uniform power sharing among its many fingers, resulting in a fairly small temperature gradient across the semiconductor die [26]. The cooling effect of the RF drive in Class A operation can also be seen.

An advantage of IR microscopy is that it can be used to determine the temperature at various locations within the packaged transistor. This is instrumental in detecting potential reliability issues with bondwires operating at elevated temperatures. As an example, the temperature profile of an array of wires carrying high current levels at 2 GHz is illustrated in Fig. 5.9. The difference in temperature between the inner and outer wires of the array can be seen clearly. All the wires have an identical cross-section and three-dimensional shape, and hence identical resistance values; therefore the high temperatures on the exterior wires indicate high current levels. This

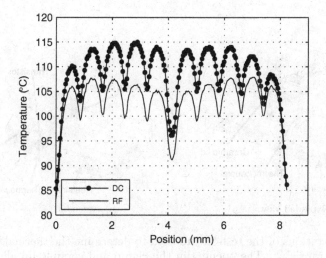

Fig. 5.8. Plot of the temperature as a function of position along scan line #1 as defined in Fig. 5.7. The temperature was measured for 2.0 GHz excitation and at DC, for 60 W power dissipation.

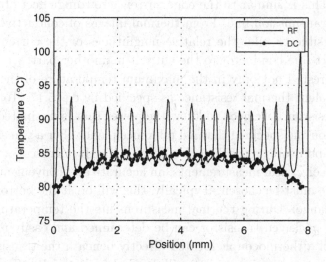

Fig. 5.9. Thermal image of an array of bondwires excited at 2.0 GHz and at DC. The temperature is measured for 60 W dissipated power along scan line #2 in Fig. 5.7. The U-shaped temperature distribution of the peak temperatures, at the wire locations, is a result of the coupling between adjacent bondwires.

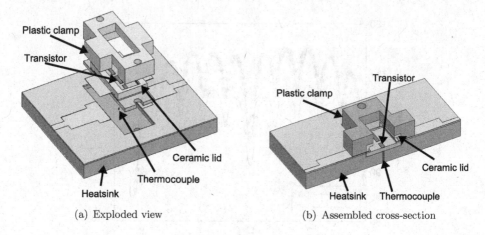

(a) Exploded view   (b) Assembled cross-section

Fig. 5.10. Illustrations of the test-fixture used to determine the thermal resistance of a packaged transistor. The opening in the clamp and ceramic lid allow the IR microscope to have an unimpeded view of the die. The thermocouple is machined into the heatsink and positioned to be beneath the active device heat-generation area.

confirms the coupled transmission line nature of the bondwire array indirectly, in which the current is forced away from the center of the array to the exterior. This is similar to the edge current crowding effect observed in a microstrip transmission line. From thermal images of conductive objects, it is often possible to infer the relative magnitudes of the current flowing within one part of a conductor to the current in another part.

We are interested not only in the maximum temperature of the die, but also the equivalent thermal resistance as specified by eq. 5.17. To compute the thermal resistance we need to know the power that is dissipated and the temperatures on the specified thermal reference planes. For a transistor, the maximum temperature on the die is found using the IR microscope. In a similar manner to electrical measurements, an accurate and convenient thermal reference plane must be chosen. Typically, this will be at the bottom side of the package flange. During thermal measurements the temperature at the bottom of the packaged transistor can be determined and easily monitored with the use of a thermocouple located directly beneath the transistor's active cell, or heat-generating area, as illustrated in Fig. 5.10. To allow an unimpeded view of the surface of the die, a hole is drilled in the ceramic lid of the transistor, or the lid is removed. For over-molded packages, the mold compound can be etched away until the surface of the die is exposed. Before mounting the transistor in the fixture, the entire device is sprayed with the high-emissivity coating, as previously mentioned.

Recent advances in IR microscopy techniques permit this technique to be used to capture rapid thermal transients. Results have been reported that enable the measurement of the temperature over a 15 µm spot size with time resolution of 1 µs [19]. This capability enables us to measure the thermal time constants that are typically found in power semiconductor devices. Historically, IR measurements were only capable of measuring time resolutions several orders of magnitude larger. This poor time resolution led to the development of the more common technique of using the electrical measurement of TSP versus time to capture the thermal transient.

5.3.2 Electrical Temperature-Sensitive Parameters

As discussed in Chapter 2, the thermal dynamics of a transistor have the effect of introducing low-frequency memory in high-power RF transistors and circuits; therefore an accurate method for measuring the transistor thermal transients is necessary. We shall describe the basic principles of the measurement of electrical TSP.

In its most fundamental form, a TSP is any parameter that can be measured by electrical means, and has a strong thermal dependence. By establishing an appropriate mapping between the electrical parameter's values and the temperature of the device, the TSP takes the place of an accurate built-in thermometer within the transistor.

Several methods have been suggested for performing thermal measurements of transistors by taking advantage of different TSPs [10, 27]. The most common TSP method presently used in the industry exploits the strong thermal dependence of a junction diode current–voltage characteristic to infer the channel temperature of the device. In GaAs MESFETs the gate Schottky diode current is a very effective TSP [28]; in Si LDMOS FETs, we use the source-to-drain junction diode reverse current.

The TSP can be calibrated by applying a small constant current through the diode and varying the temperature of the device using a controlled thermal environment such as an oven. An underlying assumption of this TSP calibration method is that the amount of dissipated power in the TSP, and associated temperature rise, is very small compared with the amount of dissipated power that occurs in the transistor under normal operating conditions. To ensure that this approximation holds, we need to make certain that the diode current used as the TSP temperature sensing mechanism is at least a couple orders of magnitude smaller than the dissipated power under normal operating conditions. Another common approach is to lay out very small testing diodes in very close proximity to the transistor's active area, for the sole purpose of using them as TSP elements. Even though the use of testing diodes is very common and straightforward, errors are introduced in this process, as the testing diode is at some distance away from the active channel of the device, which is the heat-generating area of the transistor.

The TSP measurement can be used to perform measurements under steady-state or transient conditions. We shall highlight the TSP measurement technique for the purpose of determining the transient thermal response of the transistor. The use of the FET diode characteristics in the measurement of the TSP under transient conditions is outlined as follows. First we need to bias the transistor under normal operating conditions until thermal equilibrium is reached (heating mode); then we swiftly change its bias condition to one in which a specific diode characteristic dominates. A small current is applied to the TSP diode while its voltage is measured, yielding its voltage response versus time (sampling mode). From the oven-calibrated diode voltage versus temperature characteristics, the temperature versus time can easily be computed. Notice that this method captures the transistor's cooling curve, so there is an inherent assumption that the cooling and heating profiles are the opposite of each other. In practice the measurement of the cooling curve is a lot more convenient to implement than the heating curve, which will require that the device be switched on and then switched to sampling mode, in an interactive manner. When measuring the cooling curve, the device needs to be switched from the heating mode (after thermal steady state has been achieved) to the sampling mode only once, hence making the measurement and data gathering much simpler and faster.

An alternative to the measurement of a diode current as a TSP device is to use the actual drain-to-source current thermal dependence as an indirect way to measure the temperature rise in the transistor. The drain-to-source current as a function of time can be measured with a fast data capture multimeter or a pulsed DC system. If, in addition, the drain-to-source current can be measured isothermally under pulsed conditions, and a model can be developed for it, the thermal compact model of the transistor can be generated by fitting the measured non-isothermal drain-to-source current versus time against the resulting simulated electro-thermal nonlinear current.

In contrast to optical measurement techniques, the TSP technique is incapable of discerning the two-dimensional temperature profile at the surface of the device under test, yielding only an effective average temperature. Although IR techniques are capable of obtaining the temperature gradients at the surface of the semiconductor, it is worth mentioning that the actual transistor channel is below the surface, and thus the IR techniques do not provide an exact measure of the channel temperature. On the other hand, the TSP technique does not depend on the physical location of the measurement point, therefore as the channel heats up, the selected TSP will change accordingly.

Once the thermal transients are captured, a more comprehensive thermal compact model can be developed, which includes thermal resistances and capacitances. Depending on the application, the relevant thermal time constants need to be captured. For example, if the application is a high-power RF transistor excited with a modulated signal, time constants in the order of the inverse of the baseband maximum frequency need to be characterized and modeled if one is to capture the thermal dynamical effects on the low-frequency memory effects.

5.4 Thermal Simulations

The performance of high-power RF transistors is limited by, among other things, its thermal environment. Thermal management is a very important issue that needs to be thoroughly considered during the design phase, as it represents a considerable fraction of the total system cost. In this regard, the ability to predict the maximum operating junction temperature of the transistor is driven by reliability considerations, while the ability to determine the temperature at the bottom of the package flange dictates the design of external thermal management system. In both cases, the ability to determine *a priori* the temperature signature of the transistor will greatly aid the design process.

Improved thermal performance can be achieved by separating the heat sources on the die by increasing the drain-to-drain finger spacing. This results in a decrease of the transistor's total thermal resistance. Unfortunately, if the area of the die is made too large while maintaining a given total gate periphery (by adjusting the unit gate-width or the drain-to-drain finger spacing), the electrical performance degrades. The layout of the die can be optimized with a comprehensive approach to modeling the transistor by combining electromagnetic, thermal, and physical transistor simulations [29]. Following another approach, thermal simulations can be used to generate various transistor layouts, which yield comparable thermal resistances. In turn, the electrical performance of each layout can then be measured and the die that balances electrical and thermal performance the best is selected. The selection of an optimal compromise between electrical and thermal factors is a concern of both the transistor and the power amplifier designer.

In addition to transistor layout optimization, the determination of an accurate thermal compact model is of vital importance if the power amplifier designer needs to account or predict thermal memory effects and the impact of self-heating on linearity. An approach to compact model generation that relies on measured data to extract the necessary equivalent circuit parameters is always limited to transistors or structures that are already manufactured. Furthermore, it may not be possible to determine the temperatures and thermal gradients at points that are difficult or impossible to access during the measurement process. Therefore, a robust and accurate thermal simulation process is necessary for the analysis, modeling, and design of RF power transistors.

The development of a thermal simulation process begins with studying the heat-conduction equation, eq. 5.1, where we often start by developing analytical solutions. Unfortunately, these solutions can only be determined for a few canonical geometries whose boundaries fall on consistent coordinate surfaces. Owing to the arbitrariness of structures containing intricate three-dimensional geometries and complex material properties, solutions are difficult to formulate.

Thermal modeling of the complex three-dimensional geometry of a transistor can be performed by applying standard numerical techniques to the heat-conduction equation. The essential idea is to reduce the governing equation, through approximations, to a system of linear equations that can be solved through standard matrix inversion techniques. Finite-element (FE) and finite-difference (FD) techniques are common and commercially available [30]. In each of these techniques the volume is discretized into a mesh upon which the conduction equation is solved. For complex geometries these simulations are computationally intensive. When simulating devices containing fluid flow, the FE and FD techniques are not suitable and computational fluid dynamics (CFD) are necessary.

5.4.1 Numerical Techniques

Through numerical solutions to the heat-conduction equation, it is possible to study thermal gradients across a transistor and to compute the temperature at any point within the structure. From the simulated thermal resistance matrix, the thermal resistance between any two locations is readily extractable. Simulations of the time-domain response to a time-varying input signal are frequently performed. From simulated time-domain responses, compact thermal models are extracted; this is a focus of the subsequent section.

For accurate simulations, we need to be concerned with the approximations made during the simulation. These approximations range from the
numerical method used, the density of the mesh, the boundary conditions
involved, and simplifications to the simulated geometry. A central issue for
all techniques is to obtain accurate material properties. When necessary, the
thermal properties of materials can be made to vary as a function of temperature. In the following, we present the main features of the FE and FD
algorithms, which are commonly used to solve heat conduction problems.

Simulation of an arbitrary geometry using the FE method begins by discretizing the volume into smaller volumes called *elements*. Typically these
elements are either tetrahedral or quadrilateral volumes. Through combinations of various sizes and shapes, a collection of elements, called a *mesh*, is
used to describe complex three-dimensional geometries. One of the primary
strengths of the FE method is its ability to handle irregular geometries and
boundaries. The elements are connected to points called *nodes* and any solution must be continuous along common boundaries of adjacent elements [30].

Automatic mesh generation and refinement routines are included as a
standard part of simulation packages, as it is critical to have sufficient mesh
density to approximate the problem accurately. If the mesh is too sparse,
the accuracy of the simulation is in jeopardy. Conversely, a very dense mesh
can provide accurate results but at the cost of increased computation time.
As with any numerical simulation where discretization is involved, a compromise between mesh density and computation time is required. Factors
to consider are the accuracy of the required solution, the type of problem
(static versus dynamic), and sizes of the geometries involved. An example
mesh used during a FE simulation of a transistor mounted to a package
flange is illustrated in Fig. 5.11.

After generating the mesh, any heat sources, temperatures, and boundary
conditions are assigned. The method then transforms the geometry and
boundary conditions into a system of equations, which is then solved for the
temperatures at every node within the structure.

In the FD approach, solutions to the heat-conduction equation are numerically approximated by replacing the partial derivatives with finite-
differences [31]. Unlike the FE method, a shortcoming of the FD method is
the difficulty it has with incorporating irregular geometrical boundaries [30].
One solution is to increase the mesh density resulting in a 'staircase' approximation to the irregular or curved boundary. Unfortunately, this significantly
increases the overall mesh density and reduces the efficiency of the method.
As with the FE method, a compromise between desired accuracy and simulation time must be made.

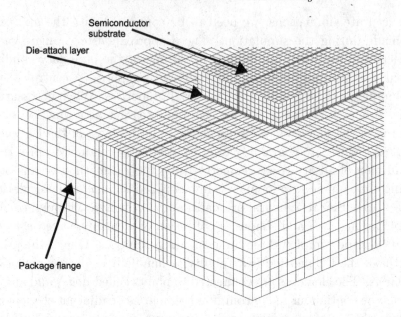

Fig. 5.11. An illustration of the mesh used in a finite-element simulation. In this example, the mesh discretizes the semiconductor die, die-attach layer and package flange.

For either approach (FE or FD), once the simulation is complete, post-processing algorithms are applied to the temperature field within the geometry to generate plots of isothermal contours or surfaces (see Section 5.4.2). The thermal resistances can be computed by accessing the nodal temperature values and the input heat-flux. For either method it is possible to derive the thermal model directly from the mesh and nodal temperature values. Each element of the mesh can be represented by a network of resistors and capacitors. A three-dimensional lattice of resistors and capacitors can be directly synthesized by extracting the equivalent circuit values for the network representing each element [5, 32, 33]. For practical problems this leads to a very large circuit containing thousands of elements. Unfortunately, this large circuit is slow to simulate and difficult to represent within a circuit simulator. Accordingly, model-order reduction techniques have been widely applied to reduce the network and yet still capture the thermal effects [33, 34].

5.4.2 A Thermal Analysis Example

As an example of the simulation techniques presented in this chapter, we present a comparison of measured and simulated results for a GaAs PHEMT transistor designed for high-power wireless infrastructure applications. This transistor can provide 50 W of output power at 2.1 GHz when operating in CW mode at its 1-dB compression point. The active area of the transistor consists of twelve active tubs, or collections of fingers that are electrically connected to one another. The transistor is fabricated on a 25 µm thick substrate with a back-side metallization and die-attach layer having a total thickness of 50 µm. The die is mounted within a standard metal-ceramic package containing a 1500 µm thick copper-tungsten flange.

Simulations were performed with ANSYS® software, and measurements of the die were taken using an IR microscope (see Fig. 5.6). Measurements were performed under DC conditions where the dissipated power ($V_{ds} \times I_{ds}$) was set to 12 W. The FE analysis was performed under the same power dissipation conditions used during the collection of the measured data.

Figure 5.12(a) shows the measured and simulated temperature profiles on the die. The top picture is taken using the IR microscope while the bottom picture is generated from simulation. The simulated and measured results are in very close agreement for the temperature variation over the entire surface of the die. To compare the measured and simulated results more closely, the temperature of the die is extracted along the horizontal lines indicated in Fig. 5.12(a) and plotted in Fig. 5.12(b). The FE simulations are in good agreement with the maximum measured temperature and the thermal gradient across the die.

5.5 Compact Models

We now turn our attention to the development of thermal compact models. To develop a model that is capable of accounting for dynamic thermal effects, we need to obtain the time-domain response of the device-under-test through either measurement or simulation. The model could also be extracted directly from the thermal impedance matrix resulting from either FE or FD simulation methods.

The thermal conductivities and capacitances of the materials used for making high-power transistors are fairly constant over the temperature range of interest. This allows us to have a linear thermal model, without any significant error. A linear model is significantly easier to extract and compute, and we can use standard techniques for its solution. A major advantage of

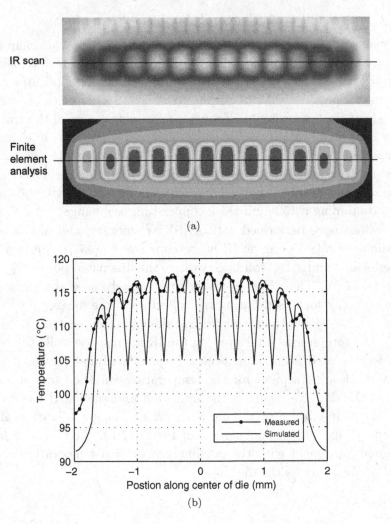

Fig. 5.12. Measured and simulated die temperature profiles of a GaAs PHEMT transistor. The top picture in (a) is a measured with and IR microscope while the bottom picture is the simulated two-dimensional temperature profile of the die. Measured and simulated temperatures along the horizontal lines indicated in (a) for the GaAs PHEMT transistor are plotted in (b).

having a linear thermal compact model that describes the thermal environment is that the temperature of the device can then be computed with any arbitrary heat excitation. This quality of a linear system is described by the so-called linear time invariant (LTI) theorem [35].

The linear thermal model can be represented by an electrical analogue. In this electrical circuit we use resistors and capacitors to model the thermal

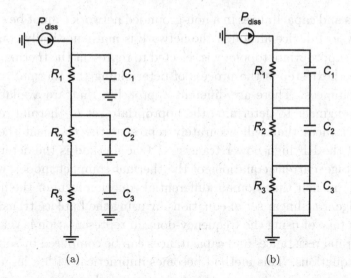

(a)   (b)

Fig. 5.13. Node and element definitions for two versions of a three-rung thermal RC network. The non-grounded capacitor thermal network is shown in (a), and the grounded capacitor network is shown in (b).

resistance and capacitance, respectively. We cannot have a thermal inductance component, as this violates the second law of thermodynamics (see Section 5.2.2). The current represents the heat-flux in the transistor, and we use a current source to represent the power dissipation in the transistor. We use a voltage source to reference a point of constant temperature in the device, such as the heatsink temperature. In this context, the ground terminal is a point of infinite heat capacity, just as the ground in an electrical circuit can sink all of the current.

The most common topology for the thermal network is a cascade of resistors and capacitors in parallel, as shown in Fig. 5.13(a). The R-C pairs are often associated with the physical structure of the transistor. The 'grounded-capacitor' thermal network shown in Fig. 5.13(b) is a physically more correct representation of the fundamental heat-flow equation.

Even though the grounded-capacitor thermal circuit is physically correct, the resistances and capacitances of a non-grounded thermal circuit can be adjusted to produce the same thermal transient response as the grounded-capacitor circuit. The thermal capacitance of a real thermal system relates the change in temperature at each position with respect to time. The non-grounded network, by adjoining neighboring capacitances instead of connecting to ground, relates the change in temperature at each position with respect to temperatures at adjoining elements in the system. The individual resistances and capacitances in a non-grounded network cannot be correlated to the physical device although the network is mathematically convenient.

Once an appropriate topology is selected to represent the thermal compact model, the next step is the process of determining the thermal resistances and capacitances. There are different approaches that are available to the modeling engineer to determine the appropriate set of thermal resistances and capacitances that will accurately represent the transient thermal behaviour of the RF high-power transistor. One method is the determination of the voltage–current equations of the thermal compact model, which can be written in their time-domain differential equation form or the frequency-domain algebraic linear set of equations by using the Laplace transform [10]. The advantage of using the frequency-domain representation is that the values of thermal resistances and capacitances can be computed by solving a set of linear equations. This method becomes impractical with a large network of resistors and capacitors, as the derivation of the voltage–current equations for all nodes in the network will be quite lengthy. Another approach is to use a circuit simulator to fit the thermal resistances and capacitances to the measured or simulated transient thermal response.

Pulsed measurement techniques, as outlined in Chapter 3, can also be used to extract the thermal resistance [36]. By exciting the transistor with short pulses with a low duty cycle, isothermal electrical measurements can be measured as a function of the ambient temperature. The thermal resistance is then extracted by comparing these measurements with the DC characteristics and the dissipated power. Furthermore, by measuring the drain current as a function of time, the dynamic thermal model can also be extracted. Thermal resistances can also be extracted using DC or small-signal measurements at different ambient temperatures [37, 38]. This is a popular method, because the measurements that are used to extract the thermal model can be combined with the measurements required for the compact electrical model extraction, making this approach practical and attractive.

5.5.1 Remarks on Building a Self-Consistent Electro-Thermal Model

A thermal model, properly implemented in a circuit simulator, is capable of computing the electrical and thermal responses of the device. The ability of an electro-thermal model to predict static or dynamic thermal effects depends on the nature of the thermal model. A model that only consists of thermal resistances is only capable of predicting a steady-state representation of the thermal environment, while a thermal model composed of resistances and capacitances is fully capable of predicting the thermal transients of the transistor.

In this chapter we have introduced all the basics concepts that are required to build the thermal part of a self-consistent electro-thermal model. But before we present how to build such a model, we need to describe the electrical nonlinear model of the transistor itself. The strategy that we will follow is to de-couple the electrical and thermal dependencies in both the measured data as well as the mathematical model construction. Once we have completed the treatment of the electrical nonlinear transistor model, we will be in a position to merge the electrical and thermal models of the FET into a self-consistent electro-thermal model. That topic will be covered in detail in the next chapter.

References

[1] Z. Radivojevic, K. Anderson, L. Bogod, M. Mahalingam, J. Rantala, and J. Wright, "Novel materials for improved quality of RF-PA in base-station applications," *IEEE Trans. Adv. Packag.*, 28, no. 4, pp. 644–649, Dec. 2005.

[2] M. Guyonnet, R. Sommet, R. Quere, and G. Bouisse, "Non-linear electro thermal model of LDMOS power transistor coupled to 3D thermal model in a circuit simulator," in *Proc. 34th European Microwave Conf.*, Amsterdam, The Netherlands, Oct. 2004, pp. 573–576.

[3] S. Wunsche, C. Clauss, P. Schwarz, and F. Wikler, "Electro-thermal circuit simulation using simulator coupling," *IEEE Trans. VLSI Syst.*, 5, no. 3, pp. 277–282, Sept. 1997.

[4] D. Denis, C. M. Snowden, and I. C. Hunter, "Coupled electrothermal, electromagnetic, and physical modeling of microwave power FETs," *IEEE Trans. Microwave Theory Tech.*, 54, no. 6, pp. 2465–2470, June 2006.

[5] J. T. Hsu and L. Vu-Quoc, "A rational formulation of thermal circuit models for electrothermal simulation - part I: finite element method," *IEEE Trans. Circuits Syst. I*, 43, no. 9, pp. 721–732, Sept. 1996.

[6] C. R. Nave, "Heat and thermodynamics," in *HyperPhysics*, Georgia State University, 2005. http://hyperphysics.phy-astr.gsu.edu/hbase/hph.html#hph

[7] G. Elert, "Thermal physics," in *The Physics Hypertext-Book*, 2006. http://hypertextbook.com/physics/thermal/

[8] J. H. Lienhard IV and J. H. Lienhard V, *A Heat Transfer Textbook*, 3rd edn. Phlogiston Press, 2006. http://web.mit.edu/lienhard/www/ahtt.html

[9] A. M. Darwish, A. J. Bayba, and H. A. Hung, "Thermal resistance calculation of AlGaN-GaN devices," *IEEE Trans. Microwave Theory Tech.*, 52, no. 11, pp. 2611–2620, Nov. 2004.

[10] J. Whang, "Thermal characterization and modeling of LDMOS FETs," Master's thesis, Massachusetts Institute of Technology, Cambridge, MA, 2000.

[11] R. Anholt, *Electrical and Thermal Characterization of MESFETs, HEMTs and HBTs*. Norwood, MA: Artech House, 1995.

[12] J. W. Sofia, *Fundamentals of Thermal Resistance Measurement*, Analysis Tech., Inc., 1995. www.analysistech.com/downloads/fundamen.pdf

[13] J. W. Sofia, "Analysis of thermal transient data with synthesized dynamic models for semiconductor devices," *IEEE Trans. Comp., Packag., Manufact. Technol. A*, 18, no. 1, pp. 39–47, Mar. 1995.

[14] D. L. Blackburn, "Temperature measurements of semiconductor devices - a review," in *20th Semiconductor Thermal Measurement and Management Symp. Dig. (SEMI-THERM)*, San Jose, CA, Mar. 2004, pp. 70–79.

[15] D. J. Channin, "Liquid-crystal technique for observing integrated circuit operation," *IEEE Trans. Electron Devices*, 21, no. 10, pp. 650–652, Oct. 1974.

[16] J. H. Park and C. C. Lee, "A new configuration of nematic liquid crystal thermography with application to GaN-based devices," *IEEE Trans. Instrum. Meas.*, 55, no. 1, pp. 273–279, Feb. 2006.

[17] M. Nishiguchi, M. Fujihara, and H. Nishizawa, "Precision comparison of surface temperature measurement techniques for GaAs IC's," *IEEE Trans. Comp., Hybrids, Manufact. Technol.*, 16, no. 5, pp. 543–549, Aug. 1993.

[18] G. C. Albright, J. A. Stump, C. Li, and H. Kaplan, "Emissivity-corrected infrared thermal pulse measurement on microscopic semiconductor targets," in *Proc. SPIE*, 4360, Nov. 2001, pp. 103–111.

[19] M. Afridi, D. Berning, A. Hefner, J. Suehle, M. Zaghloul, E. Kelly, Z. Parrilla, and C. Ellenwood, "Transient heating study of microhotplates by using a high-speed thermal imaging system," in *18th Semiconductor Thermal Measurement and Management Symp. Dig. (SEMI-THERM)*, San Jose, CA, Mar. 2002, pp. 92–98.

[20] D. Pogany, S. Bychikhin, C. Furbock, M. Litzenberger, E. Gornik, K. Esmark, and M. Stecher, "Quantitative internal thermal energy mapping of semiconductor devices under short current stress using backside laser interferometry," *IEEE Electron Device Lett.*, 49, no. 11, pp. 2070–2079, Nov. 2002.

[21] D. Pogany, V. Dubec, S. Bychikhin, C. C. Furbock, M. Litzenberger, G. Groos, M. Stecher, and E. Gornik, "Single-shot thermal energy mapping of semiconductor devices with the nano-second resolution using holographic interferometry," *IEEE Electron Device Lett.*, 23, no. 10, pp. 606–608, Oct. 2002.

[22] W. Richter, M. V. Bossche, and J.-P. Teyssier, "D 1.3.2.1 TARGET CLASSIC device annual report," TARGET, 2005.

[23] J. McDonald and G. Albright, "Microthermal imaging in the infrared," Electronics Cooling Online, 1997. http://www.electronics-cooling.com/articles/1997/jan/jan97_04.php

[24] P. W. Webb, "Thermal imaging of electronic devices with low surface emissivity," *IEE Proc. G – Circ., Dev. Syst.*, 138, pp. 390–400, June 1991.

[25] M. Mahalingam and E. Mares, *Thermal Measurement Methodology of RF Power Amplifiers*, Freescale Semiconductor, Inc., 2004. http://www.freescale.com/files/rf_if/doc/app_note/AN1955.pdf

[26] W. Batty, C. E. Christoffersen, A. J. Panks, S. David, C. M. Snowden, and M. B. Steer, "Electrothermal CAD of power devices and circuits with fully physical time-dependent compact thermal modeling of complex non-linear 3-D systems," *IEEE Trans. Comp. Packag. Technol.*, 24, no. 4, pp. 566–590, Dec. 2001.

[27] R. P. Stout, "How to use thermal data found in datasheets," 2006. http://www.onsemi.com/pub/Collateral/AND8220-D.PDF

[28] H. Fukui, "Thermal resistance of GaAs field-effect transistors," in *Int. Electron Devices Mtg. Tech. Dig.*, 26, Washington, DC, 1980, pp. 118–121.

[29] D. Denis, C. M. Snowden, and I. C. Hunter, "Design of power FETs based on coupled electro-thermal-electromagnetic modeling," in *IEEE MTT-S Int. Microwave Symp. Dig.*, Long Beach, CA, June 2005, pp. 461–464.

[30] C. A. Harper, *Electronic Packaging and Interconnection Handbook*, 4th edn. New York, NY: McGraw-Hill Companies, Inc., 2004.

[31] F. Kreith, Ed., *The CRC Handbook of Thermal Engineering*. London, UK: CRC Press, 2000.

[32] A. Ammous, S. Ghedira, B. Allard, H. Morel, and D. Renault, "Choosing a thermal model for electrothermal simulation of power semiconductor devices," *IEEE Trans. Power Electron.*, 14, no. 2, pp. 300–307, Mar. 1999.

[33] J. T. Hsu and L. Vu-Quoc, "A rational formulation of thermal circuit models for electrothermal simulation - part II: model reduction techniques," *IEEE Trans. Circuits Syst. I*, 43, no. 9, pp. 733–744, Sept. 1996.

[34] R. Sommet, D. Lopez, and R. Quere, "From 3D thermal simulation of HBT devices to their thermal model integration into circuit simulators via Ritz vectors reduction technique," in *Proc. 8th Intersociety Conference on Thermal and Thermomechanical Phenomena in Electronic Systems ITHERM*, San Diego, CA, May 2002, pp. 22–28.

[35] A. V. Oppenheim and R. W. Schafer, Eds., *Discrete-Time Signal Processing*. Englewood Cliffs, NJ: Prentice-Hall Inc., 1989.

[36] Y. Yang, Y. Woo, J. Yi, and B. Kim, "A new empirical large-signal model of Si LDMOSFETs for high-power amplifier design," *IEEE Trans. Microwave Theory Tech.*, 49, no. 9, pp. 1626–1633, Sept. 2001.

[37] R. Menozzi and A. C. Kingswood, "A new technique to measure the thermal resistance of LDMOS transistors," *IEEE Trans. Device Mat. Rel.*, 5, no. 3, pp. 515–521, Sept. 2005.

[38] B. M. Tenbroek, M. S. L. Lee, W. Redman-White, R. J. T. Bunyan, and M. J. Uren, "Self-heating effects in SOI MOSFET's and their measurement by small signal conductance techniques," *IEEE Trans. Electron Devices*, 43, no. 12, pp. 2240–2248, Dec. 1996.

6

Modeling the Active Transistor

6.1 Introduction

So far, we have described the structures and the principles of operation of field effect transistors, and how such principles and features need to be considered in the context of developing a compact model of the device for use in a circuit simulator to aid power amplifier design. We have also described in some detail how to characterize and model the transistor package and its internal passive components, in the electromagnetic, electrical, and thermal environments that surround the transistor. We shall now begin to focus our attention on the transistor itself: the active semiconductor channel where the transistor action takes place, and the gate, source, and drain electrodes that enable electrical connection to the rest of the power transistor's environment.

A typical high-power discrete LDMOS transistor die is shown in Fig. 6.1. The manifolds for the gate and drain, where the bondwires connect to the rest of the circuit, can be seen clearly. The manifolds also feed the many gate and drain fingers that form an inter-digitated pattern on the silicon surface, defining the individual transistor units that form the whole device. The source connection is made through the silicon itself to the back side of the die, making contact directly to the package, which provides the electrical ground and also a thermal reference: it is easy to make a repeatable temperature measurement at the package when operating the transistor under test. For comparison, in Fig. 6.2 we show a GaAs power transistor die; here the ground connection is made by through-vias to the back-side metallization.

The 'transistor action' in the FET occurs in the active channel that lies under the gate: this is where the current amplification happens. This part of the transistor is called the *intrinsic* device. In the real device we need additional semiconductor and metal components to connect this active region to the outside world, to get our signal in and our amplified signal out. These

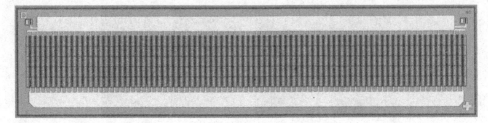

Fig. 6.1. Discrete LDMOS 80 W power transistor; this device has 164 gate fingers for a total gate width or periphery of 82 mm, courtesy of Freescale Semiconductor.

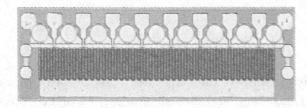

Fig. 6.2. Discrete GaAs 10 W power transistor; this device has 80 gate fingers for a total gate width or periphery of 15.2 mm, courtesy of Freescale Semiconductor.

additional components are called *extrinsic* components, or, quite often, *parasitics*. The notion of 'parasitic' can give the impression that these elements of the transistor are unwanted: while they generally tend to degrade the fundamental electrical performance, they are essential to the structure of the device, and in a careful design their negative impact is minimized. Since these parts of the structure are an integral part of the transistor, but do not contribute to the fundamental transistor action, we prefer the term extrinsic. Other considerations, such as thermal management, also play a role in the overall shape and size of the transistor die. From the modeling perspective, we need to include the electrical and thermal effects of these additional components in our model, for it to be complete. A schematic description of the transistor model is shown in Fig. 6.3. This figure illustrates the various levels or shells that we have to model and de-embed to get to the *active* transistor itself.

In our development of a complete electrical model of the transistor, we shall first describe how we can model the electrical properties of the manifolds. Second, we shall describe measurement and analysis techniques that are used to model the resistive and reactive extrinsic components. Next, for the creation of the intrinsic model of the transistor, we will measure S-parameters over the attainable gate and drain voltage bias space of the

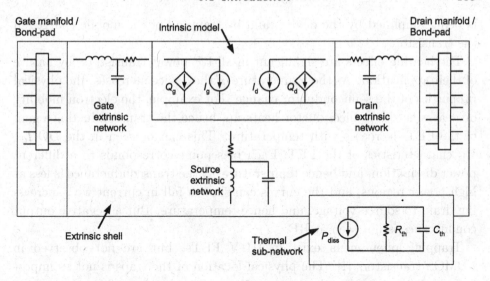

Fig. 6.3. FET model schematic diagram showing the gate and drain manifolds, the extrinsic component shell, and the intrinsic transistor model.

transistor, these measurements being taken isothermally using pulsed techniques, over a range of heatsink temperatures: this will help us to create the electro-thermal model. After de-embedding the manifolds and extrinsic components from our measurements, we are at the edge of the active transistor region. Our aim is to construct a compact electrical model of this active device, covering DC, small-signal or linear, and large-signal or nonlinear RF behaviour.

The intrinsic large-signal model that we build at this point is *quasi-static*. In other words, the model can describe the transistor's RF frequency dependence through the various reactive components in the model, such as the intrinsic capacitances at the gate and drain of the device, and the extrinsic inductance and capacitance associated with the manifold and extrinsic parts of the transistor. Further, the small-signal characteristics derived from the RF S-parameter measurements should be consistent with the small-signal parameters obtained from the DC characteristics. In practice, this quasi-static behaviour is not observed: real transistors exhibit *frequency dispersion* of their terminal behaviour. For example, the small-signal output conductance derived from the S-parameters is not the same as the slope of the DC output characteristics. This inconsistency is described as *non-quasi-static* behaviour. These non-quasi-static or frequency dispersive phenomena are caused by heating of the FET due to bias currents and the RF signal that is being amplified by the device, and by trapping of the mobile charges in the transistor.

The heating effects are significant in all RF power transistors, by nature of their application. As the temperature of the device increases, the physical properties of the semiconductor change. For example, the electron mobility decreases as the semiconductor heats up, hence the transconductance gain of the FET decreases with temperature. This can be seen in the DC I_d–V_{ds} characteristics of the FET. Each bias point corresponds to a different power dissipation, and hence temperature, so the transconductance is less at high power regions, and the curves can show a fall in current with increasing drain-to-source voltage, and hence temperature. But a negative output conductance is not seen at RF.

Trapping phenomena exist in III–V FETs, but are not observed in LDMOS transistors [1]. The physical location of the traps is not so important – it is thought that they exist at the semiconductor surface or interfaces between different regions of the transistor, such as the channel-buffer layer interface – but the characteristic trapping lifetime is the main concern. The characteristic lifetimes are in the range of milliseconds to microseconds, so the traps can only respond to electrical signals at a fairly slow rate, in the kilohertz to megahertz regime. Therefore, the traps are able to respond to DC characterization signals, but cannot react quickly enough to RF signals, and hence we observe a frequency dispersion of the FET characteristics.

We need to add parameters or components to describe these frequency dispersive effects, to complete the model of the active transistor. We will show how a sub-circuit describing the thermal environment can be added to the quasi-static electrical model so that the effects of both the static or ambient temperature as well as the dynamic electro-thermal behaviour are accounted for in a consistent manner. We use a similar approach to accommodate trapping effects into the model. The transistors that we use for model extraction are usually much smaller than the high-power die shown in Figs. 6.1 and 6.2, for a number of practical reasons. The large dies are generally unstable in a 50 Ω environment, as used for S-parameter measurements. Large devices also require high currents for accessing the whole of the I_d–V_{ds} characteristics. We perform these I–V and S-parameter measurements under pulsed conditions to obtain isothermal measurements, and such high-power pulsed current supplies and bias tees are difficult to make and expensive to purchase. Further, we perform our measurements for model extraction on-wafer, and the RF probes that are used cannot handle the high currents.

Because we use small devices for characterization and model extraction, we need to determine scaling rules for the intrinsic, extrinsic and manifold components of the transistor, for our model to be useful in the design of discrete or integrated power amplifier products. For discrete power transistors, we integrate the scaled extrinsic components and appropriate manifold models with the non-quasi-static electro-thermal model of the transistor, to obtain the *die* model, which is the basic building block for the packaged products. There are a couple of basic approaches to scaling our model up to the final device size: we can scale the intrinsic model and extrinsic components directly to the final size, and then add two-port models of the gate and drain manifolds to obtain the complete three-terminal (two-port) model of the power transistor. An alternative is to use several smaller transistor models connected between multi-port gate and drain manifolds, to obtain the full power transistor size; the small models need to be inter-connected thermally, requiring a soundly constructed thermal model. This latter approach is shown schematically in Fig. 6.4. For integrated power amplifier design, we need to provide scaling rules and guidance for the designers to be able to construct the driver and amplifier transistors on chip. These scaling rules will pertain to the extrinsic and intrinsic parts of the transistor models only; the gate and drain manifolds in an integrated design will often be custom-designed, requiring simulation of their layout using an electromagnetic field simulator – a planar simulator is usually sufficient.

In this chapter we will describe the modeling of the gate and drain manifolds and bond-pads, and their de-embedding to get to the extrinsic shell. We shall describe techniques for characterizing and de-embedding the extrinsic components of the transistor. We will then focus on the charge-based description of the active or intrinsic part of the transistor, and introduce a simple thermal model that can be coupled to the electrical model to give a dynamic electro-thermal nonlinear transistor model. During the model development, we shall comment upon the scaling issues that we need to address in order to build a compact transistor model for use in the successful design of RF power amplifiers. The complete product model, including in-package matching components and bondwires, and the model for the package itself, will be presented in Chapter 9, where various aspects of the model validation are described.

6.2 Modeling the Manifolds and Extrinsic Components

Before we can get to the details of how to construct the large-signal FET model, we need to identify and de-embed the manifold and extrinsic shells from the overall model structure shown in Fig. 6.3. While from Figs. 6.1 and 6.2 the manifolds and the metal gate and drain contacts are easy to see,

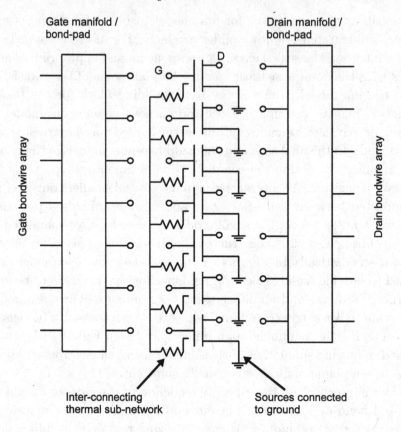

Fig. 6.4. A schematic FET model architecture for a power transistor, showing wide gate and drain manifolds connected to several individual FET models, which are connected together through a thermal network. In this way, the overall model can account for thermal distribution across a large device, and also for non-uniform feeding of current by the manifolds.

the separation of extrinsic and intrinsic parts of the transistor in electrical terms is perhaps not so clear cut. A very common way to partition the overall model is to define the extrinsic shell to be only the linear components and the intrinsic region to be the nonlinear part of the model. In fact, this division can provide a useful check on the model consistency: if the extrinsic components exhibit any voltage dependence, then this suggests that the partitioning may not be correct, and the extrinsic–intrinsic boundary may need to be adjusted. In reality, in power transistors some of the components that we might expect to find in the extrinsic shell can exhibit some voltage dependence, but it is generally more convenient from a modeling perspective to place the voltage-dependent quality in the intrinsic model, and define the extrinsic components as linear passive elements.

For high power transistors, we often need to break up the model structure even further. The manifolds at the gate and drain are physically large structures, and may behave like transmission lines. In other words, they are distributed circuit components: they have both frequency and spatial dependence of their transmission characteristics. The frequency-dispersive property can be used as the basis of the partition between the manifold structure and the extrinsic shell. The lumped components that comprise the extrinsic equivalent circuit – capacitors, inductors, and resistors – should be independent of frequency. After parameter extraction, if any of these components does show significant frequency dependence, then the partition between the extrinsic and manifold shells may be incorrectly located, according to our definition above. A frequency-dependent component may also be an indicator that our extrinsic circuit is incorrect, and perhaps a different topology is required to represent the extrinsic shell of the transistor over the desired bandwidth. However, this partition between the extrinsic and manifold regions is somewhat arbitrary, and we often use some physical part of the transistor structure to define the electrical reference planes of the manifold and extrinsic shell. For narrow-band applications this is generally sufficient. The physical size of the manifold can result in differences in *phase* in the transmission of the signal from the input to different gates in the FET (and similarly, at the drain manifold). We shall discuss these phase effects of the manifolds later.

We shall also consider that the electrical measurements made for model extraction are performed isothermally, using pulsed techniques as described in Chapter 3, for example. This simplifies matters enormously: we can treat the extrinsic and manifold components as linear, and build the electro-thermal effects into the nonlinear intrinsic model, as will be described in Section 6.5. In practical terms, the extrinsic resistances, in particular R_d and R_s, will be temperature dependent, and since the temperature is influenced by the applied voltages and currents, this means that these resistances are nonlinear functions. We choose to model these extrinsic components as linear, and build the temperature effects in later, as noted above. In the following sub-sections we will cover in detail the treatment of the metal manifolds and bond-pads in the context of the large periphery transistors typically used in RF and microwave high-power FETs. We will then describe some of the different methods used to extract the values of the extrinsic components of silicon LDMOS and GaAs FETs.

6.2.1 Metal Manifolds and Bond-pads

In discrete power transistors, the manifold metal can also serve as the bond-pad region for the bondwires that connect the transistor die to the rest of the package components. The bondwire-to-bond-pad interface is very similar to the bondwire-to-package interface: in both cases a three-dimensional wire is attached to a two-dimensional microstrip transmission line. Therefore, the method of analysis is very similar, with the differences being in the details of the metal composition and thickness of the bond-pad, and in the dielectric material, which in the die is silicon or gallium arsenide, while in the package is a plastic encapsulant or low-loss ceramic material. A detailed analysis of the bondwire-to-package interface using planar and three-dimensional electromagnetic field solvers was presented in Chapter 4.

In an integrated circuit environment, for the connections between devices on the IC, the metal manifolds connecting the gate and drain fingers of the transistors can be made much narrower, and the connection to the manifold is generally made by a simple microstrip transmission line. In essence, the treatment of the connection of this microstrip line to the manifold is the same as the bondwire-to-manifold connection. In practice, the integrated circuit designers would like to connect to the transistors using whatever metal lines they can fit into the most compact layout. This can lead to some apparent problems with model accuracy, as now the reference planes to the transistor in the IC are not the same as the ones that were used for the model extraction. The connections from the drain manifold of the output transistor of the IC to the package or matching network are made by bondwires in a similar manner to that outlined above for the discrete devices.

The metal manifold or bond-pad can be represented in the model in different ways, depending on its size and the frequency of interest. If the frequency of operation is low enough, or the manifold is small enough, it can be represented simply by a capacitance to ground. At higher frequencies, in the microwave and millimeter-wave regime, or for very large structures, the simple capacitance model is not capable of predicting the distributed nature of the manifold, and it is represented better by a transmission line model.

Essentially, what we, as modelers, have to do when modeling the manifold or bond-pad, is to make a judgment on the electrical size of the manifold; to determine whether we need a transmission-line model, and how many inputs and outputs are required. We also have to decide where the electrical reference planes are on each side of the manifold. This can be quite a challenge, especially when we are modeling very large periphery transistors that have multiple bondwire connections.

For electrically-small manifolds or bond-pads, a simple two-port representation is usually sufficient: one port connects to the bondwires, and the other port connects to the FET gate or drain finger array. The calculation of the pad capacitance is relatively straightforward: the capacitance per unit area of the metallization is a parameter that is monitored during wafer fabrication. The calculation of the pad capacitance from this parameter will need to account for fringing fields, and is usually not a simple area calculation. The alternative is to perform an electromagnetic simulation of the bond-pad to determine its S-parameters, from which the Z-parameters should yield the equivalent capacitance.

For high-power transistors that have a large gate periphery, and hence a large number of gate fingers, the concern is the phase difference between adjacent fingers or groups of fingers fed from a common point, due to the currents traveling in a transverse direction along the manifold. The two-port description of the manifold is incapable of capturing this lateral electrical delay, and a multi-port representation of the manifold is required to describe the detailed electrical behaviour completely. Typically, we would use a planar EM simulator to determine the electrical properties of the manifold; such tools are fast, accurate, and provide a very flexible simulation environment, which allows for arbitrary planar structures with arbitrary material properties.

We will now present two examples of a bondwire pad or manifold that represent these two extremes of electrical size, to illustrate the differences in behaviour, and consequently the type of manifold model that we may choose to use in constructing the model of the power transistor. The first example is an electrically small bond-pad, typical of a GaAs microwave transistor; the second example is a large bond-pad and manifold, of a size typically used in modern high-power LDMOS RF transistors.

6.2.1.1 Small Bond-pad

In this first example, we examine the electrical behaviour of a physically small bond-pad or manifold, such as may be found in a high-frequency GaAs medium power FET. The structure, illustrated in Fig. 6.5, shows a 12-finger manifold that may typically be around 200 µm in total width (depending on the source-drain pitch of the process). The port connections for the electromagnetic simulator are also shown on this figure.

As this is a small structure, we are principally interested in whether we can use a simple capacitor-to-ground model to represent the manifold. The

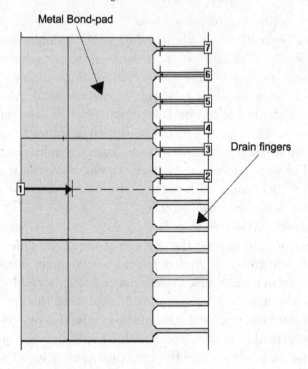

Fig. 6.5. Gallium arsenide bond-pad showing the drain fingers, and the port definitions for the EM simulation.

EM simulation has been performed as a seven-port structure, with the left-hand side of the manifold as the input, port 1, and six gates for ports 2–7: we are taking advantage of the symmetry of the structure, to simulate only half of the manifold. The multi-port S-parameters are then converted to Z-parameters to extract the equivalent circuit capacitance to ground:

$$C = \frac{-1}{\omega \, \mathrm{Im}(Z_{11})} \tag{6.1}$$

In fact, any transfer impedance Z_{k1} can be used in place of Z_{11}, in this example.

Figure 6.6 shows the extracted capacitance as a function of frequency. At low frequencies, the structure is electrically short and it can be properly represented by the simple capacitor-to-ground model. At higher frequencies, the structure begins to display some frequency dependence, indicating that the capacitor model parameters are frequency dependent, or, more practically, the manifold structure is beginning to behave like a transmission line. At some point we must decide that the simple capacitor model is no longer accurate enough. This is a modeling compromise between accuracy

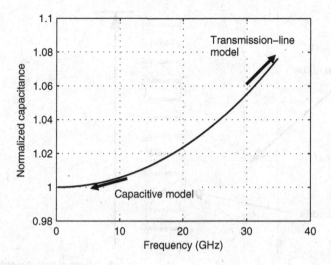

Fig. 6.6. Extracted bond-pad capacitance versus frequency of a small GaAs bond-pad. This graph is normalized to the pad capacitance at 250 MHz.

of representation of this structure, and simplicity of its implementation in the model.

6.2.1.2 Large Bond-pad and Manifold

We shall now consider the other extreme of manifold dimension: a large manifold, as is found in high power silicon LDMOS transistors. A 100 watt power transistor may have a total gate periphery of around 130 mm, and, given typical unit gate widths and source-drain spacings of modern LDMOS technology, this requires a manifold structure about 7 mm wide.

Our concern in this case is that the current flowing *laterally*, in the transverse direction along the manifold structure, will have a different phase at each of the finger feed points, and so the FET may not be optimally driven. By way of illustrating this lateral phase distribution, we shall examine the phase difference between an input at the center of the manifold, and various locations of the drain finger feeds on the output of the manifold. The structure is shown diagrammatically in Fig. 6.7: we have taken the 7 mm manifold as our model. A planar EM simulation of this structure is carried out, and we determined the difference in phase laterally from the input at center of the manifold to the various ports along the output side of the manifold. These ports do not represent gate or drain feeds, necessarily; they are simply measurement locations from which we estimate the phase.

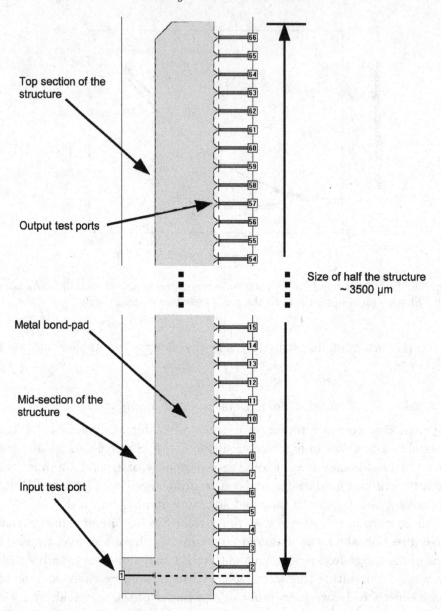

Fig. 6.7. A schematic of a large manifold structure, showing the input and output port configuration within a planar electromagnetic solver, to estimate the lateral phase delay along the manifold.

In Fig. 6.8 we show this phase difference between the center and edge of the structure. We can see that the amount of phase distribution from the center of the bond-pad to the edge is significant: even at 2 GHz we see a

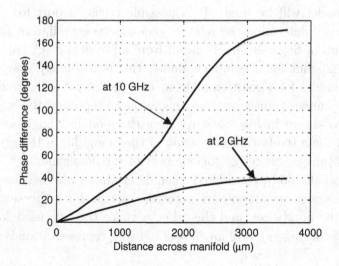

Fig. 6.8. Lateral phase in the large manifold structure, showing the phase difference between the input and output signals as a function of distance away from the center of the manifold, at 2 GHz and 10 GHz.

phase difference of over 30 degrees. At 10 GHz we can see that there are some clear physical limitations in the overall size of the gate and drain manifolds; this probably has some ramifications regarding the maximum size and hence power capability of the transistor that we can expect at these frequencies, using this simple manifold structure.

In practice, it is the designers' responsibility to decide on the level of discretization or grouping of fingers that will be fed from each bondwire in the final design of the power transistor. The decision will be based on the frequency of operation of the transistor, and the number of wires necessary to keep the phasing across the manifold to an acceptable level. The number of bondwires and the distance between them can be estimated from a simulation of the structure as indicated here. In an integrated circuit context, the concept of phase difference across the width of the manifold must be considered in the layout of the components. Feeding the power transistor uniformly across its physical width is usually not an option, and the transmission line width of the feed may be small in comparison to the manifold width. The designer must be able to determine how much phase change across the transistor can be tolerated.

Similarly, when deciding on the level of discretization or grouping of fingers during the analysis, the modeling engineer needs to make an assessment of the required level of accuracy of the model and the frequency range in which the model will be used. It is possible to use a port for each gate feed, and hence determine the relative phase between adjacent fingers, or across a group of fingers. This is useful from a physical or global modeling perspective [2], but for a compact model, this is too finely grained to be a practical model for circuit simulation and design. A compromise is to use several compact transistor model instances to represent the large power transistor (as shown in Fig. 6.4), yet here the modeling engineer still has to decide on some level of discretization of the manifold, in the grouping of several physical gates together for one port of the larger model.

Factors that should be considered when determining whether the manifold can be represented as a two-port network (one port for the side used to connect all the bondwires, and the other port for the side used to connect all the transistor fingers), or if an N-port network representation is required, include:

(i) lateral feeding effects – are all the transistor fingers excited in a uniform manner by the bondwires, or is the phase difference between fingers significant?

(ii) is the array of bondwires uniform? – in some designs the bondwires may grouped into several sets of two or three or more; and

(iii) whether different FET models are to be used for each set of fingers – the capability of having multiple FET models to represent the large periphery transistor can be useful in accounting for a non-uniform temperature profile across the die.

We note that complexity required during the modeling and simulation of a large structure at RF frequencies is similar in nature to a physically smaller structure at much higher frequencies.

6.2.1.3 Some Remarks on Manifold Modeling for Model Extraction and Model Construction

The modeling of the manifolds and bond-pads needs to be executed accurately for both model extraction and model construction objectives, though these two activities have subtly different requirements of the manifold models. We have stated earlier that the transistors that we use for model extraction are relatively small compared with the actual power transistor die. While this is essentially because of measurement limitations, the small size of these transistors allows us to use manifolds or feed structures that can be modeled accurately using a two-port or lumped-element approach. This makes the de-embedding of the manifold structures from the bias-dependent

S-parameter measurements used for the model extraction quite straight-forward. This is a very desirable quality, since the extraction of the nonlinear intrinsic model is quite difficult enough without introducing complications of multi-port structures. For model extraction, our goal is to de-embed the manifold structure from the measurements, leaving a two-port network of the extrinsic shell surrounding the intrinsic device.

For transistor model scaling purposes, we will often construct several tran-sistors with different numbers of gate and drain fingers, or a series of different unit gate widths. Again, from a practical perspective, we try and ensure that the bond-pad or manifold structure can be modeled accurately as a two-port network, for ease of de-embedding.

In model construction, we do not need to be so restrictive. Our objective in modeling the bond-pad and manifold structure is to obtain as accurate a model as is required for the application: this may also be a two-port model, or it could be an N-port model describing multiple bondwires and fingers. In the latter case, we can connect a number of nonlinear transistor models between the multi-port manifolds; each transistor model represents a number of gates and drains in the overall power transistor. These 'unit' transistor models are connected electrically through the N-port manifold models, and thermally using an equivalent thermal circuit. The overall multi-element model can then represent distributed effects such as laterally non-uniform current feeding and loading of the power transistor, or thermal gradients across the device. While such a model can be useful in aiding practical power device design, the number of coupled nonlinear devices in the circuit, for even a single power transistor, may hinder convergence in the circuit simulator, leading to long simulation times that may be unacceptable in circuit design.

6.2.2 Extrinsic Circuit Component Parameters

After de-embedding the manifold and bond-pad structures, we now need to determine and extract the extrinsic circuit components, to get access to the intrinsic transistor reference plane. Once we have de-embedded these extrin-sic components, the small-signal equivalent circuit parameters of the FET can be obtained by direct manipulation of the intrinsic plane S-parameter data. From the small-signal model parameters, we can build the large-signal model of the intrinsic FET, as will be described in Section 6.4.3.

Typically, the extrinsic element extraction methods rely on what is com-monly known as the *cold-FET* method [3]. The basic principle of this method is fairly simple: the FET is biased with the drain-to-source voltage, V_{ds}, set to zero volts. In this condition there is no electric field in the channel of the FET, and so the electrons are in the ohmic regime - they are 'cold' electrons. With V_{ds} set to zero, the controlled-current generator in the small-signal equivalent circuit model is effectively shorted out, placing the transistor in a passive condition.

The gate of the FET is then biased in order to put the transistor into either cut-off or conducting modes, in which the intrinsic small-signal equivalent circuit can be simplified sufficiently to enable unambiguous identification and measurement of the extrinsic shell components. The extrinsic parameters can be extracted using direct methods from single-frequency S-parameter measurements, or performed by optimization over a range of frequencies. The cold-FET method has been applied successfully to GaAs FETs and HEMTS [3–5], and to silicon MOSFETs [6,7].

The details of the extrinsic parameter extraction method depends on the structure adopted for the extrinsic networks. Two common arrangements are shown in Fig. 6.9, typically applied to the small-scale transistors for which this technique was originally developed. The choice of network is somewhat arbitrary, based on the particular geometry of the FET in question. Unless there is any compelling geometric feature, the network with the extrinsic capacitance outermost, Fig. 6.9(a), is usually chosen for economy of mathematical matrix inversions in the de-embedding of the extrinsic shell. This extrinsic capacitance component is generally included to represent the bond-pad capacitance of the small test FETs. In such cases, the pad capacitance is usually found from measurements of test structures, and so the pad capacitance components can be de-embedded from the extrinsic circuit easily and accurately. We have described earlier how the bond-pad and manifold layouts of power transistors can be measured and their equivalent circuits determined, and de-embedded from the measurements. We shall assume that the capacitive components of the gate and drain extrinsic networks can be subsumed into the manifolds, leaving only the series resistance and inductance components to be extracted from the cold-FET measurements. As the source of the transistor is connected directly to ground, the extrinsic source capacitance is effectively shorted. This assumption can be challenged in high-power transistor modeling: the large area of the transistor will contribute a large extrinsic capacitance - the *body* capacitance - and the resistance and inductance of the source contacts to package ground, while small, are not zero. In the equivalent circuit representation, the source of the

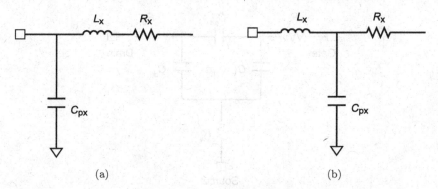

Fig. 6.9. Simple sub-networks used for the extrinsic equivalent circuits; these are connected at each port of the transistor: (a) the capacitors represent the bond-pad or manifold capacitances, (b) the capacitors represent other geometries in the FET that have a significant capacitive component, such as gate-feed to source-feed crossovers, and so forth.

active transistor is connected to ground through a series R-L and parallel capacitor network, and this can affect the frequency response and potentially the stability of the transistor.

6.2.2.1 Cold-FET Cut-Off Mode

The gate is biased below the threshold voltage, so the FET is in the cut-off condition and no conduction can take place. The intrinsic equivalent circuit for the transistor is then approximated by the capacitances C_{gs}, C_{gd}, and C_{ds} only. The S-parameters are measured at a frequency or frequencies low enough for the capacitive reactance to dominate the measurements. With our assumption that the bond-pad and manifold model includes all of the extrinsic (gate and drain) capacitance, this measurement is not strictly necessary, as we have no further extrinsic capacitance to measure and de-embed. On the other hand, if our extrinsic networks are described better by Fig. 6.9(b), we now have a measurement tool to help identify their values. If we consider that the extrinsic resistances and inductances can be neglected in this configuration, the extrinsic gate and drain capacitances are in parallel with C_{gs} and C_{ds}, respectively, so there is no way of extracting their values directly. This cold-FET measurement is often part of an investigation of the geometrical scaling of the transistor: provided that the extrinsic capacitances scale in a different way to the intrinsic capacitances, which scale directly with total gate periphery, then we have a means of identifying the extrinsic capacitances by making these cold-FET measurements over a range

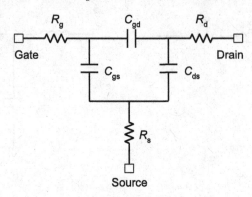

Fig. 6.10. Silicon MOSFET small signal equivalent circuit at $V_{gs} = V_{ds} = 0$ V [6].

of transistors with different numbers of gate fingers, unit gate widths, and total periphery.

Lovelace *et al.* [6] used this bias condition to extract the extrinsic series resistances in a MOSFET. They set both V_{gs} and V_{ds} to zero, yielding a completely passive structure for the FET. The equivalent circuit of the transistor reduces to the circuit shown in Fig. 6.10, in which any extrinsic inductance has been neglected. The extrinsic resistance values can be found from the real parts of the Z-parameters of the network. The FET capacitor Π-network can be converted into a T-network to illustrate this calculation. Lovelace *et al.* carried out this analysis over a range of frequencies from 1 to 10 GHz; the resistance values were determined by direct extraction at each measurement frequency, and showed good consistency, especially at the higher frequencies where the reactive component is not significantly larger than the real part of the impedance. An alternative approach would be to use the measurements at all frequencies and optimize the resistance values, using a direct extraction value as a seed. Although Lovelace *et al.* do not show any extrinsic inductance in their small-signal equivalent circuit, indicating perhaps a small device, there exists the attractive possibility that the extrinsic series resistance and inductance can be extracted from broadband S-parameter measurements made under these bias conditions, by optimizing the component values of the series R-L-C branch networks against the broadband Z-parameters derived from the S-parameters.

6.2.2.2 Cold-FET Conducting Mode

The cold-FET method was originally devised for extracting the extrinsic component values for GaAs MESFET device models [3]. In this second part of the procedure, the FET is again biased with the drain-to-source voltage

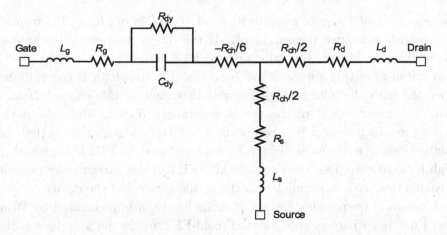

Fig. 6.11. Gallium arsenide MESFET small-signal equivalent circuit, including extrinsic components, at cold-FET bias with the gate Schottky diode in high forward bias: C_{dy} and R_{dy} represent the Schottky diode capacitance and shunt resistance, respectively, and R_{ch} is the FET channel resistance.

set to zero, but now the gate Schottky diode is driven into high forward bias. In this condition, the capacitances are all considered to be shorted by the access resistances, and the small-signal equivalent circuit becomes a network of resistors and inductors, as shown in Fig. 6.11. At high forward bias, the gate Schottky diode is dominated by the shunt resistance, R_{dy}, which is given by

$$R_{dy} = \frac{nkT}{qI_g} \tag{6.2}$$

where n is the ideality factor of the gate current characteristic at high bias, and I_g is the gate current. The Z-parameters for the network of Fig. 6.11 are

$$Z_{11} = R_g + R_s + \frac{R_{ch}}{3} + \frac{nkT}{qI_g} + j\omega\left(L_g + L_s\right) \tag{6.3}$$

$$Z_{12} = Z_{21} = R_s + \frac{R_{ch}}{2} + j\omega L_s \tag{6.4}$$

$$Z_{22} = R_d + R_s + R_{ch} + j\omega\left(L_d + L_s\right) \tag{6.5}$$

We have four unknowns, R_g, R_d, R_s, and R_{ch}, and three equations, so an additional measurement or piece of information is required for a unique solution. This could be, for example, the value of R_g from an 'end-to-end' measurement of the gate metallization, or the value of $(R_d + R_s)$ from an on-channel resistance measurement. If the device fabrication technology parameters are known, the value of R_c can be determined.

A criticism of this approach has been that the transistor is generally not operated with the gate in high forward bias, and so this characterization can yield lower values of the access resistances R_d, R_s, than the device experiences in practical bias conditions. A further observation is that this method can give low values of the access resistances in PHEMTs, where, in high forward gate bias, the parasitic MESFET of the barrier layer provides a conduction path in parallel with the actual channel of the device.

A means of overcoming these criticisms has been implemented by Wood and Root [4], where in the 'forward' cold-FET mode the gate bias is just enough to provide a conducting channel, as might be used in normal operation of the transistor in an amplifier application. In this case, the gate diode must be described by the complete model, including the capacitance, in the gate branch of the network of Fig. 6.11. The Z_{11} expression is now

$$Z_{11} = R_g + R_s + \frac{R_{ch}}{3} + j\omega \left(L_g + L_s - \frac{1}{\omega C_{dy}} \right) \qquad (6.6)$$

The RF measurements are carried out over a wide bandwidth, and the Z-parameters are optimized over frequency to obtain the values of the equivalent circuit parameters. The shunt resistance is generally neglected, as the gate is not in high forward bias, and the gate current is very small. In fact, good results can be obtained with the gate bias at zero volts, corresponding to the measurement conditions of Lovelace *et al.* The cold-FET technique can therefore be applied to LDMOS power transistors to extract the extrinsic parameters of the device model: high forward gate bias is not necessary for this technique to be applied successfully.

6.3 Scaling Considerations

We have now successfully de-embedded the manifolds and the extrinsic shell components from our carefully measured S-parameter data, and arrived at the intrinsic device reference plane, ready to begin building the nonlinear model for the power transistor. But before we dive into the nonlinear modeling, this seems an appropriate juncture to say a few words about scaling the device model. The requirements for a scalable transistor model are twofold.

First, the very large gate peripheries of modern high power RF transistors are typically in excess of 100 mm. The impedances of such transistors are extremely low, and the devices could be unstable in a 50 Ω environment. It is therefore not practical to make the S-parameter measurements for model extraction directly on these large die. Instead, we must use much smaller devices, whose peripheries are in the 1 to 10 mm range and present terminal impedances that can be measured using standard VNA equipment. The compact models that are extracted from these test devices must then be scaled to the appropriate size for the power transistor and amplifier design. The scaled models are often optimized against measured data from the large device, to obtain the last drop of accuracy.

Second, a scalable transistor model is virtually indispensable in IC design: the transistor peripheries are often adjusted by small fractions to obtain the overall performance defined in the specification, without over-engineering the part. The transistor models are therefore required to be scalable over a wide range of gate periphery, while maintaining fidelity to the transistor performance. Here, we will consider scaling effects of the transistor only – we will assume that the gate and drain manifolds are designed to suit a particular layout, be it IC or discrete device, and so scaling of the manifolds is not an issue.

'Scaling' means different things to the various people involved in the creation and usage of a device model. The modeling engineer tends to think in terms of the current flow in the transistor, and how that changes as the dimensions of the transistor are altered. This leads to very specific 'rules' about scaling, in terms of the number of (gate or drain) fingers, the total periphery, and so forth. The designer tends to think in terms of transistor performance: for example, 'If a 10 mm device can produce 5 watts of power at the P-1dB point, a 100 mm device should produce 50 watts at P-1dB.' Unfortunately for both parties, this is seldom true. This is generally attributable to factors such as: the manifold shape and structure, the number of bondwires and how they are accommodated in the modeling and design, the electromagnetic environment around the transistor and package, and, not least, the thermal environment, which will change with the die size, and also with the packaging.

Ideally, the 'design rules' would be able to accommodate all of these factors, in some way. The electrical and thermal 'scaling rules' for the transistor are essentially a subset of the design rules. The scaling properties are very much dependent upon the actual geometry and layout of the test transistor and power transistor die: these should be scaled replicas of each other. However, there are some basic guidelines that can generally be applied in the design of a set of test transistor and passive component layouts from which scaling rules can be derived. The test transistor array should include a range of gate peripheries whose DC and S-parameters can be measured using the model characterization equipment: the maximum periphery will depend on the device impedances and desired accuracy of the measurement system. The range of peripheries should include transistors of different numbers of gate fingers, and different unit gate widths. In our experience, two-finger devices do not follow the scaling rules of larger structures, and so they should not be used for scaling models.

From this array, cold-FET measurements can be used to examine the variation with number of gate fingers, unit gate width, and total gate periphery of the extrinsic circuit component values. Hence, some 'scaling rules' for the extrinsic components can be deduced. For example, the access resistances R_d and R_s usually have an inverse relationship to the total gate periphery. The gate resistance R_g generally shows a dependence on the number of gate fingers and the unit gate width, typically of the form

$$R_g^{\text{NEW}} = R_g^{\text{ORIG}} \left(\frac{W_g^{\text{NEW}}}{W_g^{\text{ORIG}}} \right) \left(\frac{\text{NF}^{\text{ORIG}}}{\text{NF}^{\text{NEW}}} \right)^2 \tag{6.7}$$

where the superscript NEW refers to the new scaled device, and ORIG the original device; W_g is the total gate periphery, and NF is the number of gate fingers. The capacitances tend to scale directly with total gate periphery, and may have a fixed offset at 'zero gate width' owing to the metallization that is required for the device structure, or bond-pad. The extrinsic gate and drain inductances tend not to have such simple scaling relationships, with mutual inductance playing a role. This can be reduced in III–V FETs using air-bridges to interconnect the source contacts. The source inductance is usually dominated by the connection to the back side of the die: using through-vias in III–V technology, and the source sinker diffusion in LDMOS. While this inductance can often be small, of the order of a few picohenries, it is in the source-to-ground branch, and can affect the stability of the power transistor. Transistor design generally tries to minimize this extrinsic component value. It is generally presumed that the intrinsic device parameters, the currents, and the charges scale directly with the total gate periphery. In the small-signal model, the capacitances scale directly with the periphery, and the resistances (in III–V FETs) scale inversely with the periphery.

In addition to the electrical scaling of the extrinsic components, there is the scaling of the thermal resistance and capacitance of the transistor to consider. The scaling of these parameters depends on how the larger transistors are constructed and laid out as die or on the IC, and implemented using the model. If we are planning to model a large power transistor die using a single transistor model, then simple scaling of the thermal parameters with area or periphery is not enough. While the thermal resistance should fall with area, the heat source is similarly distributed.

On the other hand, if we choose to model this large transistor with several paralleled models, as shown in Fig. 6.4, then the individual thermal resistances need to be coupled together to describe the lateral flow of heat, or lack thereof. It is better to carry out thermal simulations of the large structures, and hence define the values of the thermal parameters for the overall structure. The thermal coupling can then be deduced from simulation of the thermal network. This approach is also applicable in the design of RF power amplifier ICs; this can be a more difficult situation to model thermally, as the individual transistors may not be generating the same power, and therefore heat. The details of the construction of the self-consistent electro-thermal model are presented in Section 6.5.

6.4 Modeling the Intrinsic Transistor

We are now getting towards the kernel of the transistor model: the active device itself. We shall base our model derivation on measurements of S-parameters made over the attainable V_{gs}–V_{ds} bias space of the transistor. These measurements will include regions in reverse drain bias – negative V_{ds} – and below the threshold voltage; the S-parameters may also be taken over a range of RF frequencies.

When we build our model in the simulator, we will use the terminal voltages as the control inputs to the model, for both quasi-static and fully dynamic models. For the sake of clarity in the following explanations and equations, we shall presume that the measured data that we use have been electrically de-embedded, and are isothermal. By this statement, we mean that the electrical effects of the extrinsic components, manifolds, and so on, have been removed, and the measurement reference planes now lie at the intrinsic device plane. When we construct the complete model, we can, if necessary, re-reference the voltages and currents to the transistor's external terminals; this is a requirement for table look-up models, in which the table index is in the actual measured space. This re-referencing can be done at run time in the simulator. Isothermal conditions can be achieved by taking carefully controlled pulse measurements or by using some means of thermal de-embedding, as described in Chapters 3 and 5.

Once we have constructed our large-signal, nonlinear, isothermal model for the intrinsic transistor, we will add the dispersive effects to create a non-quasi-static model. These dispersive effects include electron trapping effects,

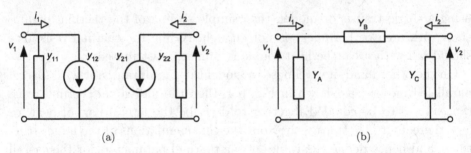

Fig. 6.12. Representations of two-port Y-parameters; (a) a completely general circuit schematic, using a controlled source to represent the transadmittances, and (b) a simplified circuit for passive elements only.

which occur in III–V semiconductor FETs, and thermal effects, which are prevalent in all power transistors. The thermal effects include the static thermal influence due to ambient temperature, and the dynamic effects of the electrical power that is dissipated as heat. This will be described in detail in Section 6.5, where we will describe a fully dynamic electro-thermal model for the power transistor.

6.4.1 A Small-Signal Model

The S-parameters represent the linearized or small-signal response of the transistor at a specified bias V_{gs}, V_{ds} and frequency, ω. This is simply the measured (and de-embedded) data, which could be stored in a three-dimensional table, indexed by the bias voltages and frequency. During simulation, the required S-parameters would be read by indexing, or interpolating to the off-grid or non-measured values. Ambient temperature could be built in by adding another dimension to the table. Index look-up models can be quite slow, especially if interpolating over several dimensions as the simulator drives to convergence.

A more compact approach is to convert the measured S-parameters to Y-parameters, using the standard rules for two-port matrix conversions see, for example, Vendelin, p. 12-13 [8]). A generic Y-parameter two-port network can be expressed in the form shown in Fig. 6.12(a), which for a passive network can be simplified to a Π-network, Fig. 6.12(b), where

$$y_{11} = Y_A + Y_B$$
$$y_{12} = -Y_B = y_{21} \qquad (6.8)$$
$$y_{22} = Y_C + Y_B$$

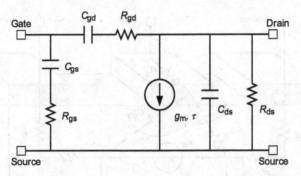

Fig. 6.13. Small-signal equivalent circuit model of the intrinsic transistor; this model is derived from Y-parameter measurements, with some admittance-to-impedance transformation in the gate-source and gate-drain branches.

In this passive network we can replace the Y_x by the corresponding circuit elements, $G_x + jC_x$, to arrive at an equivalent circuit representation that can be expressed directly in the simulator. The conductances and capacitances in this representation are assumed to be frequency-independent. A quick check on this assumption is to plot these elements against frequency. If the equivalent circuit parameters are not frequency-independent, then this may indicate a problem with the electrical or thermal de-embedding, or perhaps that the component is not bias-independent, which may occur in physically large power transistor structures, or that some dynamic phenomenon associated with the frequency dispersion effects has not been properly handled in the measurement. An example of this could be that the pulse widths are not narrow enough, or the duty cycle is too large, leading to some thermal history in the measured data. The frequency-independence of the model parameters enables us to determine their values from measurements made at a single frequency only.

Commonly, for FET models operating in the active regime, that is, above threshold and positive drain-to-source voltage (for n-channel devices), we add the controlled current source representing y_{21} to the passive network representation, as a transadmittance. This yields the equivalent circuit model as shown in Fig. 6.13, where Y_{gs} corresponds to Y_A, Y_{gd} corresponds to Y_B, Y_{ds} corresponds to Y_C, and the transadmittance is Y_m. The gate branches are generally represented as series R-C networks, as shown, and the circuit elements are often linked to some physical representation, as illustrated in Fig. 6.14; the series resistance elements mentioned above are often described as providing the charging path inside the device for the associated capacitance.

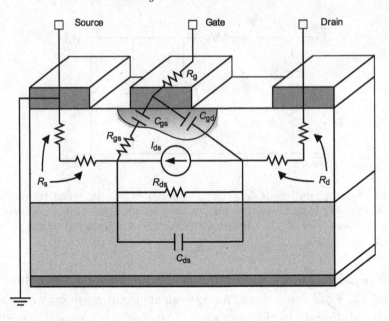

Fig. 6.14. Physical origins of the components of the equivalent circuit model of a MESFET shown in Fig. 6.13 based on [9]. © 1976 IEEE. Reprinted with permission.

Following the direct extraction method of Dambrine *et al.* [3] we can obtain the values for the equivalent circuit parameters. Starting with Y_{gd}:

$$Y_{gd} = -y_{12}$$

$$Z_{gd} = R_{gd} + \frac{1}{j\omega C_{gd}} \tag{6.9}$$

$$Y_{gd} = \frac{j\omega C_{gd}}{1 + j\omega C_{gd} R_{gd}} = \frac{(1 - j\omega C_{gd} R_{gd}) j\omega C_{gd}}{1 + \omega^2 C_{gd}^2 R_{gd}^2} \tag{6.10}$$

The *low-frequency limit* condition is usually applied, where $\omega^2 C_{gd}^2 R_{gd}^2 \ll 1$; this is typically found to be the case in practice. We can then obtain the equivalent circuit component values for the gate-drain branch:

$$C_{gd} = \frac{1}{\omega} \mathrm{Im}\left(Y_{gd}\right) = -\frac{1}{\omega} \mathrm{Im}\left(y_{12}\right) \tag{6.11}$$

$$R_{gd} = \frac{1}{\omega^2 C_{gd}^2} \mathrm{Re}\left(Y_{gd}\right) = -\frac{1}{\omega^2 C_{gd}^2} \mathrm{Re}\left(y_{12}\right) \tag{6.12}$$

where Y_{gd} is often represented as a capacitance only. In fact, R_{gd} is quite insensitive to direct extraction at a single frequency, and unphysical values can often arise, such as a negative resistance for R_{gd}. This is often indicative of errors in the de-embedding: what happens in these circumstances is that too much resistance has been accounted for in the extrinsic components, and the overall measured resistance component can only be accommodated by the value of R_{gd}. The use of very high frequencies for the determination of R_{gd} has been advocated by Kompa [10], though in practice errors in R_{gd} do not usually have a dramatic impact on the circuit behaviour of the model. It may often be easier just to leave it out.

The gate-source admittance Y_{gs} is also represented by a series R-C network in the equivalent circuit:

$$Y_{gs} = y_{11} + y_{12} \tag{6.13}$$

hence

$$C_{gs} = \frac{1}{\omega} \operatorname{Im}(y_{11} + y_{12}) \tag{6.14}$$

$$R_{gs} = \frac{1}{\omega^2 C_{gs}^2} \operatorname{Re}(y_{11} + y_{12}) \tag{6.15}$$

where again the low-frequency limit has been assumed. The drain-source admittance is a parallel R-C network, and is obtained directly from Y_{ds}:

$$Y_{ds} = R_{ds}//C_{ds} = G_{ds} + j\omega C_{ds}$$
$$= y_{22} + y_{12} \tag{6.16}$$

$$C_{ds} = \frac{1}{\omega} \operatorname{Im}(y_{22} + y_{12}) \tag{6.17}$$

$$G_{ds} = \operatorname{Re}(y_{22} + y_{12}) \tag{6.18}$$

and

$$R_{ds} = {}^1\!/G_{ds} \tag{6.19}$$

The transadmittance, Y_m, is written as a transconductance term with a delay term:

$$Y_m = y_{21} - y_{12} = g_m e^{(-j\omega\tau)} \tag{6.20}$$

$$g_m = \operatorname{Mag}(y_{21} - y_{12}) \tag{6.21}$$

$$\tau = -\frac{1}{\omega} \operatorname{Phase}(y_{21} - y_{12}) \tag{6.22}$$

The delay term accounts for the distributed effect of the charge moving along the channel of the FET during operation.

In this way, from the measured and de-embedded isothermal Y-parameters, we are able to extract the elements of an equivalent circuit model at every bias point $\{V_{\mathrm{gs}}, V_{\mathrm{ds}}\}$. This is a *bias-dependent linear* FET model, for use in small-signal AC or S-parameter simulations. We are able to perform a small-signal analysis, provided we know the gate and drain bias conditions beforehand.

In the model we can store the values for $\{C_{\mathrm{gs}}, R_{\mathrm{gs}}, C_{\mathrm{gd}}, R_{\mathrm{gd}}$ (if present), $g_{\mathrm{m}}, \tau, C_{\mathrm{ds}}, R_{\mathrm{ds}}\}$ at each bias point $\{V_{\mathrm{gs}}, V_{\mathrm{ds}}\}$ in a table. The bias points are often re-referenced to the external terminals of the transistor, that is, where the bias voltages are measured in practice, and are labeled in the simulator. The equivalent circuit values are than indexed either directly by the terminal voltages, or interpolated between the stored tabular values.

We could also create a linear model by fitting two-dimensional functions to all of the equivalent circuit values, over the measured $\{V_{\mathrm{gs}}, V_{\mathrm{ds}}\}$ space, to obtain continuous variables of the model parameters with bias. Note that this is *not* the same as having continuous variables for the model circuit parameters over an instantaneous voltage space: *it is not a large-signal model*. The reason for this is quite straightforward: the resistors and capacitors in the circuit simulator are two-terminal components. The current flow in a resistor depends on the voltage across it (commonly known as *Ohm's law*), and the charge on a capacitor likewise depends up the voltage across its terminals. If we want to make these components dependent on two circuit voltages, we need to implement them as controlled sources. In most simulators, a 'resistor' whose instantaneous value depends on two voltages can be implemented as a controlled current source. A 'capacitor' whose instantaneous value depends on two voltages must be implemented as a controlled *charge* source: this is a little more challenging, and we shall investigate this approach in the next section.

Finally, a cautionary note about the time delay element, τ, in the small-signal model. This parameter is used to model the observed time delay between a signal being applied at the gate, and the current response at the drain. This is due to the finite time taken for charge to move along the channel from gate to drain: the transit time. It is interpreted as a distributed effect, as all the charge does not move at the same rate. In the model we see that the delay is implemented as a $e^{-j\omega\tau}$ term. In a small-signal AC or harmonic-balance simulator, the frequency (or frequencies) ω is an explicit variable, chosen by the user for the analysis. In a transient or circuit envelope simulation, the variable ω is not specified for the transient analysis, and hence this model term is undefined. In general, in large-signal analysis the nonlinear model elements cannot have an explicit frequency dependence.

One way around this problem is to make the following approximation for the exponential term, using only the first term in the Taylor series expansion for e^x:

$$e^{-j\omega\tau} \approx 1 - j\omega\tau \qquad (6.23)$$

The expression for the transconductance can then be written:

$$
\begin{aligned}
g_{m0}\, e^{-j\omega\tau} &\approx g_{m0}\,(1 - j\omega\tau) \\
&= g_{m0} - j\omega g_{m0}\tau \\
&= g_{m0} + j\omega C_{m}
\end{aligned}
\qquad (6.24)
$$

The term $C_{m} = -g_{m0}\tau$ is a *transcapacitance*, it is the reactive analogue of the transconductance, in this small-signal context. In a large-signal model, the transcapacitance arises from having a capacitive element controlled by two voltages: it is required in the circuit to maintain energy conservation, as we shall demonstrate later. Using a transcapacitance component, the small-signal forward transmission (that is, S_{21}) of the FET can be modeled with reasonable accuracy in a large-signal transient analysis [11], even though a fixed delay time is being used to model what is essentially a distributed phenomenon.

However, we can adopt a more formal approach to creating a large-signal FET model than using an *ad hoc* development of the linear, small-signal equivalent circuit model. We shall develop this approach in the next few sections.

6.4.2 Historical Development of Large-Signal FET Models – a Brief Review

The first compact models for field effect transistors to appear in a circuit simulator were the SPICE Junction FET and MOSFET models [12]. These circuit models were physically based, developed from the phenomenological equations for long gate-length devices, like those outlined in Chapter 1. While these models were adequate for the low-speed silicon technology of the time, with the advent of new transistor technologies such as the GaAs MESFET, which boasted a gate-length of the order of 1 μm and transition frequency over 10 GHz, coupled with new microwave measurement techniques such as S-parameters, there was a renewed interest in the development of FET compact models. Well known examples of compact models of this era are the 'Curtice' [13] and 'Statz' or 'Raytheon' [14] models for GaAs IC FETs, the 'Curtice-cubic' [15] model for FETs used in power amplifiers, and more general-purpose models such as the 'Materka' [16],

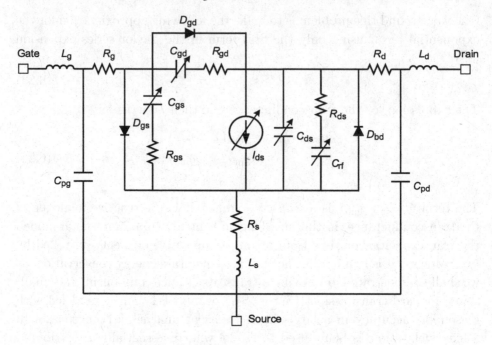

Fig. 6.15. Circuit schematic of a typical large-signal equivalent circuit model

'Parker-Skellern' [17], and 'TriQuint's Own Model' (TOM) [18]. Many of these compact models for GaAs MESFETs were extensions of the small-signal equivalent circuit model, based on the intuitive association of the circuit elements with the physical structure of the transistor, and describing the large-signal behaviour by curve-fitting the DC I_d–V_{ds} characteristics and the capacitance–voltage relationships of the transistor. And this is, in general, much the same approach as is practised today for many FET models, including those models that have been developed or adapted from the above list for newer FET technologies, such as the HEMT and PHEMT in III–V semiconductors, and, more specifically, models developed for LDMOS power FETs [19–21].

To illustrate some of the features and also the shortcomings in terms of accuracy, convergence, and so forth, from which some of these models typically suffered, we shall use a generic large-signal model schematic, shown in Fig. 6.15.

6.4.2.1 Some Remarks about Modeling the Drain Current Characteristics

Generally, the FET models mentioned above describe the DC drain current characteristic with a hyperbolic tangent function of the drain-to-source voltage, and use some polynomial or other function of the gate-to-source voltage to describe the gate control. Other parameters are added to account for finite output conductance, near- and sub-threshold behaviour, gate-leakage currents, and so forth. A generic drain current model is:

$$I_d = \beta \left(V_{gs} - V_T \right)^\gamma \tanh \left(\alpha V_{ds} \right) \left(1 + \lambda V_{ds} \right) \tag{6.25}$$

where α controls the knee region sharpness; β is related to the maximum drain current, and in some models this is expressed in terms of physical parameters such as gate oxide thickness; γ is traditionally equal to 2 – the FET 'square-law' characteristic – but in some models a different exponent is used; the λ term describes the finite output conductance; and V_T is the threshold voltage. Any or all of the model parameters α, β, γ, λ, V_T can or have been given further functional dependences on V_{ds}, temperature, and so forth, in attempts to obtain more accurate fitting of the DC curves.

The gate voltage control term, $\beta \left(V_{gs} - V_T \right)^\gamma$ in eq. 6.25, is often the focus of attention in new model developments, and a variety of functions and expressions have been proposed over the years, for example [14,15]. One particular feature of the gate voltage control expression is the reliance on the threshold voltage, V_T, which in practice is measured and defined in many different ways. This means that V_T is really more of a fitting parameter than a physical measure. In the implementation of the model in the simulator, care needs to be taken with this term as there is a discontinuity in the first derivative of this expression, at $V_{gs} = V_T$, which can lead to poor convergence, and errors at biases close to or below the threshold voltage.

In the original applications of these models, the FETs in the circuits were usually operated in Class A bias conditions, and so the errors close to threshold or pinch-off were not usually significant. In modern power amplifier applications, the transistors are operated in Class B or deep Class AB [22], to maximize the operating efficiency at maximum power output, and so model accuracy in the near-threshold region is of much greater importance.

Some models offer a refinement in the near-threshold region, by the addition of further model parameters, particularly for LDMOS FETs where the switch on characteristic can be much sharper than in III–V FETs, [20,21]. An alternative approach to obtaining a smooth and controlled I_d–V_{gs} curve in the near-threshold region is demonstrated in the *Angelov* or *Chalmers*

University model [23], in which the gate control function is a polynomial expression in $(V_{gs} - V_{pk})$, where V_{pk} is the gate voltage for maximum transconductance. This approach is claimed to account for high-order nonlinearities in the model, and this can be seen by writing out the expressions for the transconductance and higher-order derivatives,

$$\partial^n I_{ds}/\partial V_{gs}^n \tag{6.26}$$

though in practice it can be difficult to optimize for a specific order of the nonlinearity. The resulting model can be made to fit the FET output characteristics quite well, and also predict the intermodulation products [24]. Parker and Qu [25] adopt a similar approach, fitting the model to the higher-order derivatives of the transconductance.

Our discussion so far has focused on the model description of the DC I_d–V_{ds} characteristics. The basic flaw with using these DC characteristics for modeling is that RF and microwave FETs are not quasi-static. That is, the device characteristics change with frequency: this is known as *dispersion*. The output characteristics usually display the most significant dispersion. For example, in GaAs FETs and HEMTs the DC transconductance and output conductance are often quite different from the RF parameters, g_m and g_{ds} extracted from the S-parameter measurements: typically, the RF transconductance is smaller, and the output conductance higher; these properties result in a degradation of the expected performance of the transistor at RF. The dispersion is attributable to the presence of traps in the material, a feature that seems typical of III–V technologies, or to thermal effects, which, of course, are not insignificant in power transistors. A simple and well-used method of accommodating dispersion in the output conductance is to add an R-C network in parallel with the g_{ds} component, to modify the output conductance at high frequencies, as in, for example, [18, 26]. This is generally acceptable from a passive point of view, but this network lacks any influence from the input signal, and so has no output current dependence, as observed in real FETs. We shall describe techniques for modeling thermal and trap-related dispersion effects later in this chapter, in Section 6.5 and Sub-section 6.5.2.

6.4.2.2 Some Remarks about Modeling the Gate Capacitances

Whereas many of the remaining components in the large-signal equivalent circuit exhibit little or no voltage dependence, the input capacitances C_{gs} and C_{gd} can vary significantly with bias and have a discernible impact on the FET behaviour. Despite this, using the small-signal values for C_{gs} and C_{gd} at the transistor bias point can give reasonable results in small-signal and low-power amplifier applications. Additionally, in LDMOS transistors, the output capacitance C_{ds} also exhibits a significant voltage dependence. Early large-signal MESFET models, including the SPICE JFET model, incorporated a voltage dependence for the model capacitors based on the classic p-n junction or Schottky barrier capacitance–voltage relationship,

$$C\left(V_{\text{applied}}\right) = \frac{C_0}{\sqrt{\phi_{\text{bi}} - V_{\text{applied}}}} \tag{6.27}$$

where C_0 is the zero-bias capacitance, and ϕ_{bi} is the built-in voltage of the junction; the applied voltage is the gate-to-source or gate-to-drain voltage across the appropriate capacitor. An inspection of the capacitance–voltage relationships for the small-signal capacitors C_{gs} and C_{gd} indicates that the form of the relationship is very different from our expectations based on the classical equation predictions above. Examples of C_{gs} and C_{gd} plotted against V_{ds}, with V_{gs} as parameter, for a GaAs PHEMT power transistor, are shown in Fig. 6.16(a) and (b), respectively, to illustrate this effect.

In the case of the GaAs transistor, in the current saturation region, beyond the knee of the curves, the value of the capacitance C_{gd} is larger when the channel is pinched off, that is, when V_{gs} is below threshold, than when the channel is open. This is exactly counter-intuitive to our expectations from a one-dimensional analysis of the charge in the gate-drain depletion region. It is clear from the curves in Fig. 6.16(a) and (b) that the model capacitances are not simple one-dimensional capacitances, but are functions of both V_{gs} and V_{ds}. This functional dependence is the observed bias voltage dependence of the small-signal capacitances extracted from the measured (and de-embedded) bias-dependent Y-parameters. For the LDMOS transistor, a simple gate capacitance model is a parallel plate capacitor, which should be independent of voltage above threshold. From the curves in Fig. 6.17(a) and (b) it can be seen that is not so: again the capacitances are functions of both V_{gs} and V_{ds}. At higher drain voltages, the gate-drain capacitance for the LDMOS FET can be seen to approach the classical high-frequency value for beyond threshold conditions.

At this point, a simple development of a large-signal model from the small-signal parameters suggests itself: by taking only the imaginary parts of the bias-dependent small-signal Y-parameters, for clarity, we have at the gate of the FET;

$$C_{gs} = \frac{1}{\omega} \text{Im}\left(y_{11} + y_{12}\right) = C_{11} + C_{12}$$

$$C_{gd} = -\frac{1}{\omega} \text{Im}\left(y_{12}\right) = -C_{12} \tag{6.28}$$

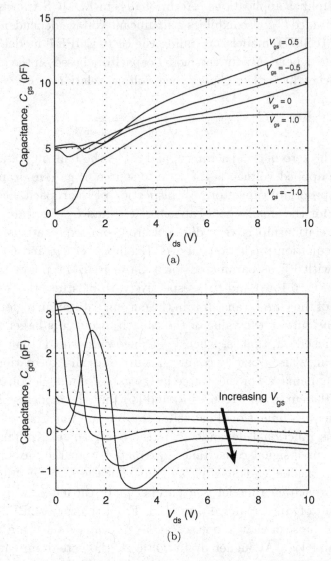

Fig. 6.16. The gate-source and gate-drain small-signal capacitances for a GaAs power PHEMT shown as functions of the applied bias voltages; (a) C_{gs} of the GaAs PHEMT, (b) C_{gd} of the GaAs PHEMT.

where the capacitances C_{ij} are now dependent on the instantaneous values of the voltages V_{gs} and V_{ds}. We can then write, for the (reactive) current at the gate,

$$I_1 = C_{gs}\left(V_{gs}, V_{ds}\right)\frac{dV_{gs}}{dt} + C_{gd}\left(V_{gs}, V_{ds}\right)\frac{dV_{gd}}{dt} \qquad (6.29)$$

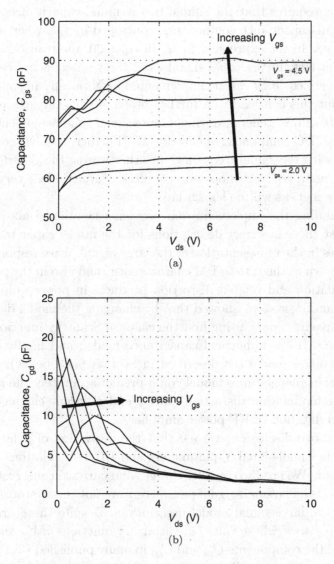

Fig. 6.17. The gate-source and gate-drain small-signal capacitances for an LDMOS power transistor, shown as functions of the applied bias voltages; (a) C_{gs} of the LDMOS FET, (b) C_{gd} of the LDMOS FET.

At first glance, this equation looks like a satisfactory description for the (reactive) port current. However, the implementation of this relationship in a compact model is more problematic. A circuit capacitor is a two terminal device that carries a charge whose value is controlled by the voltage across its terminals. Here, the capacitances are controlled by two independent voltages. This requires that the simple two-terminal capacitances are augmented by transcapacitance components, controlled by the other voltage in the dependence. In the expression for I_1 in eq. 6.29, the transcapacitances are missing; in fact, in the small-signal case, the transcapacitances are set to zero. Even so, eq. 6.29 can be implemented algebraically in some circuit simulators, but this can lead to a further problem with this expression for the current. It will, in general, during a large-signal time-dependent simulation, produce a DC component of current [27]. Further, this DC component will increase with the signal frequency. In other words, the general expression for current in eq. 6.29 does not *conserve charge*, unless very specific conditions are met, as will be shown later.

Notwithstanding the implementation problems mentioned above, it can be shown that these incorrect descriptions for the model capacitances lead to inaccuracies in the representation of the large-signal phase response of the transistor, known as the AM-to-PM characteristic, and also in the prediction of intermodulation and related distortion products in power amplifier circuits [28]. Staudinger *et al.* showed that by changing the model description of the gate capacitors *only*, using first the classical Schottky junction model, then the 'Statz' [14] symmetrical capacitance model, and finally the fully charge-conservative model (as described in Section 6.4.3 below), only the charge-conserving capacitance model could predict accurately the measured 3rd-order intermodulation distortion (IMD) and adjacent channel leakage ratio (ACLR) data for an RF power amplifier.

From the above discussion, we see that the derivation of a large-signal model from the small-signal Y-parameters and DC I–V relations is not so straightforward. We need to consider charge (and current) conservation, and dispersive effects. The small-signal parameters are not the appropriate state variables for the large-signal model description. Despite these drawbacks, 'nonlinear capacitors' whose values are arbitrary functions of V_{gs} and V_{ds} are used to model the components C_{gs} and C_{gd} in many published FET compact models. Such models can still show 'large-signal to small-signal consistency,' that is, the Y-parameters will map onto the linearized capacitances C_{gs} and C_{gd}, and the DC current I_d–V_{ds} is modeled accurately, and yet still have a non-conserving charge model under large-signal RF simulation, leading to inaccuracies in the model predictions, or even non-convergence of the simulation altogether.

6.4.2.3 Some Remarks about the Transcapacitance

We first introduced the concept of *transcapacitance* as the reactive analogue of the transconductance in the small-signal model, in Subsection 6.4.1. In

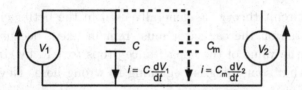

Fig. 6.18. Simple schematic circuit of a nonlinear capacitor whose value is a function of V_1 and a remote voltage V_2 [29]. © 1995 IEEE. Reprinted with permission.

the context of the small-signal model the transcapacitance is a capacitor component in the output circuit of the model, that is linearized about the bias point (V_{gs0}, V_{ds0}), and passes a reactive output current whose value is controlled by the *small-signal input voltage* v_{gs}: a *transfer capacitance*.

The capacitance components in the small-signal model are each functions of the two bias voltages, V_{gs0}, V_{ds0}. While this description is quite straightforward for the small-signal case, the behaviour of a two-terminal capacitor, whose value depends on the two signal voltages V_{gs}, V_{ds}, requires a more careful definition under large-signal conditions. Root and Hughes [11] claim that such a component can accumulate a net charge under steady-state periodic signal conditions, even though the voltage across the capacitor's two terminals returns to the same value at the end of the period. This is a violation of the concepts of conservation of charge, or conservation of energy. In a simple 'thought experiment', Snider [29] illustrates this principle with a nonlinear capacitor. We will describe this exercise briefly here.

In Fig. 6.18 we have a nonlinear two-terminal capacitor, whose value is governed by the remote voltage V_2, placed across the input voltage source V_1. The capacitor's value is (following Snider),

$$C\left(V_1, V_2\right) = 3 - V_2 \tag{6.30}$$

We will analyze this circuit over a single cycle in which the voltage V_1 steps from 1 V to 2 V, then voltage V_2 steps from 1 V to 2 V; then V_1 returns to 1 V, and finally V_2 returns to 1 V; now both voltages are back in their original conditions.

Initially, with $V_2 = 1$ V, the capacitance is 2 farads; and $V_1 = 1$ V, so the charge on the capacitor is $Q = 2$ coulombs. As the voltage V_1 is stepped to 2 V, the capacitor remains fixed at 2 F, and 2 coulombs of charge are delivered from the source V_1. Integrating the charge through the voltage source shows that 3 joules of energy have been supplied from source V_1.

Source V_1 is now held at 2 V while source V_2 is stepped from 1 V to 2 V. Since the source V_2 is not connected physically to the capacitor, then by conventional circuit theory (as generally used in the better circuit simulators), the charge on the capacitor must remain fixed, because V_1 has not changed, but the value of the capacitance drops to 1 F, by eq. (6.30). We begin to feel that something is beginning to wrong here. Nevertheless, we will press on.

The source V_1 now steps back to its initial value of 1 V. The capacitance is fixed at 1 F during this transition, so the change in charge on the capacitor is -1 coulomb, and -1.5 joules of energy are returned to the source V_1. Finally, V_2 steps back from 2 V to 1 V, the initial condition. Again the change in charge and energy are zero by conventional circuit theory – V_1 is unchanged – but the capacitance value returns to 2 F.

At the end of this cycle we have gained a net charge of 1 coulomb on the capacitor, and dissipated 1.5 joules of energy, even though this is a lossless circuit. We have apparently broken the laws of conservation of charge and energy, and yet this is how a conventional simulator would treat a circuit containing nonlinear capacitors whose values are dependent on remote voltages. Snider [29] suggests that this failure of conservation is due to not considering other possible charge and energy control mechanisms. For example, physically changing the capacitor's value requires a redistribution of the charge, and work must be done. But typically, the circuit simulator will not consider this 'external' system, and so we end up with a non-conserving system.

A means of accounting for this 'extra' charge and energy in the electrical domain is to introduce a *transcapacitance*. We add another capacitance, C_m, into the circuit in parallel with C in Fig. 6.18, and in which the reactive current is determined by

$$I = C_m \frac{dV_2}{dt} \tag{6.31}$$

The value of the transcapacitance, and the functional dependence on the voltages (V_1, V_2), is related to the value of the capacitor, C, through

$$\frac{\partial C}{\partial V_2} = \frac{\partial C_m}{\partial V_1} = -1 \tag{6.32}$$

in this example. If we now let $C_m = -V_1$, then as V_2 is stepped from 1 V to 2 V while V_1 is held at 2 V in the voltage cycle, than a charge of $C_m \cdot \Delta V_2 = -2$ coulombs is transferred from source V_1. This is exactly equal to and opposite of the charge transferred from source V_1 in the first voltage transition. If we follow the voltage steps through the complete cycle, we find that the net transfer of charge from source V_1 is zero, and the net energy dissipated in the system is zero. We now have a conservative system.

The condition described by eq. 6.32 is called the *integrability condition*, and constrains the possible values of the pair of capacitances C and C_m. We shall develop this charge conservative description later in the context of deriving our large-signal FET model.

Root and Hughes [11] claim that only one transcapacitance is necessary in the charge-based large-signal FET model. While this is mathematically correct, and indeed simpler, some modelers insist that each nonlinear capacitance (derived from the small-signal model) having a functional dependence on the two signal voltages, V_{gs} and V_{ds}, has an associated transcapacitance. This would mean that the large-signal model should have a total of six capacitances: three from the small-signal model, and three associated transcapacitances (not including, of course, our original transcapacitance proposed as the reactive analogue of the transconductance). A simple reflection on the fact that our model is a representation of a two-port network will inform us that only four unique reactive (capacitive) elements can be extracted directly from measurements of this network. These capacitances are: input capacitance, output capacitance, forward transfer capacitance, and reverse transfer capacitance, which approximately correspond to the imaginary parts of Y_{11}, Y_{22}, Y_{21}, and Y_{12}, respectively (divided by the angular frequency). These capacitances are each functions of the large-signal voltages V_{gs} and V_{ds}. The 'new' transcapacitance is the forward transfer capacitance.

6.4.3 The Large-Signal Model

We shall develop a large-signal model that uses current and charge for the state variables: such a charge-based approach has been presented by Ward and Dutton [30] and Root and Hughes [11], and further developed by Daniels *et al.* [31], Jansen *et al.* [32], and Root [33]. The use of current and charge state functions has also been used in a FET model by Werthof and Kompa [34], and in the 'Smoothie' model developed by de Vreede *et al.* [35]. We shall show that this model is charge conservative, and review the large-signal to small-signal consistency of the model. In later sections of this chapter, we shall accommodate dispersion due to thermal or trapping effects in the model.

A representation of this model is shown in Fig. 6.19 [33], in which the currents into each terminal, I_i (i = g, d, s), and the charges associated with each terminal, Q_i, are expressed in terms of the controlling voltages, V_{gs} and V_{ds}. The currents correspond to the measured currents in the model extraction process, and these terminal currents can be expressed in terms of

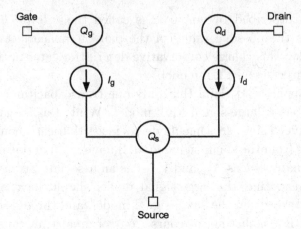

Fig. 6.19. Schematic structure of a FET model indicating terminal currents and node charge functions.

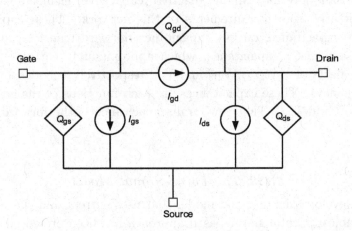

Fig. 6.20. Schematic structure of a FET model indicating branch current and charge functions.

real and capacitive components:

$$I_i^{\text{total}}(t) = I_i^{\text{cond}}\left(V_{\text{gs}}(t), V_{\text{ds}}(t)\right) + I_i^{\text{disp}}\left(V_{\text{gs}}(t), V_{\text{ds}}(t)\right)$$
$$= I_i\left(V_{\text{gs}}(t), V_{\text{ds}}(t)\right) + \frac{dQ_i\left(V_{\text{gs}}(t), V_{\text{ds}}(t)\right)}{dt} \tag{6.33}$$

While for a general FET model, a three-terminal representation can accommodate all of the operational conditions of the FET (as a switch, or in a cascode configuration, for example), our application is for RF power amplifiers, in which the source contact is generally grounded. This is convenient

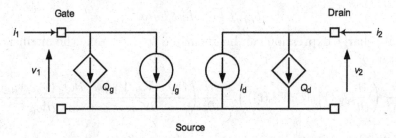

Fig. 6.21. The schematic circuit for the large-signal model, showing the voltage-controlled gate charge and gate-source current sources, and the voltage-controlled drain charge and drain-source current sources.

for model construction, and allows the controlling voltages $\{V_{gs}, V_{ds}\}$ to be measured and represented easily. Further, we can neglect the symmetry conditions which would normally apply for a general-purpose FET model. These symmetry conditions are outlined in [24].

The expressions in eq. 6.33 are perhaps more naturally cast as branch currents, represented by voltage-controlled current sources, and branch charges, represented by voltage-controlled *charge sources* [11]. This representation is shown in Fig. 6.20.

The circuit schematic of Fig. 6.20 can be transformed into the generic Y-parameter network of Fig. 6.12(a), which also applies for large-signal conditions. A redistribution of the conductive and capacitive currents in this network leads to the model structure shown in Fig. 6.21 [31, 34, 35].

In this circuit representation, the currents and charges are instantaneous functions of the controlling voltages, V_{gs}, V_{ds}. These functions are known as the *state functions*. We have

$$i_1(t) = I_g\left(V_{gs}(t), V_{ds}(t)\right) + \frac{d}{dt}Q_g\left(V_{gs}(t), V_{ds}(t)\right) \tag{6.34}$$

$$i_2(t) = I_d\left(V_{gs}(t), V_{ds}(t)\right) + \frac{d}{dt}Q_d\left(V_{gs}(t), V_{ds}(t)\right) \tag{6.35}$$

In normal operation, the conductive component of the intrinsic gate current is negligible: LDMOS transistors have an insulating oxide between gate and channel, and III–V FETs and HEMTs have reverse-biased junctions; the leakage current can be ignored, at least for the time being. That leaves us with three state variables to compute for the model: Q_g, Q_d, and I_d.

6.4.3.1 Gate Charge

The gate charge expression can be linearized to determine the intrinsic input admittance:

$$\frac{\partial}{\partial \vec{V}} \left(\frac{\partial}{\partial t} Q_g \left(V_{gs}(t), V_{ds}(t) \right) \right) = \frac{\partial}{\partial t} \left(\frac{\partial Q_g \left(V_{gs}, V_{ds} \right)}{\partial V_{gs}} + \frac{\partial Q_g \left(V_{gs}, V_{ds} \right)}{\partial V_{ds}} \right) \tag{6.36}$$

from which the admittance relations are:

$$\frac{\partial Q_g \left(V_{gs}, V_{ds} \right)}{\partial V_{gs}} = \frac{1}{\omega} \operatorname{Im} \left(Y_{11} \left(V_{gs}, V_{ds} \right) \right) \tag{6.37}$$

and

$$\frac{\partial Q_g \left(V_{gs}, V_{ds} \right)}{\partial V_{ds}} = \frac{1}{\omega} \operatorname{Im} \left(Y_{12} \left(V_{gs}, V_{ds} \right) \right) \tag{6.38}$$

If we write the admittances in terms of the bias voltage-dependent small-signal capacitances, we have the following relationships between the measured capacitances and the gate charge state variable:

$$\frac{\partial Q_g \left(V_{gs}, V_{ds} \right)}{\partial V_{gs}} = C_g \left(V_{gs}, V_{ds} \right) = C_{gs} \left(V_{gs}, V_{ds} \right) + C_{gd} \left(V_{gs}, V_{ds} \right) \tag{6.39}$$

$$\frac{\partial Q_g \left(V_{gs}, V_{ds} \right)}{\partial V_{ds}} = -C_{gd} \left(V_{gs}, V_{ds} \right) \tag{6.40}$$

The state variable $Q_g(V_{gs}, V_{ds})$ can be found by solving the pair of partial differential equations, eqs. 6.39 and 6.40 above. The necessary and sufficient conditions for a unique solution for Q_g are given by

$$\frac{\partial C_g \left(V_{gs}, V_{ds} \right)}{\partial V_{ds}} = \frac{\partial \left(-C_{gd} \left(V_{gs}, V_{ds} \right) \right)}{\partial V_{gs}} \tag{6.41}$$

which are essentially found by differentiating eq. 6.39 with respect to V_{ds} and eq. 6.40 with respect to V_{gs}. This is exactly the integrability condition introduced in eq. 6.32, which can also be expressed in integral form:

$$\oint \vec{C} \left(V_{gs}, V_{ds} \right) \cdot d\vec{V} = 0 \tag{6.42}$$

This expression states that the integral of the 'capacitance field' around a closed contour in potential is zero – it is a statement about the conservation of charge. This concept is expanded further in the next section.

6.4.3.2 Conservation Laws for Current and Charge

The *conservation of current* is essentially embodied in *Kirchhoff's current law*, which states that the net current entering (or leaving) a circuit node is zero. Generally in compact device model construction we do not need to worry about current conservation, as it is taken care of by the simulator. Later, though, we shall use current conservation to define a state function for the drain current.

The *conservation of charge* is also something that we tend to take for granted: charge can be neither created nor destroyed, and this concept is described by the *continuity equation* from electromagnetic theory [36]:

$$\nabla \cdot \mathbf{J} = -\frac{\partial \rho}{\partial t} \tag{6.43}$$

where $\mathbf{J}$ is the current density across a surface, and ρ is the charge density in the volume enclosed by the surface.

In the derivation of the state variable for the gate charge, Q_g, above, we use the notion of conservation in terms of a *conservative field* of capacitance. Again, we will use electromagnetic (or electrostatic) field theory to illustrate this concept. Probably the most familiar example of a conservative field is the *electric field*. A two-dimensional electric field in x and y is shown in Fig. 6.22; there are two arbitrary locations A and B in this field. We know that if we integrate along any path from A to B, we obtain the same potential difference:

$$\int_{A}^{B} \overrightarrow{E} \cdot \overrightarrow{dl} = \int_{A}^{B} \overrightarrow{E} \cdot \overrightarrow{dl} = V_{BA} \tag{6.44}$$
$$\text{\scriptsize contour1} \qquad \text{\scriptsize contour2}$$

Or, if we integrate around the closed contour, from A to B along contour 1 and back to A along contour 2, then we end up at the same potential:

$$\oint_{C} \overrightarrow{E} \cdot \overrightarrow{dl} = 0 \tag{6.45}$$

the electric field *conserves* potential energy.

Another consequence from vector algebra and electromagnetism is that a conservative field is also an *irrotational* field: the *curl* of the field vector is zero. For the two-dimensional electric field in our example, the curl is written as:

$$curl\left(\overrightarrow{E}\right) = \frac{\partial E_y}{\partial x} - \frac{\partial E_x}{\partial y} = 0 \tag{6.46}$$

We can see that this relation has exactly the same form as eq. 6.41, provided

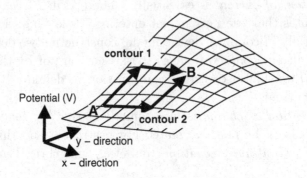

Fig. 6.22. A two-dimensional electric field in coordinates (x, y). The points A and B represent two potentials in this field, and contour 1 and contour 2 are two different paths connecting A to B.

we replace the electric field by a *capacitance field*, and the x–y coordinates become vectors in the *directions* of V_{gs} and V_{ds}, hence

$$curl_{\overrightarrow{V}}\left(\overrightarrow{C}\right) = \frac{\partial C_g\left(V_{gs}, V_{ds}\right)}{\partial V_{ds}} - \frac{\partial\left(-C_{gd}\left(V_{gs}, V_{ds}\right)\right)}{\partial V_{gs}} = 0 \qquad (6.47)$$

where

$$\overrightarrow{C} = C_g\left(V_{gs}, V_{ds}\right)\overrightarrow{u}_{V_{gs}} + C_{gd}\left(V_{gs}, V_{ds}\right)\overrightarrow{u}_{V_{ds}} \qquad (6.48)$$

$$\overrightarrow{V} = V_{gs}\overrightarrow{u}_{V_{gs}} + V_{ds}\overrightarrow{u}_{V_{ds}} \qquad (6.49)$$

and

$$C_g\left(V_{gs}, V_{ds}\right) = C_{gs}\left(V_{gs}, V_{ds}\right) + C_{gd}\left(V_{gs}, V_{ds}\right) \qquad (6.50)$$

Equation 6.47 is exactly the integrability condition mentioned earlier. And just as we did with the electric field, we can integrate along any contour between two points in the field to obtain the difference in *charge* between the two points

$$\int_{\substack{A \\ contour1}}^{B} \overrightarrow{C} \cdot \overrightarrow{dV} = \int_{\substack{B \\ contour2}}^{B} \overrightarrow{C} \cdot \overrightarrow{dV} = \Delta Q_{BA} \qquad (6.51)$$

by using the definitions for $\overrightarrow{C}, \overrightarrow{V}$ from eqs. 6.48 and 6.49. This is shown in Fig. 6.23. Just as the electric field conserves potential, here, the capacitance field *conserves charge*. A closed contour in this field would result

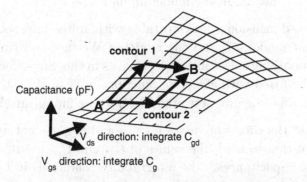

Fig. 6.23. The two-dimensional capacitance field at the gate, in coordinates (V_{gs}, V_{ds}). The points A and B represent the locations of two instantaneous signal voltages in this field, and contour 1 and contour 2 are two different paths connecting A to B. The integral along either (any) path from A to B yields the difference in charge between A and B.

in no change in the charge; in other words, charge is not being created or destroyed: charge is conserved.

6.4.3.3 Gate Charge Revisited

We now have in eq. 6.51 a recipe for calculating the gate charge state function Q_g from the measured bias-dependent Y-parameters at the gate of the FET, that is, the measured values of C_{gs} and C_{gd}. Starting from some arbitrary bias point in the $\{C_g, C_{gd}\}$ field, we can carry out the line integral in the voltage space to any other bias point, to determine the change in gate charge. By doing this over the whole measurement space of the bias-dependent gate capacitances, we build the gate charge state function of the instantaneous voltages V_{gs} and V_{ds}:

$$Q_g\left(V_{gs}, V_{ds}\right) = \int_{V_{gs0}}^{V_{gs}} [C_{gs}\left(v_{gs}, V_{ds0}\right) + C_{gd}\left(v_{gs}, V_{ds0}\right)]dv_{gs}$$

$$-\int_{V_{ds0}}^{V_{ds}} C_{gd}\left(V_{gs}, v_{ds}\right)dv_{ds} + Q_g\left(V_{gs0}, V_{ds0}\right)$$

(6.52)

where the starting point for the integration is (V_{gs0}, V_{ds0}), and the gate charge at this bias is $Q_g(V_{gs0}, V_{ds0})$, which can be set arbitrarily to zero, since we are going to compute dQ/dt to determine the current, and the time derivative of this integration constant vanishes. The gate charge $Q_g(V_{gs}, V_{ds})$ can be implemented in the model in many ways:

(i) as a two-dimensional look-up table with spline interpolations, as in the *Root* model [37] and the model of Werthof and Kompa [34], or with smoothing spline interpolations, as in the 'Smoothie' model [35];

(ii) it can be fitted with algebraic functions in (V_{gs}, V_{ds}) [32]; or

(iii) by an artificial neural network (ANN) whose inputs are (V_{gs}, V_{ds}) [38].

We note that the choice of values for C_{gs} and C_{gd} are not arbitrary. For charge conservation to hold, the values of C_{gs} and C_{gd} are related through the *curl* of the capacitances – the integrability condition. In fact, it is only when this condition is met that the model is able to fit the measured bias dependence of the Y-parameters. In practice, GaAs FETs and HEMTS, and LDMOS FETs have measured gate capacitances (Y-parameters) that do exhibit conservative charge, so this approach for the gate charge model is a good one.

The decision to use the conservative gate charge function as a basis for the large-signal model is a modeling choice, not a requirement of the simulator. Such a choice is made so that large-signal simulations of the model do not result in unphysical consequences, such as the build-up of charge. It is possible to create a non-charge-conserving model, and provided it is constructed from a charge-based function such as eq. 6.52, there should be no unbounded-charge inconsistencies in large-signal simulation, even though the bias-dependence of the small-signal capacitance will not be reproduced accurately by this model.

6.4.3.4 Drain Charge

We can apply the same method as above for calculating the drain charge state function, Q_d, except that we use the measured bias-dependent Y-parameters at the drain side of the FET in the line integral equation:

$$Q_d\left(V_{gs}, V_{ds}\right) = \int_{V_{gs0}}^{V_{gs}} [C_m\left(v_{gs}, V_{ds0}\right) - C_{gd}\left(v_{gs}, V_{ds0}\right)]dv_{gs}$$

$$+ \int_{V_{ds0}}^{V_{ds}} [C_{ds}\left(V_{gs}, v_{ds}\right) + C_{gd}\left(V_{gs}, v_{ds}\right)]dv_{ds} \qquad (6.53)$$

$$+ Q_d\left(V_{gs0}, V_{ds0}\right)$$

It turns out that the measured drain capacitances do not satisfy exactly the charge conservation conditions at the drain node. In fact, under these circumstances of a non-conserving charge, the precise form of the drain charge state function will depend upon the path chosen for the integrals in eq. 6.53. Nonetheless, imposing a charge-based description at the drain will avoid any non-physical charge build-up in large-signal simulations, even though the simulated small-signal (linearized) drain capacitances will not reproduce the measured bias dependences exactly. The discrepancy can be demonstrated by carrying out the integral for Q_d over different contours, and shows up particularly at low drain biases, around the 'knee' region of the output characteristics.

One possible reason for this departure from charge conserving at the drain node is the distributed nature of the gate-drain region in FETs, and particularly in power FETs. The partition of the measured drain-source capacitance between channel charge and the geometrically-defined capacitance, due to metallizations, for example, is somewhat arbitrary. A simple re-distribution of C_{ds} in millimeter-wave PHEMT models has been shown to produce a much better prediction of the S_{22} parameter versus frequency and bias [4, 39]. This partitioning of C_{ds} into essentially extrinsic and intrinsic components could be optimized until the integrated drain charge shows the smallest deviation from conservative behaviour.

6.4.3.5 Conservation of Energy

The conservation of energy is a basic physical principle that cannot be violated †. The relationship between charge and energy is of the general form

$$U = \int Q(V) \cdot dV \qquad (6.54)$$

We expect that the energy is conserved in our charge model of the FET, and this condition places similar constraints upon the two charges, Q_g and Q_d, as these charges place upon the capacitances attached to the gate and drain nodes. This constraint is expressed as an integrability relation for the energy, analogous to the integrability condition for the charge, eq. 6.41

$$\frac{\partial Q_g(V_{gs}, V_{ds})}{\partial V_{ds}} = \frac{\partial Q_d(V_{gs}, V_{ds})}{\partial V_{gs}} \qquad (6.55)$$

This equation states that there is a single energy function $U(V_{gs}, V_{ds})$ that is conserved by the gate and drain charges. The matrix of second partial

† Except by accident or design in simulation.

derivatives of this function U, with respect to the terminal voltages, is related to the measured capacitance functions, through

$$
\begin{bmatrix}
\dfrac{\partial^2 U}{\partial V_{gs}^2} & \dfrac{\partial^2 U}{\partial V_{gs}\partial V_{ds}} \\[3mm]
\dfrac{\partial^2 U}{\partial V_{gs}\partial V_{ds}} & \dfrac{\partial^2 U}{\partial V_{ds}^2}
\end{bmatrix}
=
\begin{bmatrix}
C_{gs} + C_{gd} & -C_{gd} \\[2mm]
-C_{gd} & C_{ds} + C_{gd}
\end{bmatrix}
\tag{6.56}
$$

A consequence of using a conservative energy function U is that the capacitance matrix must be symmetrical: the transcapacitance must be zero. But a zero transcapacitance is inconsistent with the measured bias-dependent Y-parameter data, from which we construct our charge model:

$$
\begin{bmatrix}
C_{gs} + C_{gd} & -C_{gd} \\[2mm]
C_m - C_{gd} & C_{ds} + C_{gd}
\end{bmatrix}
=
\begin{bmatrix}
\dfrac{\partial Q_g}{\partial V_{gs}} & \dfrac{\partial Q_g}{\partial V_{ds}} \\[3mm]
\dfrac{\partial Q_d}{\partial V_{gs}} & \dfrac{\partial Q_d}{\partial V_{ds}}
\end{bmatrix}
\tag{6.57}
$$

and so an energy-conserving model based on a single energy function U will not be able to predict the measured bias-dependent capacitances from which it was derived. An energy-conserving model will also be charge-conserving, but not necessarily *vice versa*; the charge-conserving model will produce a transcapacitance. However, the benefits of implementing an energy- or charge-conserving model, with the associated constraints enforced by the integrability conditions, are that the model will perform in large-signal simulation without the non-physical consequences such as unbounded charge growth.

6.4.3.6 Dispersion: FETs are Non-Quasi-Static

Before we enter a discussion about the models for the current state functions, we should address at least briefly the issue of dispersion. As we know, FETs are non-quasi-static; in other words, their terminal characteristics change with frequency. While we might expect all of the FET electrical characteristics to exhibit dispersion to some degree, the most noticeable effects are on the drain current characteristics: the transconductance and output conductance at RF are different from their DC values; the RF transconductance is lower, and the output conductance is higher. The principal causes of dispersion are (i) thermal effects, which are especially noticeable in power transistors as these devices are generally expected to operate at elevated temperatures; and (ii) charge-trapping effects, which are prevalent in III–V semiconductor FETs, but are not observed in LDMOS devices [1].

We shall account for the thermal effects on the device characteristics later, in Section 6.5 where we construct a coupling between the heat generated in the FET and the drain current characteristics that accounts for static effects arising from the ambient temperature and the quiescent bias condition, and dynamic thermal effects due to the RF electrical signal. We expect that the temperature of the transistor will also affect the gate current in junction FETs, as the reverse leakage or saturation current in Schottky junctions has a temperature sensitivity, [40]

$$J_s = A \cdot T^2 \, e^{-q\varphi_{B0}/kT} \tag{6.58}$$

The effects of charge trapping on the III–V FET and HEMT dynamics have been studied by Parker and Rathmell [41, 42] using pulsed *I–V* characterization over a range of frequencies. The FET output characteristic display quite complex dynamical behaviour over a bandwidth ranging from about 1 kHz to a few MHz. In particular, the shape of the characteristics in the 'knee' region shows droop and softening of the curves, changing with frequency. Historically this has not been a frequency regime of much interest in RF and microwave design, where the main concerns have been DC characteristics for bias circuitry, and the device response in the MHz and GHz regime for the RF signal. Nowadays in modern wireless communications systems, the signals occupy channels of several MHz width, and with the RF power amplifiers running at high output powers to maximize their efficiency. Such conditions generally correspond to the amplifier running in compression, and this nonlinear behaviour generates signal products at baseband, which can cover the frequency range over which the device parameters are changing significantly, as a result of these charge-trapping dispersion effects. This low-frequency nonlinearity can generate long-term memory effects, as described in Chapter 2. We shall describe some techniques for accommodating dispersion effects due to trapping in our model, in Section 6.5.2.

The low-frequency behaviour needs to be addressed in the construction of the model, so that we can capture the thermal and trapping dynamics, and hence be able to predict the long-term memory effects in circuit simulation.

6.4.3.7 Drain Current

We can apply a similar conservative approach to the calculation of the device currents, since the general expression for a voltage-dependent conductance is

$$I(V) = \int G(V) \cdot dV \tag{6.59}$$

The expression for the drain current is an integral equation in the measured small-signal transconductance and output conductance

$$I_d\left(V_{gs}, V_{ds}\right) = \int_{V_{gs0}}^{V_{gs}} g_m\left(v_{gs}, V_{ds0}\right)dv_{gs} + \int_{V_{ds0}}^{V_{ds}} g_{ds}\left(V_{gs}, v_{ds}\right)dv_{ds} + I_d\left(V_{gs0}, V_{ds0}\right)$$

$$(6.60)$$

If we compare this equation with its reactive analogue, eq. 6.53, we note that the gate-drain conductance term is absent. This is equivalent to setting $R_{gd} = 0$ in the small-signal model equations, which is a reasonable assumption, since this element is usually difficult to extract with any accuracy, and generally the value is small and does not make a significant contribution to the terminal characteristics. This certainly simplifies the path integrals, and the current is conserved at the terminals in any case by Kirchhoff's laws invoked in the simulator.

The drain current in eq. 6.60 is the *high-frequency* drain current, and using this for the large-signal model at RF overcomes, or avoids, the dispersion issues mentioned earlier. This is an important consideration for III–V FETs, though not so important for LDMOS.

This two-dimensional drain current function can be modeled using a similar range of techniques as described earlier for the charge functions: table look-up; multi-variate function fitting, and neural network methods.

Table models seem to be quite rare in FET modeling circles. The *Root* model [37] is probably the most widely known, constructed using the principles described above, but only available in Agilent's *Advanced Design System* (ADS) simulator. The *Root* model is process-independent, containing no material- or process-related parameters, and has been used successfully for power transistor modeling in III–V and LDMOS technologies. Its limitations are mainly related to the implementation of the spline interpolation in the simulator, although this has been overcome in the 'Smoothie' model of de Vreede [35], which uses smoothing splines. The 'Smoothie' model has also been extended to include a single-pole thermal model, for an electrothermal table model for LDMOS transistors [43]. Another means of avoiding the simple spline limitations is to use artificial neural networks to fit the two-dimensional state functions [38]. This method also provides a smoother surface with continuous higher-order derivatives, important for modeling nonlinear behaviour which is a prime concern in RF power amplifiers.

The function- or curve-fitting approach has received most attention from the FET modeling community, perhaps because relatively simple functions can give a good approximation to the I_d–V_{ds} curves, although some of the more modern FET models incorporate sophisticated function-fitting methods, including, in some cases, approximation by Volterra series [44], to try and capture nonlinearities and memory effects.

The generic I_d–V_{ds} model uses the hyperbolic tangent function $\tanh(x)$ to describe the shape of the current curves in both 'knee' and saturation regions. This function is modified by the addition of gate control and output conductance terms, of varying complexity in both implementation and parameter extraction. This generic relationship, presented in eq. 6.25, is repeated here, and we will use it to describe some interesting modifications to this curve, for power FETs in III–V and LDMOS technologies:

$$I_d = \beta \left(V_{gs} - V_T\right)^\gamma \tanh\left(\alpha V_{ds}\right)\left(1 + \lambda V_{ds}\right) \tag{6.61}$$

The parameters α, β, γ, λ, and the threshold voltage, V_T are varied to fit this equation to the measured 'high-frequency' I_d–V_{ds} characteristics obtained from the calculation of eq. 6.60. The γ parameter is often set to two as a default: the traditional FET 'square-law' characteristic. The α parameter controls the sharpness of the 'knee' region of the output current–voltage characteristics. In some implementations α is given a gate bias dependence to improve the accuracy of the model,

$$\alpha\left(V_{gs}\right) = \alpha_0 + \alpha_1 V_{gs} \tag{6.62}$$

although this can make the parameter extraction more difficult. Typically α_0 will be determined near threshold, and α_1 fitted over the I_d–V_{ds} characteristics. The λ parameter controls the slope of the output characteristics in the saturation region.

Whereas many models can give excellent predictions of the drain current in the saturation region, with more recently developed models, much greater attention has been paid to the gate control function $\beta\left(V_{gs} - V_T\right)^\gamma$, particularly in the near-threshold regime. This 'turn-on' region is of particular importance, for a number of reasons. RF power amplifiers used in wireless infrastructure applications are typically operated in deep Class AB, with the gate bias close to threshold; and high-efficiency amplifier designs are being developed, such as the Doherty architecture, which uses one transistor biased in Class AB/B and an auxiliary or peaking transistor biased in Class C, below threshold. Fager et al. [20] have shown that the detailed shape of the I_d–V_{gs} transfer characteristics near threshold has a significant effect on the magnitudes of the higher order derivatives $\partial^n I_d / \partial V_{gs}^n$, which control the level of harmonics in the output RF signal. The harmonic content, in turn, affects the level of in-band intermodulation distortion. In other words, the shape of the 'turn-on' region determines the level of distortion or nonlinearity exhibited by the transistor, so an accurate representation of the 'turn-on' characteristic is essential for our model to be able to predict the distortion components in large-signal simulation.

The venerable 'Motorola Electro-Thermal' (MET) model [19] is virtually the industry-standard LDMOS power FET model. The I_d–V_{ds} is given in eq. 6.63 below, illustrating the modifications to the gate control function compared with the simple description of eq. 6.61:

$$I_d = \beta V_{gst}^{VGEXP} \tanh\left(\frac{\alpha V_{ds}}{V_{gst}}\right)(1 + \lambda V_{ds}) \tag{6.63}$$

The control parameter V_{gst} in eq. 6.63 is given by

$$V_{gst} = VST \ln\left(1 + e^{(V_{gst1}/VST)}\right) \tag{6.64}$$

where the extractable parameter VST controls the abruptness of the 'turn-on' characteristic, and V_{gst1} is given by

$$V_{gst1} = V_{gst2} - \tfrac{1}{2}\left(V_{gst2} + \sqrt{(V_{gst2} - VK)^2 + \Delta^2} - \sqrt{VK^2 + \Delta^2}\right) \tag{6.65}$$

where V_{gst2} is essentially the gate control voltage;

$$V_{gst2} = V_{gs} - V_T \tag{6.66}$$

The extractable parameter VK is the voltage at which the gate voltage V_{gst2} saturates, and Δ is the voltage range over which V_{gst2} changes from a linear to saturated function. This is illustrated in Fig. 6.24, in which $VK = 7$ volts, and $\Delta = 1$. In the MET model, the control voltage V_{gs} is a delayed version of the gate voltage, to account for distributed effects in the gate.

Whereas the MET model can, when the model parameters are carefully optimized, predict the DC, S-parameters and large-signal behaviour, modifications to the gate control function proposed by Fager *et al.* [20] are claimed to improve the model's ability to predict distortion products for power amplifier applications. Fager *et al.* describe the transconductance and drain current transfer characteristics using four distinct regions, as shown in Fig. 6.25. They claim that by including the region of quadratic dependence of drain current on gate voltage, between the exponential sub-threshold region and the linear regime, they can incorporate additional model parameters to define the 'turn-on' abruptness. Without these additional parameters, the 'turn-on' region is too smooth, and the higher order derivatives that control the distortion are too small. Essentially, the modification that Fager *et al.*

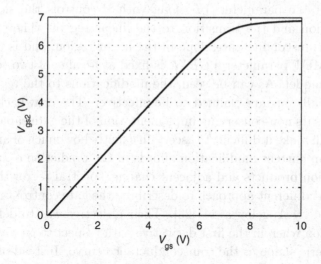

Fig. 6.24. The saturation of the gate voltage V_{gs}–V_T using the saturation voltage parameter VK and transition width.

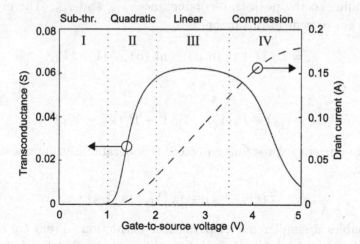

Fig. 6.25. Typical LDMOS FET transconductance and drain current as functions of the gate voltage, indicating the different operating regions [20]. ©2002 IEEE. Reprinted with permission.

make to the MET model is the inclusion of an additional parameter into the gate control function,

$$I_d = \beta \frac{V_{gst}^2}{1 + V_{gst}^p/VL} \tanh\left(\frac{\alpha V_{ds}}{V_{gst}}\right)(1 + \lambda V_{ds}) \qquad (6.67)$$

The extractable parameter *VL* along with β controls the slope of the quadratic region and the transition to the linear region. The exponent *p* modifies the transconductance slope in the linear region, and is usually set to one. The MET parameter *VGEXP* is fixed at a value of two (quadratic) in the Fager model. As can be seen, the modifications to the regular MET model are small, but are claimed to be effective. The drawback with the Fager model is its non-conserving implementation of the gate capacitor components, which make it difficult to ascertain exactly how much of an improvement the drain current modification affords in the prediction of intermodulation distortion products and adjacent channel spectral regrowth.

A somewhat different approach to describing the gate control function has been proposed by Angelov *et al.* [23]. Their objective was to define a gate control function wherein the first derivative with respect to gate voltage has the same generic shape as the transconductance curve. Instead of using the threshold voltage as a parameter in the gate control function, Angelov *et al.* define a function in terms of the measured gate voltage and drain current corresponding to the peak transconductance, V_{pk} and I_{pk}. The expression for the drain current in this model is

$$I_d = I_{pk} \left(1 + \tanh(\psi)\right) \tanh\left(\alpha V_{ds}\right) \left(1 + \lambda V_{ds}\right) \tag{6.68}$$

where

$$\psi = P_1 \left(V_{gs} - V_{pk}\right) + P_2 \left(V_{gs} - V_{pk}\right)^2 + P_3 \left(V_{gs} - V_{pk}\right)^3 + \ldots \tag{6.69}$$

This drain current expression, eq. 6.68, is separable into functions only of V_{gs} and V_{ds},

$$I_d \left(V_{gs}, V_{ds}\right) = f_A \left(V_{gs}\right) f_B \left(V_{ds}\right) \tag{6.70}$$

which enables a simpler model parameter extraction. From eq. (6.69) it can be seen that the drain current function has well defined derivatives with respect to the gate voltage. This enables the identification of these higher-order derivative terms with the order of the distortion components, in principle allowing this drain current expression to describe the nonlinear behaviour of the transistor. Validation of this premise has been carried out by Angelov *et al.* [45] using several GaAs FETs and HEMTs. An adaptation of this model in which the 'Angelov' drain current characteristic was coupled with a conservative charge model for Q_g and Q_d has been used to create a model for a III–V enhancement-mode PHEMT power transistor, designed for wireless handset applications. This model was shown to be able to predict accurately the DC, small-signal S-parameter, and large-signal behaviour, including the maximum power output and power-added efficiency points in loadpull analysis, and the adjacent and alternate channel powers [24].

The 'Angelov' model is quite widely used for modeling of III–V FETs, and appears in several commercial simulators. It has also been used for LDMOS transistor modeling, for modest size power FETs but used only at low voltages to minimize thermal dispersion [46]. A modified drain current function was used, in which the drain voltage dependence of the gate control parameters V_{pk} and P_1 are included:

$$V_{pk}(V_{ds}) = V_{pk0} + (V_{pks} - V_{pk0})(1 + \lambda V_{ds})\tanh(\alpha V_{ds}) \qquad (6.71)$$

$$P_1 = P_{1s}\left[1 + \left(\frac{P_{10}}{P_{1s}} - 1\right)\frac{1}{\cosh^2(BV_{ds})}\right] \qquad (6.72)$$

where the subscripts '0' and 's' correspond to parameter values measured or calculated at V_{ds} zero volts and in saturation, respectively, and B is a fitting parameter, related to α.

The DC and fundamental RF performance of this model was satisfactory, but the prediction of the second and third harmonic components was relatively poor. The authors suggest that this is due to dispersive effects, but it may be attributable to the smoother 'turn-on' characteristic of the 'Angelov' model, leading to poorer prediction of higher-order products, as proposed by Fager et al. [20] in their LDMOS model.

6.4.3.8 Gate Current

The controlled current source representing the gate current can also be found by the process of integrating the real parts of the input admittances over the gate and drain voltage space

$$I_g(V_{gs}, V_{ds}) = \int_{V_{gs0}}^{V_{gs}} \text{Re}\{y_{11}(v_{gs}, V_{ds0})\}dv_{gs}$$

$$+ \int_{V_{ds0}}^{V_{ds}} \text{Re}\{y_{12}(V_{gs}, v_{ds})\}dv_{ds} + I_g(V_{gs0}, V_{ds0}) \qquad (6.73)$$

Written in this form, the integral equation illustrates the formal recipe that we are following. In practice, we find that the real part of y_{12}, corresponding to R_{gd}, is difficult to extract from measured data with any accuracy: in many small-signal FET models this element is omitted altogether. And in large-signal models of the form that we are using, Fig. 6.21, the small-signal resistances R_{gs} and R_{gd} are not included, but may be re-positioned to be part of the (previously extracted) extrinsic network. If we disregard R_{gd} for the moment, then the integral term for the gate current simplifies to

$$I_g\left(V_{gs}, V_{ds}\right) = \int\limits_{V_{gs0}}^{V_{gs}} g_{gs}\left(v_{gs}, V_{ds0}\right)dv_{gs} + I_g\left(V_{gs0}, V_{ds0}\right) \qquad (6.74)$$

which is the integration of the slope conductance at the input of the transistor over the required gate voltage range. This conductance is a single-valued function, parameterized by the drain-to-source voltage. It should be fairly straightforward to model this I_g–V_{gs} relationship in the large-signal model.

If we look to the structure of the device to guide us, then in III–V MESFETs and HEMTs the gate contact is a Schottky barrier diode: the gate metal is placed directly onto the semiconductor. A simple two-terminal Schottky diode has a classical diode rectifying current–voltage characteristic. In practical III–V FETs, the semiconductor layer structure may include several heterojunctions, which will modify the simple diode characteristic, and the current flow through the diode will divide into gate-to-drain and gate-to-source current components. This line of thought leads us to the classic SPICE JFET model, which has diode instances from gate-to-drain and gate-to-source. This structure has found its way into several large-signal III–V FET and HEMT models.

In the usual bias arrangement for FET operation, the gate-drain diode is heavily reverse biased, effectively open circuit for conduction current. The gate-source diode is generally also reverse biased in quiescent conditions, although less so, and in power transistors under drive is likely to be forward biased for some part of the RF cycle. This physical view of the gate conduction in normal operation suggests that using a diode model for the gate-source controlled current source is a reasonable approach, though any drain voltage influence on this diode characteristic has to be added parametrically to the diode model. Nevertheless, an exponential current–voltage relationship is a reasonable function to use as a basis for fitting the measured gate current I_g–V_{gs} relationship, for a one-dimensional curve-fit.

The 'Smoothie' table-based model [35] uses eq. 6.73 to determine the gate current over the $\{V_{gs}, V_{ds}\}$ space, using interpolation between data points rather than function-fitting. The *Root* model does not use an integral relationship for the gate current, but uses the measured (and suitably de-embedded) gate current directly in the model table. This latter approach assumes that dispersion of the gate conduction current is negligible.

For LDMOS transistors, the gate current is zero at DC: the gate oxide provides an insulating barrier to conduction current. At high frequencies there will be a real part to the total gate current, attributable to the series resistances of the extrinsic network, the gate and source resistances R_g and R_s. After de-embedding, the conduction component of the gate current at the intrinsic gate terminal should be zero.

Conditions where there is a large gate current contribution occur in breakdown. We will need to accommodate breakdown effects in our model, as limits of the normal operating region. This is described in the next section.

6.4.3.9 Breakdown Models

RF power transistors are operated near to the limits of their electrical specifications, that is, the maximum values of voltage and current, to obtain the highest power output without compromising the reliability. The maximum voltage limit is set by the gate-drain breakdown voltage in FET devices. This breakdown voltage is generally defined in terms of a specified two-terminal current flow from gate to drain; for example, the gate-drain breakdown voltage BV_{gd} is when $I_{gate} = -I_{drain} = 1$ mA/mm of gate periphery.

There are several competing high voltage breakdown mechanisms in III–V and LDMOS FET technologies. The major factor is a high field region at the drain edge of the gate electrode, leading to avalanche breakdown, which creates high densities of electron-hole pairs in the drain drift region, and thermionic-field emission over the gate Schottky contact in III–V FETs, resulting in high gate current and eventual burn-out [47]. The avalanche mechanism in LDMOS transistors results in the generated holes being injected into the substrate, where they cause a large enough ohmic voltage drop to forward bias the source-substrate junction. This injects electrons into the base of the parasitic bipolar transistor in the LDMOS structure, resulting in high current flow. This effect is exacerbated by high resistivity substrates.

There are transistor design and fabrication techniques that can mitigate this high field region in power FETs, known as *RESURF* techniques – this is an acronym for 'reduced surface field' [48–50]. In LDMOS FETs the RESURF methods include additional dopant implants, to increase the depth of the depletion region beneath the drain drift region, thereby increasing the breakdown voltage. In both LDMOS and III–V GaAs and GaN power FET technologies, the addition of one or more electrodes between the gate and drain, closer to, and connected to the gate (or ground) metal, but isolated from the drain drift region, can also reduce the electric field crowding at the gate and produce a more uniform potential field. These electrodes are known as *field plates* [51]. The reduction of the surface field also mitigates gate-drain breakdown through the surface states, which can occur in III–V transistors. These field plate electrodes can serve an additional purpose by shielding the gate from the drain, and reducing the gate-drain feedback capacitance, and hence increasing the gain and bandwidth of the device.

For model extraction we use pulsed measurement methods that allow us to probe the breakdown regions, so that the model we derive can predict the electrical behaviour up to the edge of the safe operating region of the FET. The model must be capable of reproducing the onset of breakdown, so that the amplifier designer is able to 'push' his design envelope safely. In this respect it is often sufficient to have a qualitative model of the current–voltage characteristics for the breakdown region, and so an approximate current–voltage relationship is generally good enough.

A common technique for accommodating on-state breakdown is to model the rapid increase in the drain current by a junction diode with an artificially high turn-on voltage: the gate-drain breakdown voltage. For example, the breakdown diode characteristics are implemented in the MET model in the following way:

$$I_{\mathrm{d}}^{\mathrm{total}} = I_{\mathrm{d}0} \left(1 + K_1 \, e^{V_{\mathrm{BReff1}}}\right) \tag{6.75}$$

where V_{BReff1} is related to the drain source voltage, and K_1 is a fitting constant. The drain current $I_{\mathrm{d}0}$ may be one of the expressions for drain current described earlier, in eq. 6.61, for example. Similar exponential current characteristics have been observed in GaAs power FETs [52], and they can also be accommodated into the drain current model in this manner. The gate-drain breakdown characteristics also have a significant dependence on temperature. Again, we will use the electro-thermal modeling methods outlined in the next section.

6.4.3.10 The Large-Signal Model – a Summary

We have adopted an approach to generating the large-signal model based on current and charge conserving principles, wherein the model state functions are generated from the measured and de-embedded, isothermal, small-signal Y-parameters by integral transforms. The model structure is shown in Fig. 6.21, and the state function expressions are collected together below, for completeness:

$$Q_g (V_{gs}, V_{ds}) = \int_{V_{gs0}}^{V_{gs}} [C_{gs} (v_{gs}, V_{ds0}) + C_{gd} (v_{gs}, V_{ds0})]dv_{gs}$$

$$- \int_{V_{ds0}}^{V_{ds}} C_{gd} (V_{gs}, v_{ds})dv_{ds} + Q_g (V_{gs0}, V_{ds0}) \qquad (6.76)$$

$$Q_d (V_{gs}, V_{ds}) = \int_{V_{gs0}}^{V_{gs}} [C_m (v_{gs}, V_{ds0}) - C_{gd} (v_{gs}, V_{ds0})]dv_{gs}$$

$$+ \int_{V_{ds0}}^{V_{ds}} [C_{ds} (V_{gs}, v_{ds}) + C_{gd} (V_{gs}, v_{ds})]dv_{ds} \qquad (6.77)$$

$$+ Q_d (V_{gs0}, V_{ds0})$$

$$I_g (V_{gs}, V_{ds}) = \int_{V_{gs0}}^{V_{gs}} \text{Re}\,\{y_{11} (v_{gs}, V_{ds0})\}dv_{gs}$$

$$+ \int_{V_{ds0}}^{V_{ds}} \text{Re}\,\{y_{12} (V_{gs}, v_{ds})\}dv_{ds} + I_g (V_{gs0}, V_{ds0}) \qquad (6.78)$$

$$= \int_{V_{gs0}}^{V_{gs}} g_{gs} (v_{gs}, V_{ds0})dv_{gs} + I_g (V_{gs0}, V_{ds0})$$

$$I_d (V_{gs}, V_{ds}) = \int_{V_{gs0}}^{V_{gs}} g_m (v_{gs}, V_{ds0})dv_{gs} + \int_{V_{ds0}}^{V_{ds}} g_{ds} (V_{gs}, v_{ds})dv_{ds} + I_d (V_{gs0}, V_{ds0})$$

$$(6.79)$$

Writing the integral equation in this form retains the symmetry between real and imaginary parts of the Y-parameters, and the controlled current and charge sources. We have shown that these state functions can be implemented directly as interpolated tables in a table look-up model, or we can fit analytic functions or include artificial neural networks to calculate the charge and current surfaces.

6.5 Including Frequency Dispersive Effects in the Transistor Model

In our development of the electrical model of the transistor so far, we have been careful to ensure that the model parameters display no explicit frequency or time dependence; we have deferred any discussion of including the frequency dispersive effects in the model until now. The frequency dispersion arises from the heating in the channel of the FET, which occurs in both LDMOS and III–V power transistors, and from charge-trapping effects, which occur predominantly in III–V FETs. Both of these phenomena have been introduced earlier (Chapter 2), and we shall now explain how we can include the effects due to heating and charge-trapping in the compact FET model in a reasonably straightforward manner.

6.5.1 Temperature-Dependent Model Parameters

There are two causes of temperature change in the transistor: changing the ambient or heatsink temperature of the device; and the self-heating of the transistor attributable to the power dissipation in the FET channel, caused by the instantaneous current and voltage. These heating effects can be broadly separated as *static* and *dynamic* thermal effects, respectively.

If we consider the metal, semiconductor and dielectric materials that comprise the structure of the transistor, we see that their electrical and mechanical properties vary with temperature. The metal and semiconductor resistances increase with temperature in a fairly linear manner over the typical range of operating temperatures or channel temperatures of the FET. The materials will also expand with temperature, causing local changes in dimension, and causing changes in the stresses and strains within the materials, and at the interfaces between materials. The effects on the measured electrical characteristics of the transistor depend on the materials themselves, and the FET structure and fabrication methods.

From the modeling perspective, we can attempt to include these microscopic temperature-dependent phenomena, and construct a physically-based transistor model. The model accuracy will depend on how well these physical phenomena are described and included in the model equations. Or we can, in principle, construct a series of experiments to isolate the influence of the equivalent circuit model parameters and thence determine their temperature coefficients. This can lead to a transistor model that consists of hundreds of parameters in an attempt to describe all of the measurable phenomena, with attendant difficulties in parameter extraction, and also possible difficulties with model convergence in improperly or incompletely defined electro-thermal models [53, 54].

In contrast, our approach is to accommodate the temperature effects on the FET in the measurable or derivable terminal qualities of current and charge. We will consider both static and dynamic thermal effects on the currents and charges, and modify the isothermal models derived in previous sections in a self-consistent way.

6.5.1.1 The Manifolds and Bond-pads

The metal of the manifolds and bond-pads will have a thermal coefficient of resistivity, and so the metallic or loss resistance of this element will change with temperature. The substrate loss, which is more significant in low-resistivity LDMOS technology than the semi-insulating III–V substrates, also depend on temperature. We have also noted earlier that the manifold may be large enough in a practical power transistor to exhibit distributed effects, or frequency dispersion of the loss components, owing to its size. From a model extraction point of view, the test transistors are usually small enough that such dispersion can be neglected, and our objective is to construct a transistor model from isothermal electrical measurements, that is referenced *inside* the manifolds. This approach allows the discrete or integrated power amplifier designer the scope to build and model his own manifold and bond-pad structures. The substrate definition for the planar EM solvers used for the manifold modeling should be generated for several temperatures over the operating temperature range.

6.5.1.2 The Extrinsic Components

In our earlier analysis, we defined the elements in the extrinsic shell to be linear components, independent of the applied gate and drain voltages. In the extrinsic shell, it is only the resistive components that have temperature dependence. The inductor and capacitor elements can be considered temperature independent, to first order: the inductor losses in the metallization are subsumed into the extrinsic resistances; and any variation in capacitance due to thermo-mechanical effects, such as expansion, are negligible.

Under static thermal conditions, we can extract the values of resistance for the gate, drain and source resistances using cold-FET techniques for example, to obtain the variation of these resistances with temperature. As indicated earlier, we can then include these static temperature coefficients in the resistor definitions in the model, to mimic their temperature dependence. If we now consider the FET operation under large-signal conditions, we have self-heating of the FET due to the dissipated power in the channel, that is, the product of channel current and drain-source voltage. Therefore, the transistor temperature is dependent on the applied voltage, because the current is a dependent variable. This means that the resistors in the extrinsic shell are nonlinear components, since they are temperature-dependent and hence voltage-dependent. A voltage-dependent resistor is actually a controlled current source, and we can use these components for the extrinsic resistors in an electro-thermal model, as is done in the MET model [19].

But using a voltage-dependent resistor contradicts our earlier definition that the extrinsic elements are linear components. We shall overcome this problem by defining the extrinsic resistors to be temperature *independent*, and incorporating the thermal effects in the nonlinear intrinsic model. From the standpoint of making a compact model, this is a reasonable thing to do. Consider the drain current: if we were to use drain voltage-dependent resistors for the drain and source extrinsic resistances, and we model the intrinsic drain current source with a drain voltage-dependent current source, all of these sources are in series, and the overall terminal current can be modeled equally well with just one dependent source, in the nonlinear intrinsic part of our model, provided the thermal effects are properly captured in the drain current dependence on the voltage. Similarly, the thermal behaviour of the gate resistance can be included in the nonlinear controlled gate current source. In fact, in III–V devices, the thermal dependence of the nonlinear Schottky diode current–voltage characteristics dominates any temperature dependence of the gate metallization, and so including the thermal effects into the nonlinear gate current source is the most practical approach.

6.5.1.3 Temperature Dependence of the Drain Current

In view of the many and varied temperature-dependent effects present in the transistor, a pragmatic approach to discovering the temperature dependence of the drain current is to take isothermal measurements of the FET output characteristics over a range of ambient temperatures, and then devise a relationship between the measured current and the ambient temperature. This will yield a *static* thermal dependence. The isothermal conditions can be ensured by measuring the drain current–voltage characteristics using short pulse widths to reduce the channel self-heating. This experiment has been performed by a number of workers [55–57], and the drain current measured at a constant drain-to-source voltage, $V_{\rm ds0}$, is found to be a fairly linear function of temperature, over the typical range of operating temperatures, and can be modeled as

$$I_{\rm d}\left(T_{\rm amb}\right) = \frac{I_{\rm d0}}{1 + c\left(T_{\rm amb} - T_0\right)} \qquad (6.80)$$

where I_{d0} is the drain current measured at the reference temperature T_0, and c is the drain current static thermal coefficient, with units of K^{-1}. Following Canfield [57], we equate this coefficient to $1/T_0$, and eq. 6.80 becomes

$$I_d = \frac{I_{d0}}{1 + \dfrac{(T_{amb} - T_0)}{T_0}} = \frac{I_{d0}}{1 + \dfrac{\Delta T}{T_0}} \qquad (6.81)$$

This form of expression has also been derived by Schmale and Kompa [58], and Parker and coworkers [59, 60], in their FET and HEMT thermal model developments. Using the ambient temperature alone to account for thermal effects is not sufficient for large-signal models. The FET dissipates a considerable amount of power as heat when operating in a high power amplifier application. The instantaneous power is not, however, a good indicator of the average dissipated power, because the long time constants associated with heat diffusion and storage mean that the channel temperature cannot follow directly the instantaneous power dissipation, but is instead related to some *average power* dissipation in the device. Following the thermal analysis outlined in Chapter 5, the change in temperature can then be written in terms of this average power, related by an *effective thermal resistance* that accounts for the temperature distribution in the channel:

$$T - T_0 = R_{th} P_{avg} \qquad (6.82)$$

The effective thermal resistance, R_{th}, is not the true thermal resistance of the semiconductor, but it accounts in a practical way for the non-uniform heating in the channel, which is usually confined to the high electric field drift region at the drain edge of the gate, and any non-isotropic heat diffusion due to the FET topography and metallurgy. The effective thermal resistance can be a function of bias and temperature, particularly for III–V semiconductors and for PHEMTs. For simplicity, we will consider it to be constant; this does not change the formalism, and the results are generally acceptable. Combining eqs. 6.81 and 6.82, we can write the change in the measured drain current as a result of this average dissipated power, as:

$$I_d = \frac{I_{d0}}{1 + \dfrac{(T - T_0)}{T_0}} = \frac{I_{d0}}{1 + \dfrac{R_{th} P_{avg}}{T_0}} \qquad (6.83)$$

Parker and Root [59] write this expression as

$$I_d = \frac{I_{d0}}{1 + \delta P_{avg}} \qquad (6.84)$$

and introduce a *thermal resistance temperature coefficient*, δ, which we can see is simply R_{th}/T_0, the effective thermal resistance divided by the isothermal model extraction temperature. This coefficient is again dependent upon the detailed geometry, topography and metallurgy of the power transistor, and so is a characteristic parameter for a given device.

As stated above, the channel temperature cannot follow the instantaneous power dissipation, because the thermal time constants are very long compared with the period of the RF signal. Instead, the channel temperature is related to the average dissipated power, which in turn is derived from the instantaneous power by solving the first-order ODE describing the thermal equilibrium in the channel, expressed qualitatively as

$$\text{Heat Generation}(i_{\text{d}}, v_{\text{ds}}) = \text{Heat flow}(T - T_0) + \frac{d}{dt}(\text{Heat Storage}(T - T_0))$$

which becomes

$$P_{\text{diss}}(t) = \frac{(T - T_0)}{R_{\text{th}}} + C_{\text{th}}\frac{d}{dt}(T - T_0) \tag{6.85}$$

The instantaneous dissipated power is the product of the instantaneous drain current and drain-to-source voltage, $P_{\text{diss}}(t) = i_{\text{d}}(t).v_{\text{ds}}(t)$, and we have introduced a new model parameter C_{th}, the *thermal capacitance*.

We can relate the instantaneous power given by eq. 6.85 to the average dissipated power by substituting for $(T{-}T_0)$ from eq. 6.82, to give the following ODE, which we solve for the average dissipated power, P_{avg}

$$P_{\text{diss}}(t) = i_{\text{d}}(t)v_{\text{ds}}(t) = P_{\text{avg}} + \tau\frac{dP_{\text{avg}}}{dt} \tag{6.86}$$

The time constant, τ, is the product of the effective thermal resistance and thermal capacitance. In many FETs the value of τ is estimated to be in the MHz range, for both LDMOS and III–V technologies.

We can modify eqs. 6.83 and 6.84 so that we can accommodate the effects on the drain current characteristics of both the ambient temperature, and the dissipated electrical power in the channel due to P_{avg}.

$$i_d(T) = \frac{i_{\text{d0}}}{1 + \dfrac{R_{\text{th}}P_{\text{avg}} + (T_{\text{amb}} - T_0)}{T_0}} \tag{6.87}$$

After converting to Parker's nomenclature [59], through

(i) $\delta = R_{\text{th}}/T_0$, and
(ii) $\gamma_{\text{T}} = \delta/R_{\text{th}} = 1/T_0,$

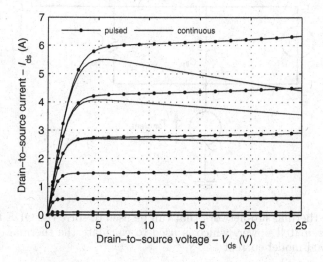

Fig. 6.26. The isothermal model drain current, and the thermally degraded drain current.

we get

$$i_d(T) = \frac{i_{d0}}{1 + \delta P_{avg} + \gamma_T(T_{amb} - T_0)} \tag{6.88}$$

By degrading the drain current in this manner we can effectively account for the dispersion in the output characteristics from DC to RF, due to thermal effects of both ambient temperature change and self-heating in the device. The isothermal current and thermally-degraded current are shown in Fig. 6.26. This formulation for the thermal dispersion has been used in a model of a GaAs E-PHEMT power transistor for handset power amplifier applications, accurately predicting the loadpull contours and distortion behaviour [24]. This model was implemented using the Symbolically-Defined Device (SDD) in ADS, and the thermal ODE was solved self-consistently with the nonlinear current equations.

A common way of implementing this model is to note that eq. 6.85 resembles a circuit theory expression for a parallel resistor-capacitor network, driven by a current source, where

(i) T-T_0 corresponds to a voltage, and
(ii) $P_{diss}(t)$ corresponds to a current.

The *thermal sub-network* is shown in Fig. 6.27; the R-C network is connected to ground through a voltage source of value $V_{source} = T_{amb}$, and the 'output voltage' corresponds to the effective channel temperature, T. This value can

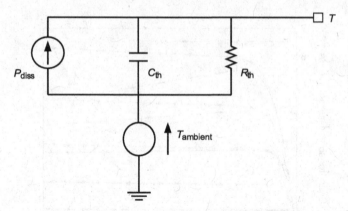

Fig. 6.27. The thermal sub-network that can be used to solve the ODE to calculate the average dissipated power, which is used to calculate the thermal degradation of the isothermal model current.

be output at a *thermal node* in the FET model, and can be used as the input signal to an external thermal sub-network representing the heatsink on which the transistor is bolted, for example. This approach to accommodating the dynamic electro-thermal effects in the transistor model is easy to implement, is quick, and is usually convergent.

For a large power device, the input drive power can be several watts, and so it is not insignificant. We should consider adding the input power at the gate to the total instantaneous dissipated power:

$$P_{\text{diss}}(t) = i_{\text{d}}(t)v_{\text{ds}}(t) + i_{\text{g}}(t)v_{\text{gs}}(t) = P_{\text{avg}} + \tau\frac{dP_{\text{avg}}}{dt} \qquad (6.89)$$

6.5.1.4 *The 'Zero Temperature Coefficient' Point in LDMOS Transistors*

An interesting feature of the thermal behaviour of MOSFETs is the occurrence in the current–voltage characteristics of a point that is independent of temperature, the so-called 'zero temperature coefficient' point (ZTC). This point arises from two competing thermal mechanisms: the decrease in the threshold voltage as the temperature rises, and the reduction in electron velocity or mobility with temperature, described by the electro-thermal model above. The ZTC is shown in the I_{d}–V_{gs} characteristics for a low-power LDMOS transistor, as shown in Fig. 6.28 [61]. The ZTC is a useful tool for RF power transistor and amplifier designers in that if the amplifier is biased close to the ZTC, where the temperature variation of the drain current is relatively small, than the amplifier has good thermal stability. The ZTC is usually found to be at a fairly low current density.

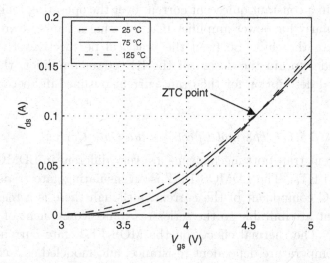

Fig. 6.28. The pulsed I_d–V_{gs} transfer characteristic measured from several different quiescent bias points, indicating the 'zero temperature coefficient (ZTC)' point. The ZTC point is seen to be at just over $I_{ds} = 0.1$ A for this LDMOS transistor.

The ZTC can be found using pulsed I–V measurements. The transistor is biased at a fixed quiescent drain current, and hence a fixed steady power dissipation and channel temperature. The drain current I_d–V_{ds} characteristics are then measured under short pulse conditions to minimize the heating of the channel due to the pulse. The ZTC can then be found by overlaying the measured current characteristics as in Fig. 6.28.

The simple electro-thermal model described above for the drain current characteristics does not accommodate changes in the threshold voltage, which are required if the ZTC effect is to be included. The changes in drain current at low currents or dissipated powers are quite small, and this can be modeled adequately by the simple electro-thermal model. Thermally induced threshold voltage shifts in MOS transistors are usually modeled by a linear change with temperature, with a small coefficient, usually of the order 1 mV/K. For completeness this can be added to the overall electro-thermal model, but care must be taken with convergence of the model. The overall change in threshold voltage over the operating temperature range of the transistor in a PA application will be quite small, around a tenth of a volt. This is a similar order to the process variation of the threshold voltage. In practical power amplifier design, it is the quiescent drain current that is the important feature, and the gate voltage is normally adjusted to produce the required quiescent current. This adjustment is often carried out to maintain a constant quiescent current over the operating temperature range, particularly for power amplifier ICs, by using a temperature sensing diode built into the chip. So from the practical perspective of using the transistor model to aid the design of the RF power amplifier, the simple electro-thermal description for the drain current outlined in Section 6.5.1.3 is sufficient.

6.5.1.5 Thermal Effects on the Gate Current

The gate current transport mechanisms are very different in LDMOS FETs and in III–V FETs. The LDMOS FET is an insulating-gate transistor, so there is no DC component of the current, although there is a real part to the RF current attributable to the series resistance component of the gate metallization. The thermal effects on the MOS FET gate current can be described a temperature-dependent resistance, and modeled as a controlled current source since the temperature of the device depends on the average dissipated power, as stated in Section 6.5.1.3. This modifies the controlled current source describing the gate current in our device model described in Section 6.4.3.8.

The gate contact of the III–V FETs is a rectifying Schottky diode, which is reverse biased in normal FET operation. In HEMTs and PHEMTs there are also heterojunctions between the different semiconductor layers that form the device, and at each of these interfaces between the layers there is a potential barrier. This layer structure modifies the ideal Schottky barrier current–voltage relationship; often this is accommodated by a large value for the 'ideality factor' in the expression for the current,

$$I_{\mathrm{g}} = I_{\mathrm{sat}}\left(T\right)\left(e^{\left[\frac{qV_b}{nkT}\right]} - 1\right) \tag{6.90}$$

where I_{sat} is the reverse saturation current, which has a strong temperature dependence, V_b is the voltage across the diode, excluding any series resistance components, and, in the exponent, n is the ideality factor. The value of n can range between one and two, according to classical p-n junction theory, depending on the dominant method of charge transport across the junction. In Schottky diodes the ideality factor is usually close to unity, but in a heterojunction FET the value can be greater than two, a non-physical value that accommodates the current transport over the many junctions between the gate contact and the channel of the FET.

As noted above, the Schottky gate contact is generally operated in reverse bias, so from eq. 6.90 the gate current will be the reverse saturation current, I_{sat}. The reverse current has an approximately $T^{\frac{3}{2}}$ temperature dependence, the current increasing with temperature. Over the typical operating temperature range of the power transistor, this can be approximated with a linear relation:

(i) at 300 K, $T^{\frac{3}{2}} \approx 5,200$;

(ii) at 385 K, $T^{\frac{3}{2}} \approx 7,700$;

(iii) at 470 K, $T^{\frac{3}{2}} \approx 10,200$.

This linear relationship can be built into the gate current controlled source, as a function of the average power dissipated, P_{avg}, which is found from the self-consistent solution of eq. (6.86), and is given by

$$I_{\text{sat}}(T) = I_{\text{sat0}} \cdot \frac{R_{\text{th}} P_{\text{avg}} + (T_{\text{amb}} - T_0)}{T_0} \tag{6.91}$$

where we also account for any change in the ambient temperature.

6.5.2 Frequency Dispersion due to Trapping Effects

The capture and release of charge carrier by traps and interface states is present in all of the III–V FET technologies that are presently used for power transistors: GaAs FET and related HEMT devices, and GaN-based heterojunction FETs. The rates for charge trapping are typically in the millisecond to microsecond regime. Therefore the traps can respond to DC measurements, but cannot react quickly enough to RF signals, which means there is some difference between the DC and RF output characteristics. This is referred to as the 'frequency dispersion' of the transconductance and output conductance of the FET.

Parker and Rathmell [42] have carried out an intensive study of the bias and frequency dependence of III–V HEMT and FET current–voltage characteristics using pulsed I–V measurements. They present an illustrative model for the I_{d}–V_{ds} characteristics in which the effective electron and hole trapping potentials are included in the expression for the control voltage:

$$I_{\text{d}} = \beta \cdot V_{\text{C}}^{\gamma} (1 - \delta P_{\text{avg}}) \tanh(\alpha V_{\text{ds}}) \tag{6.92}$$

where the parameters have the meanings defined in Section 6.5.1.3, and the drain current dispersion due to thermal effects is included. The exponent γ is typically two. The control voltage V_{C} is given by

$$V_{\text{C}} = V_{\text{gs}} - V_{\text{T}} + \sigma V_{\text{ds}} - \overline{V_{\text{E}}} + \overline{V_{\text{H}}} \tag{6.93}$$

where σ is a 'gain' parameter, and $\overline{V_{\text{E}}}, \overline{V_{\text{H}}}$ are effective electron and hole trap potentials, respectively. The average power is found from eq. (6.86), and similarly, the effective trap potentials are found dynamically from the solutions of the following first-order ODEs:

$$H \ln (f_{\text{II}} + 1) = \overrightarrow{V_{\text{H}}} + \tau_{\text{H}} \frac{1}{f_{\text{II}}} \frac{d\overline{V_{\text{H}}}}{dt} \tag{6.94}$$

$$\sigma_{\text{E}} (V_{\text{ds}} - V_{\text{gs}}) = \overline{V_{\text{E}}} + \tau_{\text{E}} \, e^{(-(V_{\text{ds}}-V_{\text{gs}})/V_{\text{B}})} \frac{d\overline{V_{\text{E}}}}{dt} \tag{6.95}$$

where f_{II} is a factor related to the impact ionization rate and is dependent on the drain voltage, and σ_{E}, H, and V_{B} are fitting constants. The effective trap potentials' rate of change is a function of bias, which accounts for the exponential term in the gate-to-drain voltage in the effective electron trap potential expression, and the impact ionization factor term in the effective hole trap potential.

These expressions can be used in the model equation for drain current, accommodating both thermal and trap-related dispersion effects by the self-consistent solution of the differential equations governed by the time constants for these phenomena, τ, τ_{E}, and τ_{H}.

These first-order differential equations can also be represented by equivalent circuits in the model, and solved self-consistently during the circuit simulation. Kallfass *et al.* [62] have implemented such an approach, using parallel resistor-inductor circuits to represent the trap-related time constants in the device. Each R-L pair is driven by a controlled current source I_{dx}, the dynamic part of the drain current, which is given by

$$\begin{aligned} I_{\text{d}}(t) &= I_{\text{d0}} + I_{\text{dx}} \, e^{-t/\tau_1} \\ I_{\text{dx}} &= I_{\text{d1}} - I_{\text{d0}} \end{aligned} \tag{6.96}$$

where I_{d1} is the drain current flowing the instant the trap effect has occurred, and I_{d0} is the drain current after the transient has subsided, and the time constant for the change in current due to the trap is τ_1.

The total drain current in the model is therefore given by a controlled current source that represents the static drain current expression for I_{d0}, as derived in Section 6.4.3, in parallel with a current source of value $V_{\text{dx}}/R_{\text{dx}}$, representing the change in current due to the trap. Additional trapping mechanisms can be included by adding further R-L sub-circuits and corresponding current sources in the drain port of the FET model. The general equivalent circuit scheme is shown in Fig. 6.29.

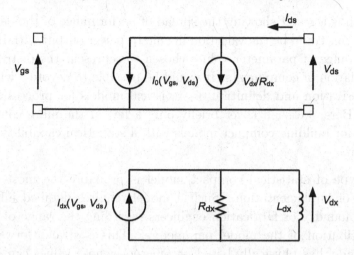

Fig. 6.29. The output part of the equivalent circuit model showing the controlled drain current source, and the additional current source of value V_{dx}/R_{dx} representing the dispersive current component. The trap-dispersion R-L sub-network is also shown.

The dynamic drain current presented in the frequency domain by Kallfass *et al.* can be re-cast in the time domain, and has a familiar appearance

$$i_d(t) = I_d^{DC} - \tau_{disp}\frac{di_d}{dt} \tag{6.97}$$

This is the drain current representation for a relaxation time differential equation description of the non-quasi-static drain current. The relaxation time τ_{disp} can represent a generic relaxation time describing all dispersion effects in the FET, and as such is a simpler version of eq. 6.86. This approach for including dispersive effects in the drain current is used in the Agilent-EEsof IC-CAP *Root* model [27].

6.6 Including Statistical Variations in the Compact Model

One of the most often-requested features of a compact model, after speed, accuracy, ease of use, and so forth, is a 'statistical capability.' By this, the circuit designers are looking for some feature in the model that can be used to predict the yield of their design in production. Most commercial RF circuit simulators include a yield prediction feature, where some of the component parameters are specified with a range of values that reflect the production tolerance or variation, and the circuit is simulated many times with the parameter values chosen from the range in a manner that reflects the distribution of values in production. The outcome of the yield or statistical simulation is a result showing the spread of performance of the design. For example, this could be the variation in gain or power output with frequency, over the range of parameter values chosen to represent the distribution of the population of components. It would be possible to devote a whole book to the derivation and definition of statistical models for process and yield control. Here, we shall cover briefly only a few of the more common approaches for building compact models with a statistical capability for yield prediction.

This type of statistical compact model is probably the most common, in terms of implementation. Such a model requires detailed information from the foundry or fabrication engineers regarding the range of variation and distribution of the model parameters. This is straightforward if the device model has physically-based parameters, whose values can be traced directly to process monitors such as sheet resistivities of the metals and semiconductors, contact resistances, and dimensional parameters such as gate-length. This is not often the case with compact FET models, where the model parameters are obtained by curve-fitting of measured data, or by function-fitting of derived functions of the data.

A statistically meaningful distribution of model parameters can be obtained by extracting many device models over several wafers and lots, over a period of time during which the fabrication and manufacturing tolerances will be exercised. The population can be generated by a continual sampling of the manufactured product over a period of time, say six months to a year or more, depending on the inherent variability of the fabrication and manufacturing processes. An alternative is to create a broad population artificially, by making and building devices specifically at the edges of the manufacturing process capability, or the *process corners*. Examples of process corners may include defining gate-lengths at the short and long extremes of the process tolerance, or building products with bondwires set at the shortest and longest lengths of the tolerance. In this way we can identify devices with the boundaries of the process. If the device or circuit is simulated over the corner-case population, then it is often assumed that this represents the limits of the process seen in production, and an estimation of manufacturing yield can be made. In practice, the corner cases are seldom reached, as they represent limits on the process capability: most devices and circuits should fall well within these boundaries. Furthermore, it is very unlikely that a device will hit more than one process corner: the joint probability of two corners occurring simultaneously in practice is very small. Including such occurrences in our boundary population will give a pessimistic estimate of the yield, which may result in over-engineering of the design.

Once we establish our population of devices, there are two approaches to generating the statistical model. First, we can extract from this population the range and distribution of the parameter variations, for all of the model parameters. We should also carry out a sensitivity analysis of the transistor performance measures on the model parameters. Generally, we would then choose a subset of the most sensitive model parameters to describe with a statistical variation in the compact model. This requires some care, as many of the performance measures depend on the interaction of several model parameters or features, and further, many of the model parameters are inter-related. For example, the figure of merit f_T is related inversely to the gate-length of the FET, and is also proportional to the ratio of transconductance and input capacitance. For the statistical compact model to be useful, then the variation in device performance with changes in the gate-length feature needs to be captured correctly in both the transconductance and gate capacitance model parameters. Moreover, the statistical variations of the parameters need to be coupled: their *joint probabilities* are important, as well as their individual variations and distributions with the population of FET models. If these factors are not considered, the 'statistical' compact model will be unable to predict properly the performance variations of the circuit, or even the FET, across this population of measured devices.

Another potential difficulty with compact models that have the statistical variations of the model parameters visible, and accessible to designers, is that the temptation to adjust a given parameter, in order to predict the device or circuit performance given some new process or device design method, is often impossible to resist. If a single model parameter is adjusted in some way, without any reference or acknowledgment that this parameter may be linked to other model parameters by the device physics or the extraction process, then the conclusions drawn from the results of such an exercise may be meaningless, or even counter-productive.

The alternative to creating a single compact model with statistically variable parameters is to use the whole population of individual models taken across the production variation. Instead of varying the model parameters randomly according to some distribution, we change the transistor model in the circuit. Often, this can be done in the simulator by sweeping over the model filename. This approach has been used successfully with table models, for example, for which there are no model parameters to vary, just the tables of measured and derived parameter values for each model.

Recently, more modern mathematical techniques have been used to find the relationship between the performance measures and the device variations over the measured population of models. In one method [63], the relationships between the measured broadband S-parameters of each transistor in the population and the extracted model parameter values are established, using a support vector machine (SVM) method. The SVM is used to handle the many inputs – the S-parameters over frequency – and the relationship between them and the most sensitive model parameters is determined by optimization. Now the parameter variations can be fed back into the compact model in the circuit, and the variation in the circuit performance with the parameter spread can then be inferred. This method is used to adjust the compact model parameters directly, and all of the joint probabilities present in the test population are included.

Another promising technique also uses neural-network-based optimization to establish a relationship between the measured S-parameters and the large-signal performance, over the population of transistors. This method is known as *neuro-space-mapping*, and has been applied to the statistical modeling of LDMOS power transistors [64]. Neural networks are placed at the input and output ports of the nominal compact model, and the neural networks are trained, or optimized, on the measured relationship between the S-parameters and the large-signal performance measures, so that the spread of the transistor behaviour over the population is captured. The circuit design can then be carried out in the usual way, using the nominal model, and then the performance spread of the circuit due to process or manufacturing variations can be impressed on the nominal behaviour, by including the neural networks in the model and re-simulating the circuit.

6.7 Closing Remarks

In this chapter we have developed a charge-conservative, self-consistent electro-thermal large-signal nonlinear model for the power transistor, based on DC and S-parameter measurements that have been de-embedded to the reference plane of the transistor die, and that are isothermal. In passing, we have shown also that a bias-dependent linear or small-signal model can be extracted directly from these isothermal S-parameters. Pulsed measurement techniques, as outlined in Chapter 3, are required for obtaining the isothermal measurements, and also for deriving the thermal model parameters. For de-embedding the manifolds and extrinsic component shells from these measurements, we have drawn on the measurement techniques outlined in Chapter 4. For the inclusion of the electro-thermal effects, we have built on the theoretical developments and analysis of Chapter 5, adding the thermal resistance and thermal capacitance to our isothermal electrical model. We have also shown how other frequency dispersive effects such as charge trapping in the channel of the FET can be included in the model framework.

References

[1] J.-M. Collantes, "Modelisation des transistors MOSFETs pour les applications RF de puissance," Ph.D. dissertation, University of Limoges, 1996.

[2] D. Denis, C. M. Snowden, and I. C. Hunter, "Coupled electrothermal, electromagnetic, and physical modeling of microwave power FETs," *IEEE Trans. Microwave Theory Tech.*, 54, no. 6, pp. 2465–70, June 2006.

[3] G. Dambrine, A. Cappy, F. Heliodore, and E. Playez, "A new method for determining the FET small-signal equivalent circuit," *IEEE Trans. Microwave Theory Tech.*, 36, no. 7, pp. 1151–59, July 1988.

[4] J. Wood and D. E. Root, "Bias-dependent linear scalable millimeter-wave FET model," *IEEE Trans. Microwave Theory Tech.*, 48, no. 12, pp. 2352–2360, Dec. 2000.

[5] M. Novotny and G. Kompa, "Unique and physically meaningful extraction of the bias-dependent series resistors of a 0.15 μm PHEMT demands extremely broadband and highly accurate measurements," in *IEEE MTT-S Int. Microwave Symp. Dig.*, San Francisco, CA, June 1996, pp. 1715–18.

[6] D. Lovelace, J. Costa, and N. Camilleri, "Extracting small-signal model parameters of silicon MOSFET transistors," in *IEEE MTT-S Int. Microwave Symp. Dig.*, San Diego, CA, May 1994, pp. 865–868.

[7] J. P. Raskin, G. Dambrine, and R. Gillon, "Direct extraction of the series equivalent circuit parmeters for the small-signal model of SOI MOSFETs," *IEEE Microwave Guided Wave Lett.*, 7, no. 12, pp. 408–410, Dec. 1997.

[8] G. D. Vendelin, *Design of Amplifiers and Oscillators by the S-parameter Method*. New York, NY: John Wiley and Sons, 1982.

[9] C. A. Leichti, "Microwave Field Effect Transistors–1976," *IEEE Trans. Microwave Theory Tech.*, 24, no. 6, pp. 279–300, June 1976.

[10] G. Kompa, "Frequency-dependent measurement error analysis and refined FET model parameter extraction including bias-dependent series resistors," in *IEEE Workshop on Experimentally Based FET Device Modelling & Related Nonlinear Circuit Design*, Kassel, Germany, July 1997, pp. 6.1–6.16.

[11] D. E. Root and B. Hughes, "Principles of nonlinear active device modeling," in *32nd ARFTG Conference Digest*, Tempe, AZ, Dec. 1988, pp. 1–24.

[12] H. Shichman and D. A. Hodges, "Modeling and simulation of insulated-gate field-effect transistor switching circuits," *IEEE J. Solid State Circuits*, SC-3, no. 3, pp. 285–289, Sept. 1968.

[13] W. R. Curtice, "A MESFET model for use in the design of GaAs integrated

circuits," *IEEE Trans. Microwave Theory Tech.*, 28, no. 5, pp. 448–456, May 1980.

[14] H. Statz, P. Newman, I. W. Smith, R. A. Pucel, and H. A. Haus, "GaAs FET device and circuit simulation in SPICE," *IEEE Trans. Electron Devices*, 34, no. 2, pp. 160–168, Feb. 1987.

[15] W. R. Curtice and M. Ettenberg, "A nonlinear GaAs FET model for use in the design of output circuits for power amplifiers," *IEEE Trans. Microwave Theory Tech.*, 33, no. 12, pp. 1383–94, Dec. 1985.

[16] A. Materka and T. Kacprzak, "Computer calculation of large-signal GaAs FET amplifier characteristics," *IEEE Trans. Microwave Theory Tech.*, 33, no. 2, pp. 129–135, Feb. 1985.

[17] A. E. Parker and D. J. Skellern, "A realistic large-signal MESFET model for SPICE," *IEEE Trans. Microwave Theory Tech.*, 45, no. 9, pp. 1563–71, Sept. 1997.

[18] A. J. MacCamant, G. D. McCormack, and D. H. Smith, "An improved GaAs MESFET model for SPICE," *IEEE Trans. Microwave Theory Tech.*, 38, no. 6, pp. 822–824, June 1990.

[19] W. R. Curtice, J. A. Plá, D. Bridges, T. Liang, and E. E. Shumate, "A new dynamic electro-thermal nonlinear model for silicon RF LDMOS FETs," in *IEEE MTT-S Int. Microwave Symp. Dig.*, Anaheim, CA, June 1999, pp. 419–422.

[20] C. Fager, J. C. Pedro, N. B. de Carvalho, and H. Zirath, "Prediction of IMD in LDMOS transistor amplifiers using a new large-signal model," *IEEE Trans. Microwave Theory Tech.*, 50, no. 12, pp. 2834–42, Dec. 2002.

[21] W. R. Curtice, L. Dunleavy, W. Clausen, and R. Pengelly, "New LDMOS model delivers powerful transistor library-Part 1: the CMC model," *High Frequency Electronics*, pp. 18–25, Oct. 2004.

[22] S. C. Cripps, *RF Power Amplifiers for Wireless Communications*, 2nd edn. Norwood, MA: Artech House, 2006.

[23] I. Angelov, H. Zirath, and N. Rorsman, "A new empirical nonlinear model for HEMT and MESFET devices," *IEEE Trans. Microwave Theory Tech.*, 40, no. 12, pp. 2258–66, Dec. 1992.

[24] J. Wood and D. E. Root, "A symmetric and thermally-de-embedded nonlinear FET model forwireless and microwave applications," in *IEEE MTT-S Int. Microwave Symp. Dig.*, Fort Worth, TX, June 2004, pp. 35–38.

[25] G. Qu and A. E. Parker, "New model extraction for predicting distortion in HEMT and MESFET circuits," *IEEE Microwave Guided Wave Lett.*, 9, no. 9, pp. 363–365, Sept. 1999.

[26] R. B. Hallgren and P. H. Litzenberg, "TOM3 capacitance model: linking large- and small-signal MESFET models in SPICE," *IEEE Trans. Microwave Theory Tech.*, 47, no. 5, pp. 556–561, May 1999.

[27] D. E. Root, "Charge modeling and conservation laws," in *Asia-Pacific Microwave Conference Workshop WS2, 'Modeling and characterization of Microwave devices and packages'*, Sydney, Australia, June 1999.

[28] J. Staudinger, M. C. de Baca, and R. Vaitkus, "An examination of several large-signal capacitance models to predict GaAs HEMT linear power amplifier performance," in *Proc. IEEE Radio and Wireless Conf. (RAWCON)*, Colorado Springs, CO, Aug. 1998, pp. 343–346.

[29] A. D. Snider, "Charge conservation and the transcapacitance element: an exposition," *IEEE Trans. Educ.*, 38, no. 4, pp. 376–379, Nov. 1995.

[30] D. E. Ward and R. W. Dutton, "A charge-oriented model for MOS transistor capacitances," *IEEE J. Solid State Circuits*, 13, no. 5, pp. 703–707, Oct. 1978.

[31] R. R. Daniels, A. T. Yang, and J. P. Harrang, "A universal large/small signal 3-terminal FET model using a nonquasi-static charge-based approach," *IEEE Trans. Electron Devices*, 40, no. 10, pp. 1723–29, Oct. 1993.

[32] P. Jansen, D. Schreurs, W. de Raedt, B. Nauwelaers, and M. van Rossum, "Consistent small-signal and large-signal extraction techniques for heterojunction FETs," *IEEE Trans. Microwave Theory Tech.*, 43, no. 1, pp. 87–93, Jan. 1995.

[33] D. E. Root, "Measurement-based mathematical active device modeling for high frequency circuit simulation," *IEICE Trans. Electronics*, E82-C, no. 6, pp. 924–936, June 1999.

[34] A. Werthof and G. Kompa, "A unified consistent DC to RF large signal FET model covering the strong dispersion effects of HEMT devices," in *Proc. 22nd European Microwave Conf.*, Espoo, Finland, Aug. 1992, pp. 1091–96.

[35] V. Cuoco, M. P. van den Heijden, and L. C. N. de Vreede, "The "Smoothie" data base model for the correct modeling of non-linear distortion in FET devices," in *IEEE MTT-S Int. Microwave Symp. Dig.*, Seattle, WA, June 2002, pp. 2149–52.

[36] B. D. Popovic, *Introductory Engineering Electromagnetics*. Reading, MA: Addison-Wesley, 1971, pp. 196–198.

[37] D. E. Root, "A measurement-based FET model improves CAE accuracy," *Microwave J.*, Sept. 1991.

[38] J. Xu, D. Gunyan, M. Iwamoto, A. Cognata, and D. E. Root, "Measurement-based non-quasi-static large-signal FET model using artificial neural networks," in *IEEE MTT-S Int. Microwave Symp. Dig.*, San Francisco, CA, June 2006, pp. 469–472.

[39] P. J. Tasker and J. Braunstein, "New MODFET small signal model required for millimeter-wave MMIC design: extraction and validation to 120 GHz," in *IEEE MTT-S Int. Microwave Symp. Dig.*, Orlando, FL, May 1995, pp. 611–614.

[40] S. M. Sze, *Physics of Semiconductor Devices*. New York, NY: John Wiley and Sons, 1969.

[41] A. E. Parker and J. G. Rathmell, "Measurement and characterization of HEMT

dynamics," *IEEE Trans. Microwave Theory Tech.*, 49, no. 11, pp. 2105–11, Nov. 2001.

[42] A. E. Parker and J. G. Rathmell, "Bias and frequency dependence of FET characteristics," *IEEE Trans. Microwave Theory Tech.*, 51, no. 2, pp. 588–592, Feb. 2003.

[43] V. Cuoco, W. C. E. Neo, M. Spirito, O. Yanson, N. Nenadovic, L. C. N. de Vreede, H. F. F. Jos, and J. N. Burghartz, "The electro-thermal "Smoothie" database model for LDMOS devices," in *IEEE MTT-S Int. Microwave Symp. Dig.*, Fort Worth, TX, June 2004, pp. 457–460.

[44] S. A. Maas and D. Neilson, "Modeling MESFETs for intermodulation analysis of mixers and amplifiers," *IEEE Trans. Microwave Theory Tech.*, 38, no. 12, pp. 1964–71, Dec. 1990.

[45] I. Angelov, H. Zirath, and N. Rorsman, "Validation of a nonlinear transistor model by power spectrum characteristics of HEMTs and MESFETs," *IEEE Trans. Microwave Theory Tech.*, 43, no. 5, pp. 1046–52, May 1995.

[46] L. Bengtsson, I. Angelov, H. Zirath, and J. Olsson, "An empirical high-frequency large-signal model for high-voltage LDMOS transistors," in *Proc. 28th European Microwave Conf.*, Amsterdam, The Netherlands, Sept. 1998, pp. 733–737.

[47] F.-C. Hsu, P.-K. Ko, S. Tam, C. Hu, and R. S. Muller, "An analytical break-down model for short-channel MOSFETs," *IEEE Trans. Electron Devices*, 29, no. 11, pp. 1735–40, Nov. 1982.

[48] J. A. Appels and H. M. J. Vaes, "HV thin layer devices (RESURF Devices)," in *Int. Electron Devices Mtg. Tech. Dig.*, New York, Dec. 1979, pp. 238–241.

[49] A. W. Ludikhuize, "A revew of RESURF technology," in *Proc. Int. Symp. Power Semiconductor Devices and Integrated Circuits*, Piscataway, NJ, May 2000, pp. 11–18.

[50] S. Karmalkar, J. Deng, M. S. Shur, and R. Gaska, "RESURF AlGaN/GaN HEMT for high voltage power switching," *IEEE Electron Devices Lett.*, 22, no. 8, pp. 373–375, Aug. 2001.

[51] S. Karmalkar, M. S. Shur, G. Simin, and M. A. Khan, "Field-plate engineering for HFETs," *IEEE Trans. Electron Devices*, 52, no. 12, pp. 2534–40, Dec. 2005.

[52] R. Menozzi, "Off-state breakdown of GaAs PHEMTs: Review and new data," *IEEE Trans. Device. Mater. Rel.*, 4, no. 1, pp. 54–62, Mar. 2004.

[53] S. A. Maas, "Ill conditioning in self-heating FET models," *IEEE Microwave and Wireless Components Letters*, 12, no. 3, pp. 88–89, Mar. 2002.

[54] A. E. Parker, "Comments on 'ill conditioning in self-heating FET models'," *IEEE Microwave & Wireless Comp. Lett.*, 12, no. 9, pp. 351–352, Sept. 2002.

[55] R. Anholt, *Electrical and Thermal Characterization of MESFETs, HEMTs, and HBTs*. Norwood, MA: Artech House, 1995.

[56] S. M. Lardizabal, A. S. Fernandez, and L. P. Dunleavy, "Temperature-dependent modeling of gallium arsenide MESFETs," *IEEE Trans. Microwave*

Theory Tech., 44, no. 3, pp. 357–363, Mar. 1996.

[57] P. C. Canfield, S. C. F. Lam, and D. J. Allstot, "Modeling of frequency and temperature effects in GaAs MESFETs," *IEEE J. Solid State Circuits*, 25, no. 1, pp. 299–306, Feb. 1990.

[58] I. Schmale and G. Kompa, "Integration of thermal effects into a table-based large-signal FET model," in *Proc. 28th European Microwave Conf.*, Amsterdam, The Netherlands, Sept. 1998, pp. 102–107.

[59] A. E. Parker and D. E. Root, "Pulse measurements quantify dispersion in PHEMTs," in *Proc. 1998 URSI Int. Symp. On Signals, Systems, and Electronics*, Pisa, Italy, Sept. 1998, pp. 444–449.

[60] A. E. Parker and J. G. Rathmell, "Broad-band characterization of FET self-heating," *IEEE Trans. Microwave Theory Tech.*, 53, no. 7, pp. 2424–29, July 2005.

[61] J.-M. Collantes, P. Bouysse, and R. Quere, "Characterising and modeling thermal behaviour of radio-frequency power LDMOS transistors," *Electron. Lett.*, 34, no. 14, pp. 1428–30, July 1998.

[62] I. Kallfass, H. Schumacher, and T. J. Brazil, "Multiple time constant modeling of dispersion dynamics in hetero field-effect transistors," *IEEE Trans. Microwave Theory Tech.*, 54, no. 6, pp. 2312–20, June 2006.

[63] H. Taher and D. Schreurs, private communication.

[64] L. Zhang, K. Bo, Q.-J. Zhang, and J. Wood, "Statistical space-mapping approach for large-signal nonlinear device modeling," in *Proc. 36th European Microwave Conf.*, Manchester, UK, Sept. 2006.

7

Function Approximation for Compact Modeling

7.1 Introduction

In the preceding chapters we have described the electrical and thermal measurement techniques for characterizing the transistor, and developed the mathematical frameworks for defining the various component pieces of the compact model, such as the package and linear matching components, the extrinsic elements, and the nonlinear intrinsic model for the high-power RF transistor. The aim of this chapter is to describe how this measured data and mathematical model can be cast into a form that can then be implemented in a compact model to run in a circuit simulator.

Before we get much further, it is perhaps appropriate to define what we mean by terms such as 'function approximation' and 'data fitting'. Strictly speaking, function approximation is taking a known but complicated function, and approximating its shape by a simpler function over some defined region and to some specified accuracy. When we 'fit' a function to measured data, we are choosing a candidate function from a large set, such as the polynomial family, and adjusting the function parameters until the data is represented with sufficient accuracy over the measured data space (and perhaps beyond). We can also think of this as using a simple function to approximate an unknown but possibly complicated function that is described by the data, at measurement points not of our choosing. We shall use the term 'function approximation' loosely to mean the fitting of measured data with a function from a given class or family of functions.

We shall, in general, in the modeling of the transistor, be dealing with discrete data, obtained by measurement or simulation, that describes the electrical and thermal behaviour of the transistor's component parts. Our task is to find ways in which this data can be represented accurately yet economically in the compact model. A common method is to find a function that 'fits' or approximates the data to some sufficient accuracy, and that can be coded into the model efficiently – by which we mean that the function is defined by a set of parameters whose values can be determined easily: this is often called *model extraction*. Sometimes the measured behaviour can be described by a closed-form equation derived from fundamental physical principles; an example of this is the current flow through a junction diode. Again, our modeling task is to determine values for the parameters in the equation.

Essentially, the modeling task is to use function approximations to reduce the measured dataset to a more compact representation, which, in itself, can be described by a small number of identifiable and extractable parameters. There is also the possibility of using the measured data directly in the model. For simple, fixed linear systems this can be quite straightforward; for example, an S-parameter block of measured data can be used as the model of a filter or matching network. For more complex systems, the number of parameters and consequently the dimension of the data space can be large and unwieldy, requiring large datasets and multi-variate interpolation methods, which can be expensive. We will generally seek to transform the measured data into a space where the representation is more compact. This is achieved either directly through the function approximations, or indirectly through a transformation of the measured data, from S-parameters to Y-parameters, for example, where the frequency can be identified as a parameter of the model.

We shall consider how the data can be presented or transformed in such ways as to make the approximating functions compact and easy to evaluate. In choosing these approximating functions we shall consider their properties of interpolation within the measured data domain, and extrapolation beyond the data domain. Our measured data will not be free from noise, and the approximating function that is chosen should be able to model the underlying shape of the data, not interpolate the measurement noise. We may need to build in some controlled extrapolatory behaviour into the approximating function or the data, so that it performs in a predictable manner outside the measured data domain, and hence enables robust convergence in the circuit simulator.

In the remainder of this chapter, we shall describe some of the features of functions and systems of equations that are used to describe transistors, and that we should bear in mind when developing our compact model. We shall then present some common methods for function approximation of the measured data that are used in transistor modeling.

7.2 Some Features of Functions and Function Approximation

The compact transistor model that we shall build is derived from measured or simulated data, obtained from real or simulated experiments that are designed to isolate the various physical behaviours in the device. The data for each experiment are discrete, and form a finite and bounded dataset. The data are also deterministic in nature, by which we mean that if the experiment is repeated, we shall get the same results, within some bounds that are governed by the noise in the measurement or the measurement system accuracy. We shall 'fit' or approximate the data using parameterized functions that mimic the data with sufficient accuracy to be of value in a compact model used for circuit design. The functional form for the approximation may be chosen *a priori*, for example, we may choose to use a polynomial to approximate a given dataset. The modeling process then becomes an exercise in finding the minimum set of polynomial parameters to fit the data. The accuracy of the approximation is determined by the error measure that is used; the choice of error measure or cost function in an optimized solution can play a significant role in the accuracy of the approximation.

To help us understand and appreciate the various aspects of function approximation, such as the choice of basis functions for fitting the data, and error measures that may be used, a brief review of the different function types that are used to describe mathematical systems will be presented [1]. A brief survey of the lexicon of terms used for describing such systems reveals that they can be classified as 'linear' or 'nonlinear', 'static' or 'dynamic', and 'time-variant' or 'time-invariant'. We shall provide a little substance to these classes, and indicate how such systems can be approximated by the appropriate functions.

A linear system is one where the output is instantly and directly proportional to the input, $u(t)$, at any time t:

$$y(t) = au(t) \tag{7.1}$$

Linear systems also possess the qualities of *superposition*, meaning that if the input signals are scaled and summed, the output signals are likewise scaled and summed, and of *decomposition*, where the output can be classified in terms of a zero-input response, or the initial conditions, and a zero-state response, due to the input signal only. Linear systems can be described mathematically in the form of ordinary differential equations (ODEs), where the zero-input and zero-state response can be associated with the complementary and particular solutions of the ODEs, respectively.

A static or instantaneous system is one where the output depends only on the system input at the time t; no past input signal information is required. In contrast, if the system output depends on previous states of the system, then the system can be described as dynamic, or as having *memory*. A simple system exhibiting dynamical behaviour is a series resistor-capacitor network. Defining the output as the voltage across the network and the input as the current, we get for the voltage

$$v(t) = Ri(t) + \frac{1}{C} \int\limits_{-\infty}^{t} i(t)dt \qquad (7.2)$$

and the response at time t depends on the integrated charge on the capacitor over all time: its *history*. We can replace the lower integral limit by some time t_0 (which may be zero) and specify the capacitor voltage at that time: initial conditions. We can write a dynamical linear system in the following general form:

$$\frac{d^n y}{dt^n} + a_{n-1}\frac{d^{n-1} y}{dt^{n-1}} + ... + a_1\frac{dy}{dt} + a_0 y = b_m\frac{d^m u}{dt^m} + b_{m-1}\frac{d^{m-1} u}{dt^{m-1}} + ... + b_1\frac{du}{dt} + b_0 u \qquad (7.3)$$

for an input $u(t)$ and output $y(t)$. If the parameters a_n, b_m are constant, the system is *stationary* or *time-invariant*. If the parameters a_n, b_m are functions of time only, then the system is *time-variant*. If the parameters a_n, b_m are functions of anything else, we have a nonlinear system.

At this point, we can say from the simple definitions above, and our observations of the electrical and thermal behaviour of the RF power transistor, that the device is a nonlinear dynamical system. This presents a significant modeling task. It is often simplified by invoking 'quasi-static' conditions: while the transistor is truly a nonlinear dynamical system, we presume that the dynamics can be considered to be invariant over the range of conditions that we want to model. This allows us to model the currents and charges in the transistor as static nonlinearities. We have chosen to adopt a compact modeling approach that enables the identification of the component parts of the device and apply appropriate modeling techniques for these parts. In other words, we have approached the modeling of the power transistor in a compartmentalized or segmented fashion, and under the quasi-static assumption.

From the perspective of compact model generation, the most important mathematical model classification is between linear and nonlinear models. We shall outline some important attributes of functions that are used during the generation of linear models of passive components, and in the development and generation of nonlinear models of active semiconductor devices. The three most important attributes of the model functions that need to be considered are interpolation, extrapolation and sensitivity to measurement errors.

For the linear components of the transistor, such as the package and the matching networks formed from the bondwires and MOS capacitors, we can adopt linear approximation techniques. In the simplest case we will transform the measured S-parameter data into another linear space such as Y-parameters, where the frequency dependence can be captured and the model can be determined by linear regression of the matrix elements over frequency. Such transformation and fitting methods are described in detail in the next section.

The active transistor is a nonlinear component: the terminal currents (and internal charges) vary in a nonlinear manner with the applied voltages and temperature. The data collected (or transformed linearly through integration or differentiation) is often multi-variate, and generally no obvious functional form for the approximation to the data suggests itself. Nonlinear function approximation is an arduous task, although there are prescriptions for simplifying the process. A nonlinear function, or dataset, can often be approximated using a set of basis functions:

$$\mathbf{y}(t) = \sum_{n=1}^{\infty} \varphi_n(a_n, t) \tag{7.4}$$

where $\mathbf{y}$ is the multi-variate output data, for example the gate charge field, and the $\varphi(t)$ are the basis functions, with the a_n being the parameters of the function elements. A fully nonlinear approximation such as this requires a nonlinear function approximation or optimization method to determine the parameters a_n that govern the approximation. Such methods include artificial neural networks (ANNs), which can be used to model the multi-dimensional current and charge fields in the nonlinear FET model [2].

Often, the nonlinear function approximation can be made simpler, still using nonlinear basis functions but by having the fitting parameters outside of the function:

$$\mathbf{y}(t) = \sum_{n=1}^{\infty} a_n \varphi_n(t) \tag{7.5}$$

This nonlinear approximation is called *linear in parameters*, and the approximation can be found by using linear techniques such as least-squares or singular value decomposition (SVD) to find the parameters of the non-linear function [3].

The dataset that is used to generate a model is both discrete, and finite in size and range. By this we mean that no matter how many data points have been measured and used during the modeling and function approximation process, there is always going to be a need to use or evaluate the model for values of the input variables that lie in between the measured or simulated points. The ability of the model function to approximate the response in between the measured data points defines its interpolation capability. Some approximation functions such as splines can be very accurate at interpolating the data, their accuracy coming from the definition of the spline function values and its derivatives at the measured points. In other words, from the choice of the error function.

The ability of an approximating function to predict the values outside the range of measured and simulated data is determined by its extrapolation characteristics. Extrapolation is of great importance in function approx-imation, and can be classified in two ways. First, we may simply want the asymptotic behaviour of the function to be well controlled, so that the function values returned for out-of-range inputs are constrained and well behaved. We can add to this requirement that the gradients or derivatives of the function outside the measurement range are also arranged to be well behaved and in fact can be arranged to guide the simulator back into the range of known data. These features are often added to the function ap-proximation after data fitting, not to provide accurate predicted data, but to assist convergence in the simulator.

Second, we may be interested in the ability of the function approximation to be predictive; in other words, to be able to provide accurate output data in regions outside the original measured data space. This could be a simple extension of the measured data space, so that the function returns values that can be validated in subsequent measurement. This could also mean that the function produces the correct result for applied signals that are not in the same signal class as those used for the transistor characterization and model extraction. The latter case is difficult to guarantee, but is a reasonable expectation if the transistor's nonlinear dynamical behaviour is described accurately by the model structure.

Measured data consists of both real information and error. Measurement error is composed of bias and noise, where bias is a systematic error or offset introduced during the gathering of the measured data, while noise is due to random errors, and stochastic processes. In many cases the task of separating the error from the actual information is quite difficult, if not impossible. Therefore, the function approximation should aim to be immune to the existence of errors. Different classes of functions, for example, splines, neural networks and polynomials react differently in the presence of non-negligible amounts of error in the measured data. Careful consideration should be taken during the function approximation process to avoid the inclusion of measurement errors. This can be as simple as the choice of the approximating function itself, or it can be determined by the choice and method of data collection – for example, using too small a voltage step in the collection of bias-dependent data can introduce thermal noise into the data.

The model that we create from the approximating functions by extracting the parameter values is intended to be used in a circuit simulator. For the design of RF and microwave circuits, harmonic balance (HB) algorithms are generally used for the simulation of the nonlinear device. The HB algorithm uses the time-domain nonlinear currents and charges to arrive at the steady-state solution of the problem, while their partial derivatives with respect to the independent variables (gate voltage, drain voltage, and temperature) are evaluated to assist the simulator in converging to a solution. The model functions therefore need to have smooth first derivatives and continuous second derivatives for good convergence properties. Additionally, we may choose to use approximating functions that are differentiable to higher orders, to enable the prediction of distortion products generated by the nonlinear behaviour of the transistor. For such predictions to be accurate, the approximating function must display high fidelity to the detailed curvature of the data itself.

Finally, we should address the complexity of the model function and the dataset that it requires. Nonlinear behaviour requires several iterations by the circuit simulator to converge to the solution; the nonlinear model function should therefore be designed to be as numerically efficient as possible, without sacrificing accuracy. The combination of compactness of the model and its accuracy can be captured in measures such as the *minimum description length*, which combine the size of the approximating function, in terms of the number of parameters, and the accuracy of the approximation to the measured data, in one metric. This measure is good for comparing models of a similar class, such as neural networks, but is not so meaningful for comparing, for example, neural network models and spline interpolators.

7.3 Practical Methods for Function Approximation

We shall now turn our attention to some practical methods for data fitting and function approximation that are used in the compact modeling of transistors. We shall illustrate these methods by some practical examples, highlighting some of the features of the approximating functions that need to be considered when building the transistor model. The techniques that we describe below do not comprise an exhaustive set of the methods that can be used in transistor modeling, but serve to illustrate some of the more common methods.

First we consider using the measured dataset itself, or transforms of these data as the basis of a model. This approach can be used to build table models, which are generally implemented in the simulator using interpolation functions, which are also outlined below. The transformation of the data can also yield a much simpler space in which to approximate the data. After this, we outline some simple function approximation methods, including polynomials, rational functions and spline interpolators, and compare the relative performance of these methods. We also describe how artificial neural networks can be used to obtain excellent approximations to multi-variate data, such as is typically found in the nonlinear modeling of the transistor's intrinsic behaviour. Finally, we describe how elementary functions can be used to build up a model that fits the complex electrical behaviour of the transistor, while using only a few fitting parameters. This approach relies heavily on the experience of the modeling engineer in choosing and applying the appropriate functions.

7.3.1 Data Representation for Compact Model Generation

Once we obtain the description of the electrical system we are going to model, the discrete data obtained from either measurement or simulation, we need to ask what is the best representation for the model to capture its behaviour within a circuit simulator. The task of generating a model is to find a method in which the data can be represented accurately and economically, then determine a mathematical expression that mimics the physical behaviour of the system. In this subsection, we discuss various representations for model development. In the sections that follow, we shall review several mathematical techniques used for function approximation.

One obvious method to generate a model or to represent the discrete data is to use these data directly. Often, measured S-parameters of a microstrip circuit or a small-signal transistor operating at a single bias are loaded directly into the circuit simulator. This enables measurement data taken from the laboratory, or results from a simulator, to be represented within the circuit simulator. For example, a small-signal transistor model that needs to be generated from measured S-parameters can be implemented with the use of a circuit component designed to read and load the measured data into the circuit simulator. The measured data can be indexed by the independent variables used during the data gathering process, for example, frequency, bias voltages or temperature. This technique finds limited application as it has significant drawbacks for modeling complex circuits. This approach is not impossible to implement but it is often impractical, largely for the size of the dataset. Also, if the data point required by the simulator was not included in the measured dataset, then multi-dimensional interpolation is required. This implementation may be acceptable for some types of system or model, but because of the large dataset required to capture the measurement space and its poor simulation efficiency, this approach is often limited in applicability. To account for the dynamic behaviour of nonlinear electro-thermal transistor models, the direct use of the measured data will not produce a practical model.

Other approaches to represent the measured data directly are often employed. These methods concentrate on developing a functional description of the input–output relationship of the device. The resulting equations are much more compact than the original dataset and the terminal behaviour of the circuit can be easily reproduced by the circuit simulator. These techniques are widely employed for the generation of models of passive circuits [4–6].

But the question remains, for both passive and nonlinear transistor model development, which is the best representation for the data? This question is non-trivial and often overlooked, as Gershenfeld articulates in his book *The Nature of Mathematical Modeling*, where he notes that one of the goals of mathematical modeling is to reduce a set of data down to a smaller set that is more independent than the original [3].

Generic tools like orthogonal transforms, Fourier transforms, wavelets, and principal component analysis (PCA) are often used to develop models throughout various fields of science and engineering. These methods transform the dataset from one specific domain into another, in order to resolve features of the physical behaviour not visible clearly in the original domain. A trivial, but instructive example, is the development of a mathematical expression for a sine wave. In the time domain the oscillatory nature of the wave and its repetitive nature is feature rich. However, by transforming the signal to the frequency domain via a Fourier transform we see a more simple representation of the data. Thus, the model can be represented by a couple of components, its magnitude and frequency, which preserves the behaviour of the original signal.

In transistor modeling, we have Kirchhoff's current and voltage laws which govern the operation of any electrical circuit. By underpinning our modeling approach on circuit theory, which must obey these physics-based governing equations, we have an advantage over models that rely only upon observational data.

Two main approaches to model development are used when equivalent circuits or circuit theory is employed. In the first approach, the discrete data are transformed from their original domain via network theory into another domain, for example a transformation from S-parameters to Z-parameters. The domain in which the data can be most easily approximated is selected. Then any function approximation technique like polynomials, splines, artificial neural networks (ANNs), or elementary functions, may be applied to represent the data directly.

Instead of transforming the data from one set of terminal parameters to another, we can use our knowledge of circuit theory and extract equivalent circuit parameters (ECPs) from the data. Most frequently, both the equivalent circuit and function approximation techniques are used in unison. Once the data has been transformed into equivalent circuit parameters, the functionality of these parameters is approximated by one of the aforementioned techniques. It is usually the case that several different equivalent circuits can be used to represent the data, and the selection criteria are often based on the difficulty in approximating the ECPs. Selection of the correct equivalent circuit removes, or reduces in the case of a distributed circuit, the ECP frequency dependency. When this model is included in the circuit simulator, the circuit parameters, are passed into the function representing ECP, the circuit simulator then determines the Y-parameters of the entire model based on the equivalent circuit parameters values returned by the approximate functions.

To illustrate these concepts, an ideal transmission line was simulated and two different transformations using terminal network parameters were used to represent the S-parameter data shown in Fig. 7.1. The transmission line is ideal, with no losses, has a characteristic impedance of 40 Ω, and an electrical length of 90 degrees at 2 GHz. One representation used an equivalent circuit corresponding to a differential section of transmission line, as derived from the Telegrapher's equations. The equivalent circuit parameters are extracted directly from the impedance matrix of a T-network, containing inductances in the series branched and a shunted capacitor. The equivalent inductance and capacitance can be extracted directly from the Z-parameters of the transmission line,

$$L_1 = \frac{\text{Im}(Z_{11} - Z_{12})}{\omega} \tag{7.6}$$

$$L_2 = \frac{\text{Im}(Z_{22} - Z_{12})}{\omega} \tag{7.7}$$

$$C = \frac{-1}{\omega \, \text{Im}(Z_{12})} \tag{7.8}$$

where ω is the radial frequency and Z_{ij} are the two-port impedance matrix elements [7]. Since the transmission line is symmetrical ($L_1 = L_2$), the total inductance of the transmission line can be expressed as $L_1 + L_2$.

An alternative representation employs the *ABCD*-parameters of a transmission line. Elementary functions within the *ABCD*-representation capture the distributed nature of the transmission line. For a lossless transmission line the *ABCD*-parameters are,

$$\begin{bmatrix} A & B \\ C & D \end{bmatrix} = \begin{bmatrix} \cos \beta l & Z_{\text{o}} \sin \beta l \\ Y_{\text{o}} \sin \beta l & \cos \beta l \end{bmatrix} \tag{7.9}$$

where l is the length of the transmission line, Z_{o} is the characteristic impedance, $Y_{\text{o}} = 1/Z_{\text{o}}$, and β is the propagation constant. Z_{o} can be extracted directly,

$$Z_{\text{o}} = \sqrt{\frac{B}{C}} \tag{7.10}$$

From β the relative permittivity of the medium can be obtained,

$$\beta = \frac{\cos^{-1}(A)}{l} \tag{7.11}$$

$$\epsilon_{\text{r}} = \left(\frac{\beta c_{\text{o}}}{\omega} \right)^2 \tag{7.12}$$

where c_{o} is the speed of light in a vacuum. Thus, for the lossless case, the characteristic impedance and the effective dielectric constant can completely represent the behaviour of the transmission line. Only these two parameters are required to represent the behaviour of the transmission line.

The S-parameters, the extracted equivalent circuit parameters, Z_{o}, and ϵ_{r} are plotted in Fig. 7.1. Examining these plots we see that Z_{o} and ϵ_{r} are constant with respect to frequency, while the inductance and capacitance are strong functions of frequency. By adopting an *ABCD*-representation of a transmission line we have removed the frequency dimension in the model,

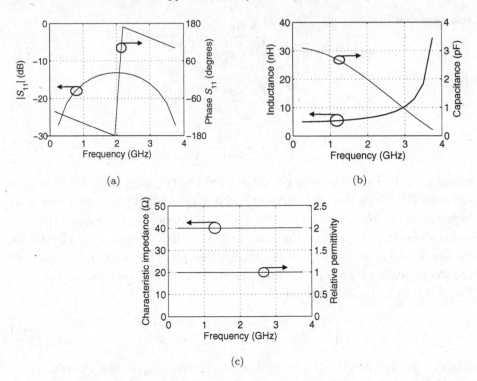

Fig. 7.1. Representations of a transmission line using (a) S-parameter data, (b) equivalent circuit parameters and (c) parameters for an *ABCD* representation, showing the progressive reduction in data dimension.

since it is incorporated in the elementary equations describing the voltage–current relationship at its ports. As a result, only a single data point is required to generate a model of the transmission line.

This pedagogical example illustrates the earlier argument that finding the best representation of a circuit can lead to a more compact representation for the model. The transformation of the output space through a set of equivalent circuits or representations can be very advantageous if the transformed functional behaviour is easier to approximate and easier to extrapolate or control. The transmission line example, presented above, illustrates both of these advantages.

In the general case, determining correct transformation can be very difficult, unless some prior knowledge about the system exists. In the case of transistor models and transmission line elements, the prior knowledge takes the form of existing equivalent circuits. So it is possible to transform the discrete data obtained from either measurement or simulation into equivalent circuit parameters. If a good equivalent circuit is selected, the resulting functionality of the equivalent circuit parameters will be easier to approximate than the original data. If the model and equivalent circuit are selected *a priori* the number of measurements or simulations required to generate the discrete data can be minimized thereby reducing the effort required to characterize the circuit.

7.3.2 Polynomial Functions

A polynomial is a power series with constant coefficients and integer exponents, and is of the following general form:

$$y(x) = a_0 + a_1 x + a_2 x^2 + ... + a_n x^n \qquad (7.13)$$

Polynomial functions are probably the most frequently used method for the approximation of arbitrary datasets. They have a simple structure; they are linear in the coefficient parameters, which means that the linear least-squares method can be used to determine the coefficients and hence obtain the polynomial approximation to the data. Polynomial functions are also easy to evaluate, and they are continuous and smooth. These properties make polynomials attractive for approximation of many classes of problems, but in their simplicity lay several important limitations. In particular, the

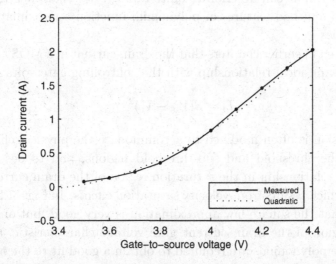

Fig. 7.2. Measured drain current versus gate-to-source voltage data in the vicinity of the threshold voltage, and an approximation by a quadratic polynomial, or 'square-law' relationship, reflecting the model expression of eq. 7.14.

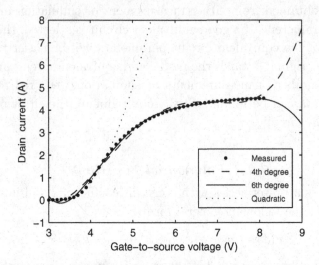

Fig. 7.3. Measured drain current versus gate-to-source voltage data, and approximations using even-order polynomials, over the whole of the measured dataset. The square-law approximation is quite poor over this data range.

interpolation and extrapolation properties of polynomials are quite poor. For example, high-degree polynomials can provide a very small error approximation to the data, but can have an oscillatory response in between the data points used to for the fit. Outside the range of values of the data, the quality of the fit can deteriorate quite rapidly. We shall illustrate some of these good and bad features of polynomial function approximation with a few examples.

We have seen in earlier chapters that the drain current in a MOS transistor has the following ideal relationship with the controlling gate voltage:

$$I_d = \beta(V_{gs} - V_t)^2 \tag{7.14}$$

This expression is often modified to accommodate the physical effects and nuances in the threshold and sub-threshold regions, and to describe the nearly linear relationship in the saturation region of the drain current characteristics, which results from velocity saturation effects. In Figs. 7.2 and 7.3 we can see that the square-law approximation is very good, but only in the threshold region of the drain current–gate voltage characteristic, and that higher-degree polynomials are required to obtain a good fit to the measured data over the whole range of gate voltage. Figures 7.3 and 7.4 show how the approximation generally improves as the polynomial degree is increased. From these figures we can also see that the polynomial functions can diverge

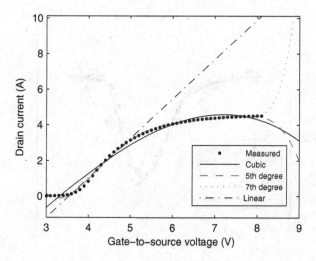

Fig. 7.4. Measured drain current versus gate-to-source voltage data, and approximations using even-order polynomials, over the whole of the measured dataset.

dramatically from the data outside the range of the fit, demonstrating the poor extrapolation property of polynomial function fitting. The divergence increases with polynomial degree. We can also detect some oscillatory response *within* the data range for some of these polynomial approximations. Nevertheless, polynomial approximations are easy to implement and quick to evaluate in many simulators and mathematical packages, making them popular, at least for initial investigations of data fitting and function approximation.

7.3.3 Rational Functions

A rational function is simply the ratio of two polynomials:

$$y = \frac{a_0 + a_1 x + a_2 x^2 + ... + a_{n-1} x^{n-1} + a_n x^n}{b_0 + b_1 x + b_2 x^2 + ... + b_{m-1} x^{m-1} + b_m x^m} \tag{7.15}$$

where n and m are non-negative integers that define the degrees of the numerator and denominator polynomials, respectively. The numerator and denominator define the zeros and poles of the model function. The constant parameter b_0 is usually set to one (by division through the numerator and denominator), reducing the number of parameters to be fitted in the model by one. Rational function models are also known as Pade approximations. The rational function family includes the polynomials, which comprise the set when the denominator is a zero-order polynomial, that is, a constant.

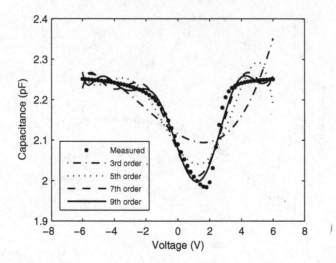

Fig. 7.5. A MOS capacitor C–V data fitted with 3rd, 5th, 7th, and 9th-degree polynomials.

Rational function models offer several advantages over the more simple polynomial models. They can take on a much wider range of shapes compared with polynomial expressions, which means that they can often be used to model structures with high curvatures, using only moderate degrees in the numerator and denominator, resulting in fewer coefficients overall than a comparable polynomial fit. Rational function approximations also offer superior interpolation properties than polynomials, and are typically smoother and less oscillatory. Further, the extrapolation characteristics of rational function models are better than polynomials; their asymptotic behaviour can be tailored to model the desired dataset outside the domain of interest, through

$$y\big|_{\lim x \to \infty} \approx \frac{a_n}{b_m} x^{n-m} \tag{7.16}$$

This ability to adjust the fit of the model both within the data domain and in extrapolation, enables the user to obtain good agreement with a particular theoretical behaviour.

These features are illustrated in Figs. 7.5, 7.6 and 7.7, in which polynomial and rational function models have been determined for a MOS capacitance–voltage relationship. The capacitance–voltage curve has a particularly sharp change in curvature at the threshold voltage, which requires a very high-degree polynomial to fit the shape of this curve accurately: a smooth fit is obtained with a 19th-degree polynomial. The lower-degree polynomials

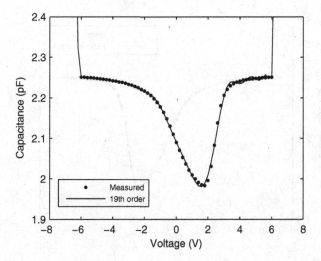

Fig. 7.6. The same MOS capacitor C–V data fitted with a 19th-degree polynomial, showing acceptable accuracy in the data domain, but poor extrapolation behaviour.

yield oscillatory behaviour in the data domain, impairing the interpolation capability. The extrapolation of the high-degree polynomial is also poor: outside the data domain the model function can 'take off' very sharply. In contrast, the rational function model produces a smooth fit within the data domain, and the extrapolation is also well behaved, as it is controlled by the asymptotic properties of the model. This rational function is sixth-degree in both numerator and denominator – a *sixth/sixth model* – or thirteen fitted parameters in total, fewer in number than required for the polynomial model.

Rational functions, like polynomials, are generally easy to evaluate, although there are some potential difficulties that can arise. Unconstrained rational function fitting can, at times, result in undesired vertical asymptotes owing to roots in the denominator polynomial. If the function evaluation is close to one of these roots of the denominator polynomial, that is, a pole of the model, then a large error in interpolation could occur in this region. It should be borne in mind that when using a rational function approximation, poles are always introduced into the model, even though there may have been none in the original data or function.

In building a rational function model, the question of how to choose the degrees of the numerator and denominator for a given data shape needs to be addressed. Some simple observations for the asymptotic behaviour can be used as a guide:

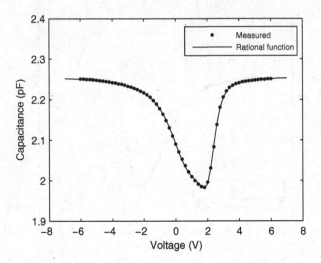

Fig. 7.7. The same MOS capacitor C–V data fitted with a 6th/6th rational function, showing good model fit in the data domain, and excellent extrapolation behaviour.

   (i) if the numerator and denominator are of the same degree, $n = m$, then the horizontal asymptote is given by a_n/b_m;
  (ii) if $n < m$, then the horizontal asymptote is $y=0$; and
 (iii) if $n > m$, then the function tends to infinity at large x.

From these simple observations, a rational function model can be built and tested. The starting values for the parameters can be found from a simple linear fit, using p points from the data set for the p parameters in the model. For example, a quadratic/cubic rational function model

$$y = \frac{a_0 + a_1 x + a_2 x^2}{1 + b_1 x + b_2 x^2 + b_3 x^3} \qquad (7.17)$$

has six parameters, so we need six data points; choosing six pairs of $\{x, y\}$ from the dataset, and re-arranging eq. 7.17 for each pair of points, we get six linear equations in the parameters, a_n and b_m

$$y(1) = a_0 + a_1 x(1) + a_2 x(1)^2 - b_1 x(1) y(1) - b_2 x(1)^2 y(1) - b_3 x(1)^3 y(1)$$

$$\vdots$$

$$y(6) = a_0 + a_1 x(6) + a_2 x(6)^2 - b_1 x(6) y(6) - b_2 x(6)^2 y(6) - b_3 x(6)^3 y(6)$$
$$(7.18)$$

The coefficients can be estimated using a linear least-squares algorithm, and used as the starting values in the rational function approximation for the full dataset. As with polynomial function fitting, the degrees of the numerator and denominator polynomials need to be increased until a suitable model error is obtained.

In summary, rational functions offer a flexible data fitting method, and the asymptotic properties can be controlled, resulting in good extrapolation behaviour. Rational function models are particularly appropriate for modeling networks, functions, and data structures in the frequency domain, where the pole-zero representation is quite natural.

7.3.4 Splines

We have seen that the polynomial and rational functions are capable of producing an accurate approximation to data, but to get this accuracy over the whole dataset requires us to use high-degree polynomials, which can introduce problems with ringing and poor extrapolation. This is because these methods are global function approximations, whereas the detail in the data is often local. We have also seen that low-degree approximations can yield a good fit over a limited range of the data: the square-law approximation to the drain current–gate voltage relationship in the neighborhood of the threshold voltage is a good example of this, as shown in Fig. 7.2. By using polynomials to fit the data over small intervals, and stitching these local functions together with suitable matching conditions at the interval boundaries, we can obtain an accurate fit to the whole dataset, while at the same time using only low-degree polynomial functions. These *piecewise polynomial* approximations are known as *splines*.

The *cubic spline* fits the data values with a cubic polynomial that is smooth in the first derivative and continuous in the second derivative, both within the interval and at the interval boundaries. The cubic spline can be developed from a consideration of interpolation in the interval between two adjacent points in the dataset, $\{x_j, x_{j+1}\}$. Linear interpolation essentially fits the data with a piecewise linear function; the slope is constant in the interval, but discontinuous at the interval boundaries or data points, and the second derivative is not defined. A cubic interpolating function has a second derivative that will vary linearly across the interval, and so meets the requirement for continuity. The first derivative of the cubic polynomial is then arranged to be equal to the first derivative calculated at the end of the previous interval, at x_j; the first derivative of the cubic function in the next interval is then calculated to be equal to the value of the cubic function in the interval $\{x_j, x_{j+1}\}$ at x_{j+1}. The value of the cubic polynomial itself at the end points of the interval is made equal to zero, and so does not affect the data values at these points, $\{y_j, y_{j+1}\}$. From these constraints, and the

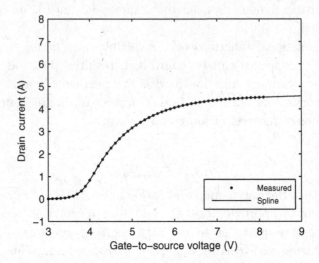

Fig. 7.8. Measured drain current versus gate-to-source voltage data, and a cubic spline interpolation that passes through each measured data point, giving an excellent fit to the data.

boundary conditions at the end points of the dataset, a cubic polynomial for each interval can be designed, such that the whole dataset can be interpolated, with the interpolating function passing through each point in the dataset. This is a *natural* cubic spline, and an example of interpolation using such an approach is shown in Fig. 7.8.

An attractive feature of this form of interpolation or approximation is that the splines for the entire curve can be calculated using only three values for each data point on the curve, the nearest neighbours $\{y_{j-1}, y_j, \text{ and } y_{j+1}\}$. The approximation is therefore a local one. The data can be written as a tri-diagonal sparse matrix, and hence calculation of the splines can be performed efficiently using linear techniques. Splines are generally used for interpolation rather than function fitting, and are often implemented in the circuit simulator where they can be used to interpolate values from tabulated data. The smooth and continuous higher-order derivatives of the cubic spline meet the simulator requirements for function evaluation, and the spline interpolation is fast and accurate.

An alternative formulation of splines is known as *B-splines*: the 'B' refers to the *Bernstein* polynomials, which form the basis functions of the spline interpolator. The 'B' form can be written as a linear combination of splines

$$\sum_{n=1}^{N} B_{n,k} a_n \tag{7.19}$$

where $B_{n,k}$ is the nth B-spline of *order* k for a monotonic sequence of *knots* or breaks $(t_n...t_{n+k})$. $B_{n,k}$ is a polynomial of degree less than k, whose value is zero outside the interval $(t_n...t_{n+k})$. The B-spline need not go through all of the data points, but uses the data points or knots to determine the best interpolation or approximation of the whole dataset. This is a more general form of writing the spline, and again, being linear in parameters and having small support (that is, the B-spline is zero outside the interval), can be solved using linear techniques.

Spline approximation can be extended to two or more dimensions, by the tensor product of the univariate splines, for example

$$\sum_{n=1}^{N} \sum_{m=1}^{M} B_{n,k}(x) B_{m,j}(y) a_{n,m} \qquad (7.20)$$

An alternative multi-dimensional spline formulation is the *thin-plate* spline, which uses a radially symmetric function as the basis function for the data approximation; for example, in two dimensions

$$f(x) = \sum_{j=1}^{n-3} a_j \Psi (x - c_j) + \text{linear terms} \qquad (7.21)$$

where x is a vector of points in the data space, the a_j are linear coefficients, and c_j are values of x which are the centers of the radially-symmetric functions in the data space. The radial basis function Ψ is often a Gaussian function. An important feature of this function is that it is defined everywhere in the data space, and so there is no need to track interval boundaries. On the other hand, it is desirable for the radial function to fall to a negligible value for a large value of $(x - c_j)$, so that the influence of the basis function is restricted to being local. Thin-plate spline approximation has been used to good effect in the approximation of loadpull measurement data for device validation [8]. For multi-variate function approximation, however, a more general approach is to use artificial neural networks, of which the thin-plate spline is but one specific example.

7.3.5 Artificial Neural Networks

Artificial neural networks have become a popular method for multi-variate nonlinear function approximation or data fitting. The reasons behind this popularity include such factors as: ease-of-use – there are many general-purpose and application-specific software tools available, for example, the Matlab® Neural Network toolbox [9]; wide applicability – the technique has been developed for many technical fields; and the potential for high accuracy of function approximation, which is often realized in practice. This last feature is supported by a theoretical proof stating that a single-layer neural network can, in principle, approximate any nonlinear function to an arbitrary degree of accuracy, given certain provisos [10]: the *universal approximation theorem*. Neural networks have a regular structure that enables rapid evaluation in matrix-oriented software, and hence they are suitable for implementation in circuit-based simulators [11].

A neural network is a general, parameterized, nonlinear mapping between a set of input variables and set of output variables (or data). As we have seen, such mappings can be created using polynomials or rational functions, provided that there are enough terms in the approximation, that is, the polynomial is of sufficiently high degree. For a polynomial of degree M, the number of free parameters for a space having dimensionality d, is proportional to d^M [12]. Therefore, to achieve a high accuracy in a multiple-dimension space, a large number of free parameters must be determined. The greater the number of dimensions, the more free parameters are required. This means an increased complexity of the polynomial, increased memory requirement for coefficient storage, and reduced computational efficiency of the function evaluation. This problem is known as the *curse of dimensionality* [3].

One of the most significant advantages of neural networks as a function approximation technique lies in the manner in which they deal with this problem of scaling with dimensionality. Since the functions of the neural network are adapted as part of the training process, the number of functions has to be increased only as the problem complexity increases, not simply as the dimensionality of the problem grows. For neural networks, the number of free parameters typically grows only linearly or quadratically with the dimensionality of the input space [12]. Therefore, for multi-variate function approximation, neural networks have significant advantages over other techniques, as they permit a compact representation of a multi-variate function, requiring minimal storage of coefficients and being very efficient to evaluate.

An artificial neural network is a system of interconnected nodes, called *neurons*. The function of each neuron is to take a weighted sum of all of its inputs, the *induced local field*, which is then operated on by some transfer function, called an *activation* function, to produce an output signal. The activation function can be a linear function, or a nonlinear function that is smooth and differentiable. A schematic representation of a neuron is

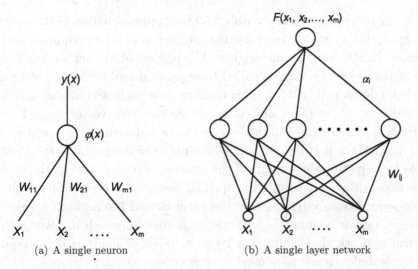

Fig. 7.9. Schematic representations of a single neuron and a single layer neural network.

illustrated in Fig. 7.9(a). The mathematical description of the neuron is [13]:

$$y_i = \phi(v_j(n)) \tag{7.22}$$

where $\phi(\cdot)$ is the activation function, and $v_j(n)$ is the induced local field of neuron j, defined by

$$v_j(n) = \sum_{i=1}^{m} w_{ji}(n)x_i(n) + b_i \tag{7.23}$$

Here m is the total number of inputs into neuron j and b_i is a bias or offset signal applied to neuron j. The w_{ji} is the weight of the link that connects the output of the jth neuron to the input of the ith neuron.

The neurons are typically arranged in layers. A simple *single hidden layer* neural network is shown in Fig. 7.9(b). The 'hidden' refers to the layer of neurons in the middle, which has no direct connection to the outside world. For nonlinear function approximations, the hidden layer has a nonlinear activation function, and the output layer, a single neuron in this example, has a linear activation function. The output of the single layer network, $F(\overline{x})$, where $\overline{x} = [x_1, x_2, \ldots, x_m]$, is given by,

$$F(\overline{x}) = \sum_{i=0}^{n} \alpha_i \phi \left(\sum_{i=0}^{m} w_{ji}(n)y_i(n) + b_i \right) \tag{7.24}$$

where m is the number of neurons and n is the number of input variables.

The key to obtaining an accurate function approximation is the accurate estimation of the free parameters: the number of neurons required, and the values of the interconnecting weights. The process of finding the free parameters of the neural network is called *training*, and many training techniques have been developed [12–15]. This training process is a nonlinear optimization problem, where the goal is to minimize the error vector formed by the difference between $F(\overline{x})$ and the function to be approximated, f, or the measured data. For a network with a given number of neurons, the error is minimized by adjusting the weights of the interconnections between the neurons in the network w_{ji}. To obtain the optimal network, in one approach, hidden neurons are successively added to the network and the network trained until the best error estimate is obtained. Another approach involves training a neural network that contains a larger number of neurons than required; pruning techniques are then used to remove unnecessary neurons [16]. One of the goals of the training is to obtain a network that is capable of 'generalization': the ability to predict the correct output for input data that was not used in the training dataset. This is essentially what we require from the function approximation of discrete measured data. Training techniques for improving the generalization of a neural network include 'early-stopping' using cross-validation, in which different subsets of the measured data are used for training and validation, with the validation being carried out periodically during the training to assess whether a suitable minimum in the error surface has been reached [12, 13, 17, 18].

As noted earlier, the universal approximation theorem for neural networks has helped to establish neural networks as a significant method in the field of function approximation. This theorem states that a network with a single hidden layer of neurons is capable of approximating any given function, with any degree of accuracy, provided the activation functions are non-constant, bounded and monotonically increasing [10]. As this theorem is only an existence theorem, it does not predict the optimal number of neurons, nor does it indicate whether a single layer of neurons is better than a many-layered structure in terms of the function approximation.

A known problem with a neural network containing only a single hidden layer is that the neurons tend to interact with each other globally: modifying the weights to make the approximation better at one point in the data will degrade it at another. This global interaction can be overcome through the use of two hidden layers. Neurons in the first layer tend to partition the space into sub-domains and other neurons learn to approximate the function locally. The second layer fits more global features. That is, the outputs of the first layer tend to operate only on localized areas of the

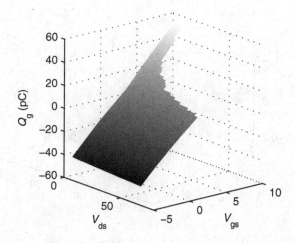

Fig. 7.10. Neural network approximation of the gate charge surface for an LDMOS transistor.

function range and output zero elsewhere. The second layer combines these localized regions [13].

We shall demonstrate the capability of neural networks as function approximators in nonlinear transistor modeling by using as an example the conservative gate charge field described in Chapter 6. The gate charge is a function of both the gate and drain voltages. Its value at a given set of instantaneous gate and drain voltages is calculated by performing a contour integral of the total small-signal gate capacitance along the gate voltage and the small-signal gate-to-drain capacitance along the drain voltage, from some bias point, as described in eq. 6.52. The gate charge is obtained from the contour integrals at every point in the V_{gs}–V_{ds} bias space at which the S-parameters were measured to give the small-signal equivalent circuit capacitance parameters.

The gate charge surface is an observational result: we have no *a priori* simple functional description for this surface. To approximate the charge surface we have used a two-hidden-layer network, with seven neurons in each layer. The hyperbolic tangent was used for activation function in each hidden layer, and the output layer used a linear function.

The fitted gate charge data for an LDMOS transistor is shown in Fig. 7.10. In this figure, the charge surface appears to be almost flat, of a constant slope in the V_{gs} direction, indicating an approximately voltage-*independent* gate

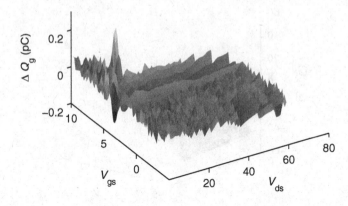

Fig. 7.11. The error surface of the difference between the neural network approximation to the gate charge and the charge derived from the measured RF data.

capacitance. For a MOS transistor, this is, to first order, a reasonable expectation: the gate capacitance is approximately equal to the parallel-plate oxide capacitance, with the semiconductor 'electrode' being attributable to the channel electrons in inversion, and to the depletion charge below threshold. A careful examination of the gate charge plot reveals slightly different slopes in inversion and below threshold, some small curvature along the drain voltage axis, and some structure in the active region, though this detail is extremely small. To verify that the neural network approximation to the gate charge data is accurate, we compare the model function with the data obtained from integration of the measured capacitances. The difference between the data and the neural network approximation is shown in Fig. 7.11, and shows only random errors at below 1% relative error. An alternative view of this difference is presented in Fig. 7.12. Here we show the gate charge values for the neural network approximation and the data for various values of the gate voltage, over the range of the drain voltage. We obtain a 'dots-on-lines' graph, which indicates that the neural network has returned the correct value of the charge. These two figures show that the neural network is capable of approximating this data very accurately.

While the comparison between model and data shows that the neural network can approximate the discrete charge dataset, of more practical value if this model is to be used in the context of a circuit simulator, is the ability of the neural network to approximate the capacitance functions. In Fig. 7.13, the line gradient of the neural network approximation is compared with the

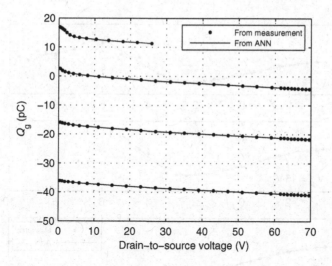

Fig. 7.12. Slices of the gate charge surface over the drain-to-source voltage, for several values of gate voltage, showing good agreement between the data and the neural network approximation.

numerical derivative of the charge surface. The derivatives are taken with respect to the gate-to-source voltage, and will therefore yield the total gate capacitance, which is dominated by the gate-source capacitance in this MOS transistor. The derived capacitance from the neural network agrees with the small-signal capacitance derived from the S-parameter measurements for this 7.2 mm gate-width test device. The smoothness of these curves illustrates the smooth function approximation capability of the neural network, and its suitability for this class of nonlinear modeling.

One drawback that has been observed with neural network function approximation is that they can have poor extrapolation properties if the training has not been performed with the objective of building good generalization properties [19]. The extrapolation behaviour tends to degrade as the number of dimensions or inputs to the network increase. One technique for improving the generalization and extrapolation is to incorporate prior knowledge of the function to be fitted. The reasoning for this is that for inputs outside the data range, the priors will dominate. A practical example of this is the knowledge-based neural network (KBNN). Within a KBNN, some of the neurons are coded with an analytical approximation for the function to be approximated, and these are combined with the regular neural network. This 'knowledge' provides extra information about the target function or data to be fitted, and can help reduce the amount of training data required,

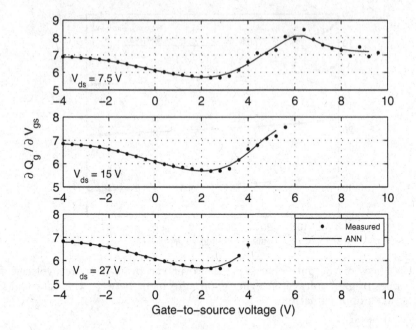

Fig. 7.13. Q_g derivatives – composite plot

and improve the accuracy of the function approximation, both within the dataset, and in extrapolation. There are several successful examples of using knowledge-based neural networks for device modeling [11, 20–23].

7.3.6 Elementary Functions

In the previous sections we have shown several function approximation techniques that are used commonly during the modeling of RF and microwave structures and devices. All of them are quite flexible and generic and do not require a detailed knowledge of the characteristics and underlying physics of the problem to approximate the data successfully. As previously demonstrated, this flexibility does come at a price, for example, poor interpolation and extrapolation qualities, large number of parameters and added complexity, among others.

At times the modeling engineer is presented with the opportunity to use certain elementary functions that, by their nature, more closely resemble the physical properties or behaviour of the structure or device being modeled. Elementary functions are algebraic combinations of fundamental functions, like exponents, trigonometric functions, and so forth. Within the domain of empirical models, the use of elementary functions can provide a wide range of advantages. This is especially true when there is some information about the physical properties of the problem, which can hint at the generic functionality that describes the relationship between the inputs and the outputs being modeled.

The selection and development of an appropriate elementary function can improve its extrapolation qualities outside the fitted space. In addition, if the elementary function is developed using physical principles governing the behaviour of the device or structure, the model parameters used in the function could represent specific physical parameters or properties, which could allow the user to understand and adjust the model for changes of that particular physical property.

The usage of elementary function approximations brings a new set of challenges and issues that need to be addressed. The development of an appropriate function does not typically follow a clear and automated process, and at times it could be slow and difficult, bordering into a craftsmanship activity. Because elementary functions are developed to approximate the behaviour of a specific set of data, structure or device, they do not tend to be generic or transportable. For example, if a function is developed to fit accurately the nonlinear current of a silicon LDMOS device, the equation and the fitting flexibility of its model parameters might not provide the required level of accuracy to reproduce faithfully the nonlinear drain current of a gallium arsenide PHEMT device.

To capture highly nonlinear behaviour, the complexity and the number of model parameters of the elementary function tends to increase to accommodate the requirements imposed by the data. This often leads to a non-trivial fitting or parameter extraction process to arrive at a meaningful and robust solution. This means that the modeling engineer's job is not completed simply by finding a good functional expression, but a model extraction procedure needs to be developed as well.

During the modeling of the nonlinear drain-to-source current of FETs, the use of the hyperbolic tangent as a basis for elementary function development is quite common [24–27]. As an example, we will now show the nonlinear drain current equation used in the MET LDMOS model [28], which is an electro-thermal model that can account for dynamic self-heating effects and was specifically tailored to model high power RF LDMOS transistors used in base-station, digital broadcast, land mobile and subscriber applications. The MET LDMOS is an empirical large-signal nonlinear model, which contains a drain current description that is single-piece and continuously differentiable, and includes static and dynamic thermal dependencies.

The model is capable of accurately representing the drain-to-source current–voltage characteristics and their derivatives at any bias point and operating temperature. This single continuously-differentiable drain current equation models the sub-threshold, triode, high current saturation and drain-source breakdown regions of operation.

The forward bias drain-to-source current equation is given by the following set of equations:

$$V_{gst1} = V_{gs} - (VTO + (GAMMA * V_{ds})) \tag{7.25}$$

$$V_{gst2} = V_{gst1} - \frac{1}{2}(V_{gst1} + \sqrt{(V_{gst1} - VK)^2 + DELTA^2} \\ - \sqrt{VK^2 + DELTA^2}) \tag{7.26}$$

$$V_{gst} = VST * \ln\left(e^{\frac{V_{gst2}}{VST}} + 1\right) \tag{7.27}$$

$$VBReff = \frac{VBR}{2}(1 + \tanh[M1 - V_{gst} * M2]) \tag{7.28}$$

$$VBReff = \frac{VBR}{2}(1 + \tanh[M1 - V_{gst} * M2]) \tag{7.29}$$

$$I_{ds} = (BETA)\left(V_{gst}{}^{VGEXP}\right)(1 + LAMBDA * V_{ds}) \\ \tanh\left[\frac{V_{ds} * ALPHA}{V_{gst}}\right]\left(1 + K1 * e^{VBReff1}\right) \tag{7.30}$$

The last equation combines a hyperbolic tangent function, which provides good fitting flexibility in the linear and saturation regions, with an exponential function, which is used to describe the breakdown characteristics of the transistor. To provide an accurate description of the drain current dependency with the gate-to-source voltage, several expressions are used to map the measured voltage to a different space. This provides a logarithmic change in drain current in the sub-threshold region and a monotonic drain current saturation (transconductance going to zero) at high voltages.

An example of the quality of fit that can be achieved with this kind of equation for a high-voltage LDMOS device is shown in Fig. 7.14. This complex elementary equation provides very good extrapolation and interpolation characteristics, while keeping the number of fitting parameters to only fifteen. This number of parameters is small compared with the amount of parameters that will be required if more generic function approximations

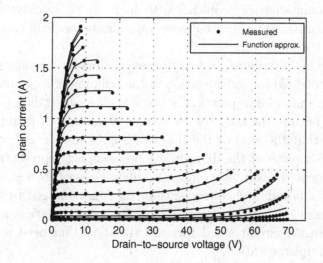

Fig. 7.14. Measured and fitted drain current of a Si LDMOS transistor using eq. 7.30

techniques are used. In addition, several of the model parameters in the equations have meaningful correlations to physical parameters of the device, for example, VTO is the threshold voltage, VST is the sub-threshold current slope and VBR is the drain-to-source breakdown voltage of the transistor at zero gate-to-source voltage.

The rest of the parameters are used to improve the fitting quality in the different regions of the data. For example, $GAMMA$ is used to adjust for VTO changes with drain-to-source voltage; the parameter $LAMBDA$ is used to control the output conductance in the saturation region; and $ALPHA$ controls the output conductance in the linear region. The VK and $DELTA$ parameters are used to adjust for the transconductance collapse at high gate-to-source voltages which yield to a maximum drain-to-source current or I_{max}. The rest of the parameters provide fine adjustments to the overall shape of the elementary function approximation.

7.4 Conclusions

In this chapter we have presented some of the different approaches to function approximation of the measured data used in building compact transistor models. There are various degrees of complexity of the functions available for data fitting, and we have illustrated the degrees of freedom that the modeling engineer has in choosing an appropriate function for a given set of data. For example, artificial neural networks are more suitable for the function approximation of problems with high dimensionality, whereas a polynomial approximation may be more appropriate for fitting a curve that is a function of a single variable.

Some of the key attributes of function approximation to measured data have been highlighted for consideration in the building of a compact model. These include issues of interpolation between the measured data points, and extrapolation beyond the range of the measured dataset. It is important to keep in mind that the goal of the function approximation is to provide a continuous description of the data, in a more compact representation that can be implemented easily in a circuit simulator; issues such as speed and accuracy of evaluation, and convergence of the function need to be considered. In the next chapter we describe how these mathematical models are implemented in the circuit simulator, and the steps that need to be taken to verify this implementation.

References

[1] J. Wood, "Volterra methods for behavioral modeling," ch. 2, in *Fundamentals of Nonlinear Behavioral Modeling for RF and Microwave Design*, J. Wood and D. E. Root, Eds. Norwood, MA: Artech House, 2005.

[2] J. Xu, D. Gunyan, M. Iwamoto, A. Cognata, and D. E. Root, "Measurement-based non-quasi-static large-signal FET model using artificial neural networks," in *IEEE MTT-S Int. Microwave Symp. Dig.*, San Francisco, CA, June 2006, pp. 469–472.

[3] N. Gershenfeld, *The Nature of Mathematical Modeling*. Cambridge, UK: Cambridge University Press, 1999.

[4] B. Gustavsen and A. Semlyen, "Rational approximation of frequency domain responses by vector fitting," *IEEE Trans. Power Delivery*, 14, no. 3, pp. 1052–1061, July 1999.

[5] J. De Geest, T. Dhaene, N. Faché, and D. De Zutter, "Adaptive CAD-model building algorithm for general planar microwave structures," *IEEE Trans. Microwave Theory Tech.*, 47, no. 9, pp. 1801–1809, Sept. 1999.

[6] R. Lehmensiek and P. Meyer, "Creating accurate multivariate rational interpolation models of microwave circuits by using efficient adaptive sampling to minimize the number of computational electromagnetic analyses," *IEEE Trans. Microwave Theory Tech.*, 49, no. 8, pp. 1419–1430, Aug. 2001.

[7] D. M. Pozar, *Microwave Engineering*, 2nd edn. New York, NY: John Wiley & Sons, 1998.

[8] P. Hart, J. Wood, B. Noori, and P. Aaen, "Improving loadpull measurement time by intelligent measurement interpolation and surface modeling techniques," in *67th ARFTG Conference Dig.*, San Francisco, CA, June 2006.

[9] *Matlab Neural-Network Toolbox*, The Mathworks, Inc.

[10] G. Cybenko, "Approximation by superposition of sigmoidal functions," *Math. Control, Signals Systems*, no. 2, pp. 303–314, 1989.

[11] Q.-J. Zhang and K. C. Gupta, *Neural Networks for RF and Microwave Design.* Norwood, MA: Artech House, 2000.

[12] C. M. Bishop, *Neural Networks for Pattern Recognition.* New York, NY: Oxford University Press, 1995.

[13] S. Haykin, *Neural Networks: a Comprehensive Foundation*, 2nd edn. Upper Saddle River, NJ: Prentice Hall, 1995.

[14] A. H. Zaabab, "Device and circuit-level modeling using neural networks with faster training based on network sparsity," *IEEE Trans. Microwave Theory Tech.*, 45, pp. 1696–1704, Oct. 1999.

[15] L. Prechelt, "A quantitative study of experimental evaluations of neural network learning algorithms: current research practice," *Neural Networks*, 9, no. 3, pp. 457–462, 1996.

[16] L. Prechelt, "Connection pruning with static and adaptive pruning schedules," *Neurocomputing*, 9, pp. 49–61, 1997.

[17] L. Prechelt, "Early stopping–but when?" in *Neural Networks: Tricks of the Trade.* London, UK: Springer-Verlag, 1998, pp. 55–69.

[18] I. Rivals and L. Personnaz, "Neural-network construction and selection in nonlinear modeling," *IEEE Trans. Neural Networks*, 14, no. 4, pp. 804–819, July 2003.

[19] J. Xu, M. C. E. Yagoub, R. Ding, and Q.-J. Zhang, "Robust neural based microwave modeling and design using advanced model extrapolation," in *IEEE MTT-S Int. Microwave Symp. Dig.*, Fort Worth, TX, June 2004, pp. 1549–52.

[20] P. M. Watson and K. C. Gupta, "EM-ANN models for microstrip vias and interconnects in multi-layer circuits," *IEEE Trans. Microwave Theory Tech.*, 44, no. 12, pp. 2495–2503, Dec. 1996.

[21] F. Wang and Q.-J. Zhang, "Knowledge based neural network models for microwave design," *IEEE Trans. Microwave Theory Tech.*, 45, no. 12, pp. 2333–2343, Dec. 1997.

[22] J. W. Bandler, R. M. Biernacki, S. H. Chen, P. A. Grobelny, and R. H. Hemmers, "Space mapping technique for electromagnetic optimization," *IEEE Trans. Microwave Theory Tech.*, 42, no. 12, pp. 2536–2544, Dec. 1994.

[23] J. Gao, L. Zhang, J. Xu, and Q.-J. Zhang, "Nonlinear HEMT modeling using artificial neural network technique," in *IEEE MTT-S Int. Microwave Symp. Dig.*, Long Beach, CA, June 2005.

[24] W. R. Curtice and M. Ettenberg, "A nonlinear GaAs FET model for use in the design of output circuits for power amplifiers," *IEEE Trans. Microwave Theory Tech.*, 33, no. 12, pp. 1383–94, Dec. 1985.

[25] A. Materka and T. Kacprzak, "Computer calculation of large-signal GaAs FET amplifier characteristics," *IEEE Trans. Microwave Theory Tech.*, 33, no. 2, pp. 129–135, Feb. 1985.

[26] A. E. Parker and D. J. Skellern, "A realistic large-signal MESFET model for SPICE," *IEEE Trans. Microwave Theory Tech.*, 45, no. 9, pp. 1563–71, Sept. 1997.

[27] I. Angelov, H. Zirath, and N. Rorsman, "A new empirical nonlinear model for HEMT and MESFET devices," *IEEE Trans. Microwave Theory Tech.*, 40, no. 12, pp. 2258–66, Dec. 1992.

[28] W. R. Curtice, J. A. Plá, D. Bridges, T. Liang, and E. E. Shumate, "A new dynamic electro-thermal nonlinear model for silicon RF LDMOS FETs," in *IEEE MTT-S Int. Microwave Symp. Dig.*, Anaheim, CA, June 1999, pp. 419–422.

8

Model Implementation in CAD Tools

8.1 Introduction

The ultimate goal for a modeling engineer is to see the model that he has developed be used successfully by circuit designers in their CAD simulations of circuits and design of new products. All of the hard work in the characterization, construction, and mathematical analysis of the transistor to produce the compact model is only of value if it enables the circuit designer to be more confident in the design, and to produce better circuits. The design cycle is often speeded-up when better models are used. These goals and outcomes follow from the 'Law of Simulation and Modeling' that states: 'a model is (mostly) useless unless it is embedded in a simulator.' A successful model is one that is used, and is used in the design of successful products.

The *implementation phase* is therefore an essential element of any model development. The practical implementation in the CAD tool can take many forms, from an equivalent circuit composed of basic circuit elements already available in the simulator, to programming of the mathematical expressions in a high-level software language. To create a successful model, the modeling engineer needs to be involved in the detail of the implementation process: it is fully a part of the model development and deployment. The basic knowledge of how the simulator operates, its methods of solving the circuit equations and arriving at convergence, is invaluable in the implementation of a model that is accurate, fast, and easy to use.

For the sake of simplicity, we shall refer to the design software packages delivered by electronic design automation (EDA) vendors for the computer-aided design of circuits and ICs as 'CAD tools'. Information about simulators and CAD tools can be found in the user guides and books that describe how to use the simulator, its commands, and so forth. Such sources are usually more concerned with providing the user with the means to run a simulation successfully, rather than with describing its inner workings. On the other hand, the treatises on how to design and construct the algorithms for solving the matrix equations in the simulator are often too focused to be of assistance for model development. A useful text that describes how the various simulation tools work, arrive at convergence, and how to get the best from the simulator, is the book by Kundert [1].

In this chapter, we shall present some of the aspects of model implementation that the modeling engineer should consider in building the compact model into the CAD tool. We shall focus on the general principles and tasks that need to be undertaken in building the mathematical model of the transistor into a simulator in order to illustrate the requirements that need to be satisfied for the construction of a successful compact model in the target CAD tool. In particular, we shall describe some of the methods for implementing the model in the simulator, and highlight the need for verification of the model implementation, indicating some examples of the tests that can be carried out. We shall not embark upon an exhaustive catalogue of techniques of how the model can be implemented, nor a line-by-line description of the software that might be used. This approach would be too specific and detailed to be of general utility.

We shall also consider aspects of model portability between simulators, of intellectual property (IP) protection, and the building of libraries or design kits to support RFIC and PCB-level design. We shall begin with a brief review of the various classes of simulator, including descriptions of convergence requirements, and how the compact model interacts with the simulator.

8.2 An Overview of the Various Classes of Simulator

Circuit simulators became important in the 1960s and 1970s with the growth of the market for integrated circuits. As the circuits became larger and more complicated, hand-design became impossible, so simulation became an essential part of the design process. It is also necessary to evaluate the design before committing to die-level manufacturing, since it is not possible to adjust or 'tune' a mass-produced monolithic integrated circuit after manufacture. Many of today's circuit simulators have their roots in the SPICE simulator [2] developed at the University of California at Berkeley. SPICE is basically a time-domain simulator, calculating the circuit response at successive time-steps; it has DC and small-signal AC simulation capabilities included. Many of the commercial CAD tools employ proprietary numerical techniques in their simulators, and for RF and microwave applications, the suite of simulators often includes nonlinear frequency domain solvers.

The SPICE simulator has its origins as a class project of Prof. Ron Rohrer at Berkeley. The first SPICE simulator was written by Nagel and Pederson [2], and was released in 1972, with SPICE2 following in 1975: this became the *de facto* standard for analogue circuit simulation. It owed its popularity partly to the fact that it included passive and active device models as part of the simulator, which was unusual at the time. The Gummel–Poon integral charge control model for the bipolar transistor [3] and the Shichman–Hodges quadratic FET model for the MOSFET and JFET devices [4] were the state-of-the-art models that were included. The SPICE simulator was also robust, accurate, and easy to use. Much of the development of SPICE focused on accurate and efficient numerical methods for solving the circuit equations, nonlinear equation solution techniques, sparse matrix techniques, and semiconductor device compact modeling [5]. Mostly, though, one suspects that SPICE was popular because it was free.

The 1980s saw many developments in the field of circuit simulators. Several IC manufacturers developed their own in-house variants of SPICE. This was because circuit simulation was an essential part of the design of the IC, and having a sophisticated simulation tool was seen as a competitive advantage. Commercial CAD vendors also began to appear, with CAD packages offering a SPICE-like simulator, often with proprietary algorithmic developments, along with other features such as schematic-entry circuit elements, component layout, schematic and layout verification, and so forth, integrated into a single software package. SPICE3 was released by the Berkeley group in the late 1980s, offering improved software architecture, though with substantially the same algorithms.

At this time the first RF and microwave CAD simulators became available. These tools focused on S-parameter and noise simulation, in addition to DC simulation, as many microwave devices were described by their frequency-dependent S-parameters or small-signal equivalent circuit models. These tools were quite successful, essentially offering automated Smith chart capability, and usually coming with a library of frequency-domain transmission-line and related distributed components, which SPICE and similar time-domain packages do not model so well.

Probably the most far-reaching development in the 1980s was the harmonic balance simulation technique. This method computes the steady-state solution of nonlinear circuits in the frequency domain, and was targeted for use in microwave circuit simulation, where the traditional time-domain techniques can require millions of time-steps before the steady-state condition is reached, making the circuit simulation an expensive exercise. The harmonic balance simulator, called *Spectre*, was developed by Kundert while studying in the Berkeley group [6]. This simulator was adopted by Hewlett Packard for their *Microwave Nonlinear Simulator* (MNS), which was a significant part of the MDS™ CAD tool, and remains a central feature of Agilent's ADS™, and is also used by Cadence™. All of the microwave CAD vendors now offer harmonic balance as one of the main algorithms in their software.

At the International Microwave Symposium in 1996, the *circuit envelope* simulation technique was announced separately by Sharrit [7] and Ngoya [8], with Hewlett Packard filing a contemporary patent [9]. The envelope simulator is directed at the efficient simulation of circuits driven by RF signals with analogue or digital modulation: the RF response is calculated using a harmonic balance simulator, and the modulation signal is analyzed using a time-domain simulator that is stepped at a rate appropriate to the modulation frequency. This overcomes the drawbacks for this class of problems of traditional time-domain simulation, where the step-size must be on the RF timescale, and harmonic balance, which would require a large number of tones to describe the signal.

Another 1990s simulator development was the release of the system-level simulator from the Berkeley group, called *Ptolemy*, which has also been adopted by Agilent in their suite of RF simulators. Ptolemy is a time-stepping simulator, running at the data clock rate; it can also be thought of as a simulation controller, calling the ADS RF simulators such as 'Envelope', and also managing co-simulation with the Mathworks Matlab™ simulator for digital signal processing (DSP). Compact models can be used in Ptolemy through the system call to the RF simulator.

A recent interest has been in the development of the analogue hardware description language, Verilog-A, as a simulator-independent tool for model development and deployment. This has implications for compact model portability between simulators, and is discussed in Section 8.7.

8.2.1 Circuit Simulation Basics

The simplest type of simulation is the steady-state analysis of the DC operating conditions of the circuit. The DC analysis is an important component of the simulator. Repeated applications of the DC analysis for different input signals enable the DC characteristics of a device model to be traced. A DC analysis is performed before a small-signal frequency-domain analysis or harmonic balance simulation is carried out, to establish the operating point of the circuit and the devices in it. A DC analysis is also performed at every time-step in a transient simulation.

We shall use the DC analysis to illustrate some of the basic features of circuit simulation, and hence some of the requirements of the compact model implementation. The basic purpose of the simulator is to solve Kirchhoff's equations for the circuit; for SPICE-derived simulators (and this is virtually all of them), this is done by setting up a nodal admittance matrix description of the circuit netlist and solving Kirchhoff's current equation,

$$\mathbf{I} = f(\mathbf{V}) \tag{8.1}$$

where the individual branch currents are determined from the node voltages at the ends of the branch. The function relating the branch current to the voltage can be a simple linear proportional relationship, as for a resistor; a differential relationship describing a capacitor; an integral relationship describing an inductor, although in practice an alternative technique is used, and is outlined later; or a nonlinear function such as those we have derived for the compact transistor model. The general nonlinear branch current equation is

$$i\left(v(t)\right) + \frac{d}{dt} q\left(v(t)\right) + j(t) = 0 \tag{8.2}$$

with some initial condition $v(0) = V_0$.

In DC analysis the currents and voltages are steady with time; their time derivatives are therefore zero: capacitors behave as open circuits, and inductors as short circuits. Further, linear elements are replaced by their real conductances at zero frequency: this includes transmission line elements. The equations for DC are derived from eq. 8.2 by setting the time derivative to zero, and the input signal $j(t)$ is constant. The resulting DC equations are a set of nonlinear algebraic equations, that is, no derivatives or integrals. Solving a large system of nonlinear algebraic equations can be a challenge, and the general method is to use iterative techniques. A common technique used in circuit simulators is the Newton–Raphson algorithm, which solves equations that can be written in the form

$$f\left(\bar{v}\right) = 0 \tag{8.3}$$

Newton's method takes an initial guess and linearizes the equations about this point, solves the linearized equations to find a new solution, which then becomes the initial guess for the next iteration. This is illustrated in Fig. 8.1 with a simple example, and the algorithm can be written for the $(k+1)^{\text{th}}$ iteration as

$$v^{(k+1)} = v^{(k)} - \mathbf{J}^{-1}\left(v^{(k)}\right) f\left(v^{(k)}\right) \tag{8.4}$$

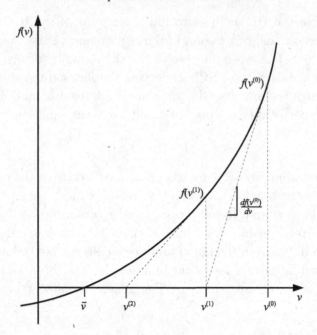

Fig. 8.1. Newton's method applied to finding $f(v) = 0$, starting with the initial guess, $v^{(0)}$, and linearizing the function to estimate the next guess, $v^{(1)}$, and so on until the solution is found.

where J is the *Jacobian* of the function f

$$J(v) = \frac{d}{dv} f(v) \qquad (8.5)$$

For a circuit comprising N nodes, then v and $f(v)$ are N-vectors, and the Jacobian is an $N \times N$ matrix. This will usually be a sparse matrix, as circuit connectivity is usually local, and matrix-inversion routines that are quick to execute have been developed for sparse matrices [10].

The Newton–Raphson method is guaranteed to converge if the function f is continuously differentiable, and the initial guess is close enough to the solution (among certain other requirements). The choice of the function f is made by the modeling engineer, and it is his responsibility to ensure that the model function has a continuous first derivative, and that its second derivative exists, to satisfy the first convergence criterion. Of course, the second criterion cannot be guaranteed, although there are several ways of getting a better initial condition for the simulation; these are often documented in the user guide for the particular simulator.

The iteration of the Newton method proceeds until the approximate solution satisfies two convergence criteria, at which point the solution is considered to be sufficiently close to the 'correct' value. The first convergence condition specifies that the Kirchhoff's current law should be satisfied within a certain tolerance:

$$\left| f\left(v^{(k)}\right) \right| < \varepsilon_R \tag{8.6}$$

This is known as the *Residue* criterion. The second convergence condition specifies that the difference between the solutions of successive iterations is small:

$$\left| v^{(k)} - v^{(k-1)} \right| < \varepsilon_U \tag{8.7}$$

This is the *Update* criterion. If both criteria are satisfied, than we can consider $v^{(k)}$ to be a valid solution. The residue criterion is important when the impedance is small: a large current change occurs for only small changes in the node voltage. On the other hand, the update criterion is important for large impedances, where the change in current is small for large changes in node voltage. In practice, the simple limits ε_R and ε_U comprise absolute and relative components. In this way, the convergence criteria scale with the voltages and currents.

While the Newton method is still widely used for its simplicity and general robustness of convergence, modern simulators also have a selection of sophisticated optimization or function minimization algorithms that can be used for better convergence behaviour. These optimizers include functions such as conjugate gradient, 'minimax' routines, and so forth, which may have advantages for certain circuit configurations and signals.

The numerical efficiency of the optimization or minimization algorithm is also a consideration. The optimizer function may need to be evaluated several times before convergence is reached. A computationally inefficient implementation in the model increases resulting in a long simulation time. Furthermore, if multiple instances of a transistor model are used within the same circuit, the problem is exacerbated. The computational efficiency of the implementation is often dictated by decisions made during the model generation process.

In the illustrations above of how the simulator works, we have used a classical nodal analysis description for the circuit. This is fine for most passive components and voltage-controlled devices, but for current-controlled elements the functions and Jacobian entries are difficult to express and cumbersome to manipulate. To overcome these problems, *modified nodal analysis* (MNA) was developed. In this representation, the 'drive' signal vector can contain currents and voltages, and the responses likewise. By arranging the admittance matrix rows, the current-controlled components can be grouped together, and the resulting matrix is sparse. The algebraic equations can be solved using standard techniques indicated earlier to arrive at the converged solution for the branch currents and node voltages in the circuit.

8.2.2 AC Simulation

The AC simulation is a small-signal frequency-domain simulation carried out on the circuit after it has been linearized about the quiescent point that is found from the DC analysis. The linearization is a Taylor series expansion about the quiescent conditions, which in simple form means that the branch current function $i(v)$ is replaced by the small-signal conductance, and the charge function $q(v)$ is replaced by a capacitance, as shown below

$$g_{AC} = \left.\frac{di}{dv}\right|_{v_{DC}} \tag{8.8}$$

$$C_{AC} = \left.\frac{dq}{dv}\right|_{v_{DC}} \tag{8.9}$$

Hence the small-signal AC current can be written

$$i_{AC} = (g_{AC} + j\omega C_{AC})\, v_{AC} \tag{8.10}$$

where a single-frequency sinusoidal excitation is used, allowing the time derivative to be written in the frequency domain. This is an expression of the small-signal Y-parameters. The AC simulation calculates the linear properties of the circuit, and usually we are interested in measures such as the transfer function relation, or the two port (or multi-port) network parameters, such as the S-parameters of the circuit.

8.2.3 Transient Simulation

The transient simulation is a time-stepping simulation in which a Newton–Raphson iterative solution is found at each time-step. The spacing of the time points is governed by the highest frequency expected at any node, which in turn is determined by the rate of change of the node voltage. The first time derivative at a given node voltage is found from the present and previous values of the node voltage using curve-fitting or integration routines. The time-step is adjusted to keep the derivative small, and so the initial guess for the node voltage at the next time-step is close to the present value.

Transient simulations can take a long time to arrive at a suitable solution, especially when we are concerned with steady-state large-signal solutions for microwave circuits. This may require many time-steps before the transient solution has died out. The solution to this problem is the harmonic balance simulation.

8.2.4 Harmonic Balance

Harmonic balance is a frequency-domain analysis used for the simulation of the steady-state behaviour of nonlinear circuits and systems. The technique is based on the fact that the circuit response to a sinusoidal excitation can be described to a given accuracy by a Fourier series with a finite number of coefficients. The truncated Fourier series representation for a single-tone excitation at frequency ω_0 is

$$V(\omega) = \sum_{k=0}^{M} V_k \, e^{(jk\omega_0 t)} \tag{8.11}$$

where V_k is the (complex) Fourier coefficient of the voltage at the kth harmonic, and there are M tones in the approximation. The zero-th component is the response at DC. The circuit currents are found by iterative solution of the system of algebraic equations for all of the harmonic components. Compared with the DC case, there are now M times the number of equations to solve. This is still more efficient than a transient solution, because the solution of the system of $M \times N$ algebraic nonlinear equations gives the steady-state response directly.

The harmonic balance analysis can accommodate multi-tone excitations and hence can simulate intermodulation distortion in nonlinear circuits such as power amplifiers. The total number of coefficients that need to be solved in the Fourier analysis can increase rapidly with the number of tones considered. The matrix size can be an issue for many tones, and sparse matrix and 'Krylov' sub-space solution methods need to be used. Such techniques can be less robust and accurate in terms of convergence than the direct iterative methods, although significantly more efficient in terms of memory usage.

The frequency-domain description can also handle transmission line and other distributed component definitions more easily than transient analysis, as there is no closed-form solution for these devices in the time-domain representation. On the other hand, nonlinear device models are evaluated in the time domain, which is the natural domain for nonlinear description, and then converted to the frequency domain at each iteration using the discrete Fourier transform (DFT). Fast DFT algorithms have been developed for many signal-processing applications, and are adapted to circuit simulation, making this domain transformation very efficient.

In common implementations of the harmonic balance simulator, the circuit is first partitioned into linear and nonlinear components, and then a DC simulation is carried out to establish the operating point of the circuit. The linear component network is then simulated at all frequencies using the small-signal AC simulator, which is typically very fast. Only the nonlinear components need to be analyzed using the harmonic balance simulator. Details of the harmonic balance algorithm and its implementation have been presented by Kundert *et al.* [1,6] and Rizzoli *et al.* [11].

8.2.5 Envelope Simulation

The circuit envelope simulator is a combination of a harmonic balance simulator and a time-domain simulator. It is designed for the analysis of RF communications circuits, where the stimulus is represented by one or more carriers having time-varying modulation envelopes. The harmonic balance analysis of the RF carriers is performed, and then the time-domain analysis is carried out on the envelope signal. The envelope analysis is shown schematically in Fig. 8.2.

This analysis technique offers significant advantages over both harmonic balance and transient analysis, when these techniques are applied individually to modulated RF signals. First, fewer tones and harmonics are required than for a harmonic balance analysis, since the modulating tones are captured in the envelope. This reduces the memory requirement and the number of algebraic equations that need to be solved for the harmonic balance component of the envelope analysis. It is found that fewer harmonics are required to simulate accurately a given nonlinear device using envelope analysis. Second, the time-step for the time-domain part of the analysis is determined by the bandwidth of the modulation signal, and not of the RF carrier, so significantly fewer time points are needed to arrive at the converged, steady-state solution.

The formulation of the current and voltage variables for envelope simulation in the MNA is very similar to that for harmonic balance, shown in eq. 8.11, except that the Fourier coefficients are now functions of time, at the modulation or envelope timescale. In envelope simulation, the models

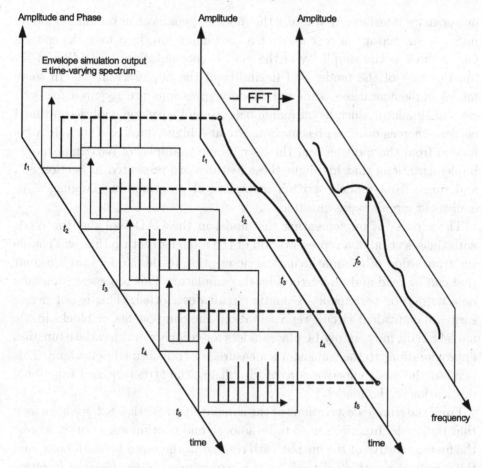

Fig. 8.2. A modulated signal and its time-varying spectrum simulated using the envelope technique, with time-domain and frequency-domain representations also shown.

are required to return a current at each harmonic tone in the harmonic balance part of the simulation, and an additional current term for the transient envelope current caused by the time-varying change in envelope voltage at each tone. This extra term is the rate of change of charge at the envelope timescale. This behaviour is accommodated by a charge-control model, such as we have described in Chapter 6.

8.3 Overview of the Model Implementation Process

Here we shall review some of the basic steps and considerations that are taken in creating a model in a CAD tool. The target CAD tool may have one or more interfaces that assist the modeling engineer in building a model into the simulator. The choice of a particular interface may depend on factors such as the simplicity of the model representation, protection of IP, functionality of the model and flexibility of the implementation. In some model implementations, different model expressions are required for each class of simulator, such as harmonic balance and envelope analyses outlined earlier, whereas other representations are at a higher level and such detail is hidden from the modeler and the user. Some examples of particular model implementations that highlight these features are presented in Section 8.5, and range from models written in high-level programming languages, to equivalent circuit representations.

The process of implementing the model in the CAD tool then proceeds with the selection of a representation of the model structure that will enable the translation of the mathematical description of the model into a format that can be read and understood by the simulator. This representation may be a netlist, or perhaps a schematic circuit description. This is not necessarily an equivalent circuit representation; the components or blocks in the model circuit may simply be placeholders for complex multi-variate function approximations to the transistor's measured or transformed behaviour. This is often the most effective way of visualizing the structure and functional description of the model.

From the simple descriptions of the simulators in Section 8.2, we have seen that the model functions need to be smooth and continuous, in other words, the first derivative of the output with respect to the input is continuous, and the second derivative exists. This is a requirement of the Newton iterative method for convergence of the model, and is also required to define the small-signal admittances for the linear simulation. The choice of model functions for data approximation is large, some examples being described in Chapter 7, and so this should not be a limitation. In table-based models, the simulator itself interpolates between data points, often using cubic splines, ensuring local continuity.

Some of the model function approximations have many parameters that are also functions of the input variables; the approximation of the FET drain current using elementary functions is a good example. These functions should also be continuous in the first derivative for the Newton optimizer. The model Jacobian matrix is formed from these first derivatives of the model functions. In practice, the algebraic construction of the Jacobian for the model is probably the most time-consuming and error-prone part of the model implementation in the simulator. Some simulators use more sophisticated optimizers that require continuous second derivatives also, for the Hessian matrix, although the Hessian can often be approximated using only the Jacobian matrix in the iteration [12].

For time-domain simulations, the derivatives with respect to time also need to be calculated. This can be carried out at run-time in the simulator, by using difference equations or function fitting of the present and previous time point data. The voltages and current state variables of the model must be recorded for some history in the simulator: the data are expressed as vectors of the present values, first- and higher-order time derivatives, and often a vector of time-delayed state variables.

The modeling engineer must ensure that the implementation of the model functions does not lead to numerical instability. Examples of numerical problems that can affect the numerical stability and hence the robustness of the convergence process include: fractional expressions with denominators that limit toward zero, variables or exponentials raised to high powers, unintended negative roots, and so forth.

For example, the following function, *softexp* can be used in lieu of the regular exponential function in the simulator, preventing its value from becoming too large, reaching the simulator maximum floating point value, or causing numerical instabilities:

$$softexp(x) = \begin{cases} e^x \text{ for } x < MaxExp \\ (x + 1 - MaxExp)\, e^{MaxExp} \text{ for } x \geq MaxExp \end{cases} \tag{8.12}$$

An optimal value of *MaxExp* may vary from simulator to simulator, but a useful starting value, from our experience, is around 100.

The modeling engineer must also consider the asymptotic behaviour of the model functions as implemented in the simulator. The iterative methods may select a point beyond the validated data range of the function during the progress towards convergence. It may be necessary to condition the model function so that its extrapolation is well-behaved, and guides the Newton method to a safer approximation.

Passive component models are often implemented as equivalent circuit descriptions, or using network parameters in the frequency domain, which can be very convenient. Recalling that a DC analysis is always carried out before an S-parameter simulation, the DC properties of the passive model must be defined. Unless values for the S-parameters at DC are provided in the model, the simulator will extrapolate the high-frequency S-parameter data to DC and use the real part for the DC component value. This can lead to errors.

Passive component models must also obey the circuit theory requirements of passivity and causality. Passivity can usually be ensured by the presence of a dissipative component – a resistor – in the passive model. Passive models exhibiting gain are possible if implemented incorrectly [13]: this can happen if the S-parameters or function approximations to them are used directly, without investigating the resistive behaviour, especially at DC. Causality means that the output is uniquely determined by the present and past inputs, and that the model equations do not change with time. For passive models, causality is usually assured if the network transfer parameters are functions of the input signals only.

In the active device model, causality is often addressed by using a time-delayed gate voltage as a control input to the drain current source. This ensures that the output signal is a function of the past values of the input. As we have seen in Chapter 6, the time delay is an expression for the transcapacitance. Further, using a charge-conservative implementation for the active device, by integrating the transcapacitance to yield the drain charge, will produce a causal model.

The large-signal model architecture that we have adopted, using controlled current and charge sources for the nonlinear elements, matches the simulator analyses very well. The DC behaviour is convergent, providing a stable base for small-signal analysis. The charge description is suitable for time-domain, harmonic balance and also envelope analyses. The convergence properties lie in the details of the model functions used to approximate these controlled sources.

Finally, we presume that some basic software engineering is applied in the creation of the model in the CAD tool, for example, in the range checking of the parameter values that are input by the user. This reduces the risk of having convergence problems due to unrealistic model parameter sets.

8.4 Model Verification Process

The verification process ensures that the model is implemented correctly and is yielding the expected results. At this stage of the model implementation we are not ascertaining if the model and the specific model parameter set are producing realistic or valid results when compared with measured data. The goal at this stage is only to gain confidence that no errors were made during the implementation of the model in the simulator.

One of the first steps in the model verification process is to check that all the model parameters are being properly read by the simulator. This can be done in the debug stage, or by the simple but crude expedient of printing or otherwise outputting the parameter values.

The verification process is much simpler for linear or passive devices than for nonlinear models. Once the model of the passive component is implemented in the simulator, some of the basic circuit characteristics of a passive device should be verified, such as causality and passivity. Passivity can be checked by small-signal frequency-domain analysis, and looking for gain at any frequency. Causality can be checked in the time domain, using a step response. A DC analysis will determine whether the correct DC behaviour has been captured in the model. As noted earlier, this is important for models of passive components that have been implemented in the frequency domain, where extrapolation of the high-frequency response to DC may result in incorrect results, non-passivity, or even non-convergence. The modeling engineer should also check for the asymptotic behaviour by running tests outside the nominal data domain.

The verification of the model implementation of a large-signal model is considerably more complex than the linear model counterpart. As part of the verification process, there are several tests that should be performed to ensure that the model is self-consistent and implemented correctly.

One such test is the small-signal to large-signal consistency of the model. This can be accomplished by performing linear small-signal S-parameter simulations, and comparing the results with a large-signal simulation in a setup that mimics the block diagram of a vector signal analyzer, with a very small incident power to ensure that the device is being operated in the small signal regime. Under these conditions, the incident, transmitted and reflected powers can be computed, and from their ratios, the two-port S-parameters can be calculated. If the model is properly implemented and there are no issues with the simulator, the small-signal and large-signal results should match within the numerical accuracy of the linear and nonlinear solvers in the simulator.

This test is not a test for the charge-conserving property or otherwise of the model. When used in a large-signal simulation, models that are not charge-conserving can produce an unphysical DC component, which increases with the signal frequency [14]. This characteristic of non-charge-conserving models can be used to verify the implementation of a model that has a charge-conserving formulation, by verifying the absence of the aforementioned DC component.

For an electro-thermal model, there are consistency checks that can be carried out to ensure that the electrical and thermal components of the model produce consistent results. For an LDMOS device, one check is to verify that a zero temperature coefficient point in the gate voltage–drain current characteristics can be obtained. If the model is capable of predicting dynamic self-heating effects, the modeling engineer can verify that the temperature rise is computed correctly, based on the simulated power dissipation of the device and the thermal resistance that is used during the simulation.

An essential part of the verification process is to evaluate the model performance in all of the different simulation types: DC, AC or small-signal S-parameters, large-signal harmonic balance, transient (time-domain), and envelope. This is to ensure that proper convergence, and asymptotic behaviour is obtained. This is conveniently done using a suite of test cases, where the results of the simulation are known through analytical solution, or are at least predictable.

Often, the models need to be implemented in several CAD packages or different operating systems. Again, once the basic implementation of the model has been verified, a set of test cases can be established that can be used to compare the simulation results in the different simulators or operating systems.

8.5 Types of Model Implementation

The format of the model implementation in the CAD tool can range from an equivalent circuit representation using standard built-in components, to programming the nonlinear functions in a high-level language. Here we shall outline some of these types of implementation, and indicate some typical examples of their use. We shall also describe briefly some examples of nonlinear model prototyping.

8.5.1 Equivalent Circuit Models

This approach is popular for linear models, both passive and active. The circuit topology is generally determined by the small-signal network parameters that are used, but usually has a basis in some physical interpretation of the circuit elements. For example, a linear FET model may be derived from the terminal Y-parameters, that have been transformed from the measured S-parameters at a given bias condition. The two-port Y-parameters yield four real and four imaginary components that can be associated with lumped-component resistors and capacitors in a straightforward manner. An example of the linear FET model is shown in Fig 8.3.

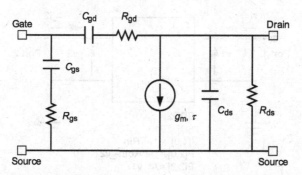

Fig. 8.3. Small-signal equivalent circuit model of the intrinsic transistor; this model is derived from Y-parameters, with some admittance-to-impedance transformation in the gate-source and gate-drain branches.

The values of the lumped component parameters can be fixed, or they can be defined by functions of the control signals such as the bias voltages, temperature, and so forth. An alternative to the function approximation is to read the parameter values from a table that is indexed by the control signals. In this case the simulator will interpolate between these measured data points at run-time.

Equivalent circuit representations are also used for modeling passive components. This can be quite straightforward, although it can become more complicated when distributed effects need to be accommodated, and the parasitics need to be included in the model. The CAD tool's built-in transmission-line components may not be appropriate for the structures under consideration. This can be the case for passive components used in silicon RFIC power amplifiers, for example, where the loss in the silicon substrate is significant, and not included in the standard models. In these circumstances using one or more network parameter components, such as S-, Z-, or Y-parameter blocks, may be a better choice. With these models, the variation of the parameter values with frequency, dimensions of the structure, and so forth, can be included in the network variables using equations derived from either the function approximation of the measured data, or from physical principles. As mentioned earlier, care with the function approximation needs to be taken to ensure that passivity and causality are maintained.

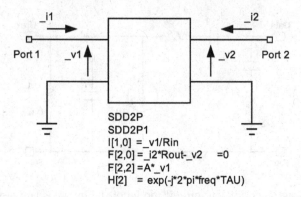

Fig. 8.4. Example of a two-port symbolically-defined device schematic component describing a voltage-controlled voltage source amplifier, using a time-delayed input signal, showing the port equations and an example of a weighting function.

8.5.2 The Symbolically-Defined Device

The symbolically-defined device (SDD) is an equation-based component that is well suited to the rapid prototyping of nonlinear device models. It is a component available in Agilent ADS.

The SDD is a multi-port component that can be placed directly into the circuit schematic in the simulator. A schematic symbol for an SDD is shown in Fig. 8.4. The current at a given port is defined using an equation, or *constitutive relationship*, that is expressed in the time domain. These equations can be functions of the port voltages or currents, their time derivatives or delayed versions of the port voltages or currents, or of external signals that can be referenced. They can be explicit or implicit expressions. For example, in an n-port SDD, the explicit representation of the current at the kth port is given by:

$$i_k = f(v_1, v_2, ..., v_n) \tag{8.13}$$

and the current defined by an implicit expression at the kth port is

$$f_k(v_1, v_2, ..., v_n, i_1, i_2, ..., i_n) = 0 \tag{8.14}$$

The explicit equation is a voltage-controlled expression, which is how model equations are often written. The explicit form is used in standard nodal analysis, and is very efficient to execute in the simulator, as the expression can be solved directly, and no new variables are created. In contrast, the implicit representation requires modified nodal analysis for its solution, adding an extra branch variable, i_k, and current equation to the set of algebraic equations that needs to be solved.

The weighting functions H[m] are used to defined time derivatives or time delays. The weighting function H[1] is the first time derivative, and it is a built-in function. A time delay can be constructed by defining the following weighting function:

$$H\,[2] = e^{(-j\omega\tau)} \tag{8.15}$$

where ω is the frequency and τ is the delay.

The equations used in the SDD can be the function approximations for the model functions described in earlier chapters, using the device gate and drain voltages as the controlling signals, for example. The expressions should be constructed with the usual considerations for simulator convergence in mind: the functions must be continuous in the voltages and currents, and should be differentiable with respect to the voltage and current. Additionally, the derivatives should be continuous, to aid convergence. An advantage of using the SDD as a vehicle for nonlinear model implementation is that the Jacobian is calculated within the component itself, during simulation. This avoids the lengthy time involved and potential for errors that arise when the modeling engineer has to create the Jacobian for the model by hand. The development and prototyping of the model is therefore significantly faster when using the SDD; often, this is also the final form of the model for distribution. The numerical efficiency of SDD devices is just slightly worse than similar models implemented in a high-level computer programming language.

Several practical examples of how to use an SDD component for nonlinear model implementation are provided in the Agilent ADS *User Guide: User-Defined Models* [15].

8.5.3 *The Frequency-Domain Defined Device*

The frequency-domain defined device (FDD) is an equation-based component that enables the construction of nonlinear device models in the frequency domain. This device is designed to work with the envelope simulator, and therefore enables models to be created quickly and efficiently for devices that are used in RF communications circuit and systems simulations. The FDD is a multi-port component that can be placed directly into the circuit schematic in the simulator. It is a component available in Agilent's ADS.

In the FDD, the spectral values of the current and voltage at a given port are written as algebraic functions of the spectral voltages and currents at other ports. This means that the voltages and currents are expressed as functions of frequency, as well as the other port signals. Delayed signals can also be included, allowing the propagation delays of signals in an RF power transmitter, for example, to be accommodated in the model description. The FDD is designed for the development of nonlinear behavioural models that are defined in the frequency domain, and several sub-circuit examples are provided in the Agilent ADS *User Guide: User-Defined Models* [15]. Since the FDD only functions in the frequency-domain, and the transistor models that we are developing for the design of RF power amplifiers and circuits will be used in the full range of available simulations, we shall not pursue the FDD for transistor models any further here.

8.5.4 High-Level Computer Language Model Implementation

So far we have discussed model implementation methods that do not require a formal computer programming structure. All modern CAD tools allow for the implementation of models using a high-level computer programming language, commonly ANSI-C. This is often the native language of the simulator implementation itself. This type of model implementation is the most complex but also the most flexible and powerful. The computer code can be written to manipulate the model's response, depending on its parameters and stimulus, and to provide enhanced function approximation control to improve the convergence and overall robustness of the model.

Models implemented with a high-level computer language can be either linear or nonlinear in nature, although the most common usage is in the implementation of nonlinear transistor models. There is no inherent limitation to the number of ports or terminals that the model must have in this kind of model implementation. Since the access to the simulator functions is more direct compared with schematic entry model implementations (equivalent circuit, SDD, FDD, and so forth), the modeling engineer has the ability to use the high-level computer language model implementation to gain access to state variables that determine the model's response currents and charges.

The details of how to write the C-code are different for each CAD tool package, and can be found in the written documentation that is provided by the CAD vendor. For the most part, once the model code is written, it is compiled and linked with the supplied object code that forms the simulator, or it is compiled into a dynamic link library file, which is loaded at the CAD tool boot time. Usually the model code will include the appropriate header files, which contain macros, type definitions, and interface function declarations. Once the newly developed models have been implemented in the CAD tool package environment, they can then be used in the same way as built-in elements.

At this stage of the model implementation process, the model formulation, list of model parameters, and model topology should already be complete and available. If the model is nonlinear, the partial derivatives of the non-linear functions with respect to the control voltages must be available for the definition of the Jacobian. The first step is to decide which of the model parameters need to be entered from the schematic. Error trapping code should be included to prevent the entry of model parameters that do not make physical sense or that could cause convergence problems during the simulation of the model. The symbol for the model that will be used in the schematic capture also needs to be constructed. The modeling engineer must decide where to insert the newly created model within the existing library and schematic palette structure of the CAD package. Typically, the newly created model becomes part of an existing custom library of models, facilitating its eventual distribution to the end-users.

More recently, CAD tools have been introduced with graphical user interface facilities that streamline the creation of the necessary computer code required to implement these linear and nonlinear models, simplifying the overall model implementation process.

8.5.5 *Model Implementation using Verilog-A*

As is often the case in the software world, new standardized higher-level languages are developed that allow a programmer to work at a higher layer of abstraction and generate the necessary computer code more efficiently. To address these issues for analogue modeling, an analogue hardware description language (AHDL) called Verilog-A has been developed that promises a portable, robust, and efficient platform for compact model development. To simplify the model implementation process, Verilog-A also includes the automatic differentiation of the functions describing the nonlinear currents and charges of the model, which are required to compute the Jacobian during solution of the nonlinear problem.

The use of this language allows the modeling engineer to concentrate on model development and model extraction activities, rather than focusing on underlying simulator-specific implementation details. In the past few years this language has emerged as the leading candidate for a new compact model development environment, and is beginning to be widely adopted [16]. The functional representation is very flexible, enabling complex expressions to be included in the model description easily. On the other hand, the ability to read file-based data in table form is presently quite limited. New language capabilities for Verilog-A are still being proposed [17]; Verilog-A offers some advantages in some aspects, and limitations in others. The development of Verilog-A is overseen by the Verilog-AMS Working Group [18], and updated standards are released on a regular basis.

The Verilog-A model implementation should be transparent to the end-user, as the resulting model will possess the same capabilities and qualities as a model that has been implemented using a high-level computer programming language, such as C-code. It is clear that a Verilog-A model implementation provides distinct advantages to the modeling engineer, but it also provides the model user with the flexibility to be able to change an existing model without having to delve into the details of C-code programming.

Verilog-A models are loaded in the simulator in a similar manner as C-coded models. For the most part, the loading process is automatic and the model user does not need to be concerned with it. Verilog-A models have the same attributes as other models with regard to the ability of the model engineer to distribute it. Nonlinear transistor models implemented in Verilog-A require that the nonlinear currents and charges be described in a similar fashion as in a C-code implementation. As with SDD and FDD model implementations, the speed penalty imposed by Verilog-A models compared with high-level computer language implementations is quite small.

8.6 Building a Model Library

After extracting, generating, and validating a model the only remaining task is its distribution. Models are often a collection of files, containing the parameters, equations, schematic, and potentially, layout descriptions. To aid in the distribution and delivery, most CAD tools provide an infrastructure for defining a collection, or library, of models and their automatic installation. Without the defined infrastructure, as specified by the CAD vendor, it may be impossible to install multiple libraries, as their installations may conflict with each other.

When delivering a library, details pertaining to IP must be examined. It is often the case that the model formulation, layout information, or even the model parameters themselves are confidential. If necessary, this information can be encoded, a feature often provided by the CAD vendor or developed by the modeling engineer. Encoding allows the designer to use the models without having access to confidential details. In many cases there is no need to hide the transistor model parameters as they have no proprietary value. Such an encoding process may find further use when there is a need to deliver a compact model of an entire circuit. For a 50 Ω RFIC power amplifier, the need to protect the IP associated with the circuit topology of the IC necessitates the encoding of the circuit. The encoding of such a model needs to be done in a way that protects the owner of the IP while providing the end-user with access to an accurate model of the circuit. One simple method of protecting IP is to generate an S-parameter representation of large portions of the circuit. If properly implemented, the S-parameters can be freely distributed with little risk.

The model library may include layout instances of transistors, associated with the compact model. This is more usual when the model is being used in the context of an IC design process. As the integration level increases, and the number of components in the circuit being designed becomes larger, the need for some kind of layout-versus-schematic (LVS) consistency check becomes important. For small RF or microwave circuits, full synchronization between layout and schematic can be achieved without the use of an LVS tool. Special care must be taken to ensure that any specific LVS requirements are taken into account during the development of the library. To complement a library of models that contains both electrical and layout component definitions, the delivery of a design rule checker (DRC) provides the design engineer with the ability to ensure that the proposed circuit complies with all the manufacturing rules of the given technology.

Once the decision is made to release a library of models, the librarian needs to assess the important but often overlooked issues related to customer support. To facilitate the overall process of installing the library, clear and concise installation instructions or help files need to be made available to the library users. Libraries of models are sometimes composed of many elements with their appropriate options and definitions (in both schematic and layout), so a manual describing each element of the library is often necessary. In addition, the comparisons of model versus measured data for a given model should be made available, in the form of a validation report, which ultimately will provided needed information of the model to the design engineer.

When planning for the release and support of a library, the number of computer platforms and operating systems that need to be supported must be carefully considered. The larger the number, the more effort will need to be expended on library verification. The more portable the library is between computer platforms, the less testing and support will be required. The modeling engineer should establish a procedure for library verification, quality control and release that is simple to implement and maintain.

8.7 Portability of Models and Future Trends

A layer of complexity that may be added to the modeling engineer's job is the implementation of models and libraries in multiple design environments and computer operating systems. This is especially true when referring to a library that is released openly to support a diverse customer base, which will in general be using more than one design environment.

The portability of a model is influenced by two distinct but equally necessary layers of model implementation in a CAD package: the user interface (UI), and the linkage to the algorithms coded in the simulator. The UI layer is used to display the model parameters in the schematic view of the design tool, to define ports, and to provide linkage to a layout representation of the transistor if required. Some CAD package vendors develop their own programming language and functions for the implementation of models in the UI layer, while others adopt standard UI tools such as Tcl/Tk. On the other hand, the implementation layer uses standard programming languages, but the method for interfacing with the simulator is different from one CAD tool package to another. In the simulator implementation layer, all the nonlinear equations that describe the model and the partial derivatives required for convergence are implemented. This layer also interfaces with the UI layer to pass all the necessary model parameters and to return any error messages encountered during the solution of the model.

The support of multiple CAD tools is challenging for the modeling engineer because the UI and simulator implementation layers are unique to each design tool, and to implement a model the engineer must become fluent in the layers of each design tool they intend to support. The simplification of these tasks is at the essence of the drive to improve model portability. Today's market conditions are such that models and libraries are generally not portable between simulator environments, and the model distribution is a more difficult task to accomplish.

To complete the task of implementing models in several circuit simulators, the models must be coded by hand in each environment and extensively tested. Depending on the complexity of the model and the ease of implementation of the given design tool, this endeavor may take many months of non-value-added activities, which instead could be dedicated to improve the inherent accuracy limitations of the models.

One method of overcoming this problem would be to write the model in a language or format that is non-specific to a given simulator, but that can be imported by the target CAD tools. The analogue hardware description language Verilog-A has been adopted by many modeling engineers, as it appears to be a portable, robust, and efficient platform for compact model development. Even though the usage of Verilog-A shows significant promise, there are still some model implementation issues that need to be resolved, as outlined in Section 8.5.5. The adoption of Verilog-A as a standard model development interface has been slower for simulators targeted for microwave design, than for low-frequency analogue IC design.

To allow a model developer access to the benefits of Verilog-A, hybrid flows have emerged, where programs are employed to take the Verilog-A code and translate it to specific simulators. One example of a public-domain tool is ADMS, which employs XML document-type definition (DTD) to translate directly to the simulation environment of choice [19]. Once the XML DTD is complete, then models are written once and automatically implemented into the various simulators. With this type of hybrid flow, models can be developed efficiently within Verilog-A and the time-consuming and error-prone task of porting the code to multiple CAD packages is completed using an automatic translator.

It is not yet clear if Verilog-A will become the standard modeling language for compact microwave transistor models. Alternative approaches, such as the one described above using ADMS, will continue to gain acceptance until the usage of Verilog-A or another programming environment becomes prevalent. It is clear that the pressure for improving model portability will only increase as the usage of models becomes more widespread and the need to have models and model libraries in multiple commercially available CAD packages continues.

References

[1] K. S. Kundert, *The Designer's Guide to SPICE & SPECTRE*. New York, NY: Springer, 1995.

[2] L. W. Nagel and D. O. Pederson, "SPICE (Simulation Program with Integrated Circuit Emphasis)," University of California, Berkeley, ERL Memo No. ERL M382, Tech. Rep., Apr. 1973.

[3] H. K. Gummel and H. C. Poon, "An integral charge-control model of the bipolar transistor," *Bell System Tech. J.*, 49, pp. 827–852, May 1970.

[4] H. Shichman and D. A. Hodges, "Modeling and simulation of insulated-gate field-effect transistor switching circuits," *IEEE J. Solid State Circuits*, SC-3, no. 3, pp. 285–289, Sept. 1968.

[5] D. O. Pederson, "A historical review of circuit simulation," *IEEE Trans. Circuits Syst.*, 31, no. 1, pp. 103–111, Jan. 1984.

[6] K. S. Kundert and A. Sangiovanni-Vincentelli, "Simulation of nonlinear circuits in the frequency domain," *IEEE Trans. Computer-Aided Design*, 5, no. 4,

pp. 521–535, Oct. 1986.

[7] D. Sharrit, "New method of analysis for communication systems," in *IEEE MTT-S Int. Microwave Symp. Workshop, 'Nonlinear CAD'*, San Francisco, CA, June 1996.

[8] E. Ngoya and R. Larcheveque, "Envelop transient analysis: a new method for the transient and steady-state analysis of microwave communication circuits and systems," in *IEEE MTT-S Int. Microwave Symp. Dig.*, San Francisco, CA, June 1996, pp. 1365–1368.

[9] D. Sharrit, "Method for simulating a circuit," U.S. Patent 5 588 142, May 12, 1995.

[10] W. H. Press, S. A. Teukolsky, W. T. Vetterling, and B. P. Flannery, *Numerical Recipes in C*. Cambridge, UK: Cambridge University Press, 1992.

[11] V. Rizzoli, C. Cecchetti, A. Lipparini, and F. Mastri, "General-purpose harmonic balance analysis of nonlinear microwave circuits under multitone excitation," *IEEE Trans. Microwave Theory Tech.*, 36, no. 12, pp. 1650–1660, Dec. 1988.

[12] *Matlab Optimization Toolbox*, The Mathworks, Inc.

[13] T. Weller, L. Dunleavy, and W. Clause, "Avoid frequency extrapolation errors," *Microwaves & RF*, pp. 98–104, Sept. 2002.

[14] D. E. Root, "Charge modeling and conservation laws," in *Asia-Pacific Microwave Conference Workshop WS2, 'Modeling and characterization of Microwave devices and packages'*, Sydney, Australia, June 1999.

[15] *ADS User Guide: User-Defined Models*, Agilent Technologies, Agilent EEsof EDA.

[16] K. S. Kundert and O. Zinke, *The Designer's Guide to Verilog-AMS*. New York, NY: Springer, 2004.

[17] L. Lemaitre, G. Coram, C. McAndrew, and K. Kundert, "Extensions to Verilog-A to support compact device modeling," in *Proc. IEEE Int. Workshop on Behavioral Modeling and Simulation (BMAS)*, San Jose, CA, Oct. 2003.

[18] http://www.verilog.org/

[19] L. Lemaitre, C. McAndrew, and S. Hamm, "ADMS automatic device model synthesizer," in *Proc. IEEE Custom IC Conf.*, Orlando, FL, May 2002.

9

Model Validation

9.1 Introduction

Model validation is the process of determining the degree to which a model is an accurate depiction of the real world from the perspective of the intended use of the model. Model validation has different purposes depending on the person's perspective. For example, for the modeling engineer the main purpose of model validation is to guide the development and refinement of a model, while for the user of the model, its validation provides a confidence level for the accuracy and limitations of the model. A solid and comprehensive model validation exercise results in increased confidence in assumptions behind the construction of the model and a higher level of assurance of its predictive capabilities outside the validation domain.

There is a subtle but important difference between the validation and verification of a model [1–3]. The verification of the model is the process by which the implementation of the model in the CAD package is demonstrated to be consistent with the equations and topology of the model, and to ensure that the model produces the expected results. In other words, the verification ensures that the model was properly implemented in the circuit simulator, as has been outlined in Section 8.4. On the other hand, model validation is the process by which the model simulation results are compared with an independent set of data not used during the model extraction. In essence, the model validation provides confidence and guidance on the predictive qualities of the model. An example of model validation is the comparison of model versus measured loadpull contours for a model in which only DC–IV characteristics and small-signal S-parameters were used during the model extraction.

Before the model validation process can begin, the modeling engineer must complete the model verification phase. For linear models, the model verification is fairly simple, but for nonlinear transistor models, this phase of the model development is often quite complex and lengthy.

A compromise has to be made between the cost of validating the model and the desired level of confidence required for the application. To validate a model over the entire possible range of usage is not practical, therefore a judicious approach should be taken to arrive at the necessary and sufficient criteria for model validation. The model validation is carried out by performing specific measurements and model comparisons that will produce the highest level of confidence in the model with the minimum level of work.

A common misconception is to equate model validation with model tuning. If an *ad hoc* fit of the measured data is obtained by tuning a model, it does not imply that the model was validated. A tuned or adjusted model can still be simulated with an independent set of model validation data, as long as it provides the required level of confidence on the predictive capabilities of the model.

In this chapter we will elaborate on the concept of model uncertainty and the possible sources of error during the model extraction and validation process. This will be followed by several validation examples of linear and nonlinear models. The first example will demonstrate the validation of the passive matching networks of a high-power RF transistor by comparing the modeled and simulated responses of a complex structure used within a 140W 2.1 GHz air-cavity ceramic packaged transistor. Four additional examples will be presented to illustrate different aspects of nonlinear transistor model validation. The second example of the section is the validation of an unmatched silicon LDMOS die utilizing a loadpull system. The third example is a GaAs PHEMT power transistor used for wireless handset applications, and the validation includes DC and loadpull comparisons between measurement and model predictions. The fourth example will combine aspects of the first and second examples, that is, the linear model of internal matching components used in commercial high-power RF transistors and the nonlinear transistor model, in the validation of the product model for a commercially available packaged RF transistor. The last example in the chapter will present an alternative approach of model validation, by using time-domain data instead of the more common and traditional methods based on frequency-domain data and measurement techniques.

9.2 Model Uncertainty and Sources of Error

From the 'Laws of Simulation and Modeling' mentioned earlier, we should bear in mind that, 'All models are inaccurate, it is just a matter of degree.'

Deciding the correctness of a model during the validation phase is somewhat subjective. The modeling engineer usually makes a judgment on the quality of the model based on the results obtained during the model validation phase of the model development process. Some objectivity is also introduced to quantify the quality of a model during the model validation, which usually entails the use of some type of statistical test or mathematical procedure that describes the differences between the simulated model response and measured data.

Inevitably, once we decide to use measured data to compare the results of the model, and, therefore, make a judgment on its accuracy, we need to address the issue of model uncertainty and the different sources of errors introduced during the development, extraction, and validation of the model. To understand the sources of error and uncertainty, we need to ascertain the validity of the data. By this we mean that we need to ensure that the data used during the different phases of model development are adequate and correct. We can express this truism in other words: there is no substitute for good data.

By definition models are simplifications of the physical world. It is the job of the modeling engineer to develop a mathematical framework from theories that capture the essence of the behaviour being modeled. The limits and assumptions that were used to derive the model are a source of model inaccuracy. An example of this is a nonlinear model that is constructed around electrical device physics principles alone. If such a model does not account for self-heating effects, the limitations of the theory used to describe it will introduce significant sources of error.

Another aspect of model construction that introduces inaccuracies and uncertainties to models is the lack of an absolutely correct model structure. This problem is quite common in over-determined systems, in which a vast number of model topologies could generate a comparable quality of fit of the measured data used during the model development and extraction process. Simplification of the problem by using geometry or boundary conditions during the model construction is also another source of model uncertainty.

With today's proliferation of commercially available circuit simulators, further sources of model inaccuracy and uncertainty are those introduced by the differences in the algorithms used to solve the linear and nonlinear problems. For example, different implementations of numerical methods and techniques to solve nonlinear set of equations often result in differences in robustness of the convergence, and slight differences in the converged solution. In addition, inherent (or inevitable) limitations of the numerical methods also introduce errors to the final solution of the problem, and, if not properly addressed, can compromise the overall accuracy of the model. Further examples of these limitations are the different convergence settings used during large-signal analysis or the number of harmonics used during a harmonic balance simulation, all of which vary from simulator to simulator, and generally affect convergence and stability of the steady-state solution of the large-signal problem.

When generating simulated data during the model validation process it is very important that simplifications of the physical environment during simulation do not affect the results. An example of introducing a difference between measured and simulated data is ignoring harmonic terminations during the loadpull model validation procedure. It is well known that the performance of a transistor is highly dependent on the impedance terminations at the fundamental frequency of operation and its harmonics. Moreover, the low frequency impedance presented to the input and output of the transistor will have a significant effect on the thermal and electrical dynamics of the device, and hence has the potential for severe consequences to the transistor distortion characteristics. Therefore, to compare the predictive capability of models, the impedances presented by the loadpull system must be properly captured and duplicated in the simulator [4]. The measurement thermal boundary conditions must also be replicated correctly during the extraction and validation phases of the model development. Any simplifications here will lead to inconsistencies between the model predictions and the validation measurements.

A very common technique used to simplify the modeling of very complex structures or devices is to partition the model into its linear and nonlinear parts. Additional segmentation of the problem to reduce its overall complexity can also introduce inaccuracies into the model [5]. There are several issues to consider when implementing a segmentation approach to the simulation of a large complex circuit. As mentioned in Chapter 4, a set of procedures that allow the individual circuit components to be separated from one another must be devised. The separation procedure must not perturb the behaviour of the components; that is, the component when analyzed by itself must operate in the same way as it does within the larger, more complex structure. The critical issue with the segmentation approach is that the planes at which the circuit is divided must be carefully selected such that the field configurations on either side of the plane match. If the field configurations are not the same at the segmented plane, then a discontinuity is artificially introduced during the analysis.

An often-overlooked aspect of the validity and consistency of the data is the issue of process variations and how they affect the selection of an appropriate device-under-test to generate, extract and validate a model. During the model extraction, the particular part used might not be representative of the actual product population distribution, and adjustments may need to be made to account for the variation, in an attempt to arrive at a nominal model; that is, a model that represents a typical device in the population. Often, the circuit designer would like the model to be able to accommodate the variations in the fabrication process, in order to give some indication of the yield of the design. The device models must be extracted, or other measurements made, over the range of the process, to enable some statistical variation of the device parameters to be included in the model. Some comments on statistical modeling are found in Chapter 6.

Once the development of a new transistor model is complete, the modeling engineer is often required to generate many transistor models for library development. Automatic extraction can be used to reduce library development time. In many instances, models that contain numerous model parameters suffer from non-deterministic model parameter extraction techniques and procedures. The uncertainty in the value of a given model parameter can relate directly to the level of confidence in the model.

9.3 Validation Criteria for Power Amplifier Design

So far we have presented the basic definitions of model validation, model verification and *ad hoc* models, followed by a discussion of the different sources of uncertainty and error encountered during modeling of the transistor. In this section, we will present model validation from the perspectives of the power amplifier designer and of the modeling engineer. The goodness and usefulness of a model must be assessed from the the perspective of the user; the accuracy of any model must be defined in the context in which it is used.

9.3.1 The Power Amplifier Designer Perspective

The power amplifier designer is presented with the task of weighing different options that trade off multiple device metrics. For example, the designer is often faced with a compromise between linearity and efficiency, and therefore needs to have access to models that can help decide on the best approach that will allow him to make such choices quickly, efficiently, and with confidence. To expect a perfectly accurate model is unrealistic, and inevitably the question of value needs to be addressed: does the use of the model provide an added value to the design process?

An important feature that a model must possess, for it to be a useful PA design tool, is the ability to predict the source and load impedance values that will produce a certain level of large-signal performance. This capability allows the designer to make the necessary trade-off of performance in the impedance space.

The capability of a model will affect the way it is used. A very simple model should be capable of predicting the correct trends over the specified independent variables, that is, be able to duplicate the device's measured behaviour over the range of frequency, quiescent conditions, power level, and so forth. A different scenario facing the user of a model is the troubleshooting of power amplifier performance issues. In this case the designer is using the model to investigate ways to resolve problems. Having access to models at this stage can provide an in-depth view of the problem. Models can provide an advantage over direct measurements: they can be used to display information at reference planes that are impossible to achieve in real life. Examples of this are time-domain waveforms at multiple locations in the circuit, and the dynamic trajectory of the RF signal at the intrinsic reference plane of the transistor.

As the model accuracy improves and the usefulness of the model increases, the designer can take advantage of the model to perform simulations more relevant to the power amplifier design. In this case, the model is being used to make specific design trade-offs, which include, among many others, the design of the bias network used to improve the PA performance, the determination of the small-signal frequency response of the circuit, fundamental impedance termination for output power, gain and efficiency under a single- and two-tone excitation, and performance simulations that will allow the designer to make decisions about the sizing of the transistors and the PA line-up budget. More advanced simulations related to power amplifier design can also be performed, for example, determination of distortion characteristics under multi-carrier or digitally modulated signals, the effect of harmonic termination on transistor performance, low frequency terminations and their influence on video bandwidth and distortion characteristics, and the incorporation of electro-thermal phenomena.

As we continue to move up the ladder of model complexity and usefulness, the model can be used to aid understanding and help the designer make system-level design decisions, such as: including different power amplifier linearization techniques, considering thermal constraints, bias and signal-level temperature compensation, and comparing the benefits of different high-efficiency circuit architectures.

Another important factor that the power amplifier designer uses to evaluate the usefulness of a model is its ability to account for statistical process variations of the transistor. Having access to a statistically-based model allows the designer not only to design a power amplifier that meets the design targets, but also to determine the factors, conditions and sources of variation which could influence the yield and manufacturing variations of the product. With this information, changes can be made to the power amplifier design to minimize the effects of process and manufacturing variations, and ensure the centering of the design.

To a lesser extent the speed of simulation of a model will influence the way it can be used. Small-signal analysis is the most common method of simulation; it is easily understood and fast, and hence it lends itself to optimization and frequent analysis. As the complexity of the analysis goes up, so will the simulation time. If it is too long, circuit designers tend to analyze only the final design.

An important issue to the model user is the timeliness of the delivery of the model. Very accurate and sophisticated models delivered outside the design window provide little help. On the other hand, a model that is not as precise but is made available at the right time has the potential to aid the design process.

A recent trend in power amplifier design is to use advanced circuit architectures for high-efficiency power amplifiers, including Doherty circuits, class F, bias-modulation, and so forth. The high level of complexity of these high-efficiency PA architectures require a more complicated design process, which presents challenges for and demands on transistor models. Like traditional PA architectures and modes of operation (such as class A, AB, and so forth), these new high-efficiency architectures PA require models that are capable of predicting the saturated power of the device when tuned for maximum output power, and that the transistor model works well not only at the fundamental frequency used in the design, but also at low frequency and up to multiple harmonics.

9.3.2 Model Validation from the Modeler's Perspective

The viewpoint of the modeling engineer can be somewhat different from that of his customer, the design engineer, when it comes to validating a compact model. The concepts of *validation* – 'are we doing the right thing?'

– and *verification* – 'are we doing the thing right?' – can take on different meanings. Sometimes, the validation phase for the modeler is no more than finding out what the compact model will be used for, to guide his judgment as to what form the model needs to take. Whereas this is an important specification or requirement of the model, as we have discussed already, there is more to validation than simply capturing the requirements. The verification phase, as noted earlier, comprises a set of tests that the modeling engineer will perform during and after the construction and implementation of the model, to assure himself that the architecture of the model is sufficient for the design tasks, and the implementation of the various functions in the simulator is correct.

The specifications describing how the model will be used in design help the modeling engineer to decide how complex the model needs to be. For instance, if the design is a linear circuit, then a small-signal AC or S-parameter simulation will suffice, and the model needs only to mimic the measured S-parameters of the transistor. The model structure could then be an equivalent circuit model, requiring the extraction of the circuit parameter values, or it could be a table of measured data, indexed appropriately during simulation. The modeling engineer then needs to verify that the model has the correct equivalent circuit topology and the correct parameter values, or that the table returns the correct S-parameters when indexed by the simulator frequency and the bias voltages. From the power amplifier designer's viewpoint, this verification may also be sufficient validation of the model in use.

In contrast, when the circuit is a large-signal, nonlinear design, such as a power amplifier, the model must be capable of predicting the nonlinear behaviour that the circuit designer specifies. Again, this specification will guide the modeling engineer's choice of model structure, to accommodate the requirements. We discussed the various model architecture options that are available to the modeling engineer in more detail in Chapter 2, but in general terms, we can choose from physically-based models, such as the BSIM MOSFET model, through equivalent circuit models, which form the basis of most compact models, to 'black-box' or behavioural models. All of these model structures are capable of predicting large-signal transistor behaviour to a greater or lesser degree of accuracy.

The physically-based models can contain many parameters, making model extraction an involved process, and the job of verification of the model implementation very arduous. The black-box models can be very simple, and hence straightforward to implement and verify, but they will generally only mimic the measured input-output relationship, or data of a similar form or class. So, unless these measured data are of a similar form that is specified by the designer, it is doubtful whether such a model will provide value to the designer.

As modeling engineers, we have adopted a more pragmatic view of verification and validation when constructing compact models for power amplifier design. We capture the designers' requirements of the model, which often include scaling considerations, thermal behaviour, and so forth, in addition to the usual demands for accurate prediction of large-signal behaviour at various load conditions when driven by digitally modulated signals. Clearly, we cannot hope to cover every possible contingency. Instead, we focus on describing the linear, thermal, and nonlinear parts of the packaged transistor by models that are based in fundamental circuit theory and thermal physics, so that the compact model is capable of accommodating a wide range of electrical and thermal stimuli, and can therefore be used in a variety of circuit configurations. The model implementation in the simulator is verified using test cases for the mathematical algorithms, and by comparing simulation with a set of measurements of the form used to generate the model: bias-dependent S-parameters, over a limited voltage and frequency range. The validation of the model is specific to the designer's requirements, and often includes high-power loadpull measurements using a variety of test signals, including single-tone, two-tone, and digitally modulated excitations. In this way we can exercise the model using signals that were not used in the extraction, and that are of value to the designer.

9.3.3 Model Validation Criteria

As we mentioned in the previous sub-sections, the designer can use the model in many different ways, each one with a different level of expectation on the accuracy of the model, depending on the complexity of the analysis and how the model will be used. The model validation criteria will, in general, be determined by the application. For example, if the model is only to be used for predicting small-signal behaviour, then the model will be validated against a set of S-parameter measurements. If the model is to be used for predicting large-signal performance, then the model validation procedure needs to be broader in scope, including loadpull measurements, for example, as outlined above. The broader the model scope, the greater the range of the validation criteria to establish the accuracy of the model.

9.4 Validation of the Model Against Measurements

In this section we will describe several examples of model validation pertinent to high power RF and microwave transistors. The examples will range from the model validation of linear networks, an unmatched LDMOS transistor, a GaAs PHEMT power amplifier, an internally matched 900 MHz 160 W LDMOS packaged transistor and a novel way to validate nonlinear transistor models using time-domain measurements instead of the traditional frequency-domain techniques.

9.4.1 Passive Components Within a Packaged High-Power Transistor

As an example of the simulation techniques presented in Chapter 4, we generate a model of the package and matching networks of an air-cavity internally matched transistor, designed for use within W-CDMA base-stations operating in a frequency range of 2.11–2.17 GHz [6]. The high-power transistor is capable of outputting 30 W (average) when operating with a W-CDMA signal and 140 W of output power when operating in continuous-wave mode (1–dB compression point).

The intricate matching circuitry formed by the bondwire and MOS capacitors is illustrated in Fig. 9.1. Rows of parallel bondwires form arrays which interconnect the dies, MOS capacitors and the package. The packaged transistor contains three active dies each having a gate periphery of approximately 80 mm. The matching network consists of the package, six MOS capacitors, a capacitor integrated inside the window-frame and 189 bondwires. All bondwires have a diameter of 50 μm and are made of an aluminium alloy. The matching networks for this transistor are manufactured by specifying the values of the MOS capacitors and by specifying the number of bondwires, controlling the three-dimensional shapes of the bondwires, and the distance between neighboring wires.

Two types of MOS capacitor are employed in this device; a 22 pF capacitor is used to create part of the matching network on the gate side of the transistor and a 350 pF capacitor is used on the drain side. The 350 pF capacitor forms a part of the matching network whose topology is termed *shunt-L*. There are two arrays of wires connected to the drain of the transistor, one establishing a connection to the 350 pF capacitor and the other to the package leadframe. The section that connects to the 350 pF capacitor is designed such that the array of bondwires attached to the drain resonates out the drain-source capacitance (C_{ds}). The value of the MOS capacitor on the drain side must be large enough so that within the operating bandwidth,

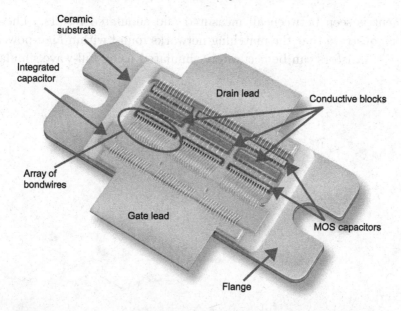

Ceramic substrate

Integrated capacitor

Drain lead

Conductive blocks

Array of bondwires

Gate lead

MOS capacitors

Flange

Fig. 9.1. A photograph of the test-package containing the package and matching networks [6]. © 2006 IEEE. Reprinted with permission.

the shunt wires are effectively shorted to ground, while still blocking the flow of DC current.

To compare simulated and measured results, a special version of the transistor was manufactured that contained highly conductive metal blocks in place of the transistor die, as shown in Fig. 9.1. This transforms the packaged device into a resonant circuit, which is very sensitive to the loss and reactance of the circuit components. Thus, the vector network analyzer can be used to measure the S-parameters of the passive structure directly.

The task of generating a model for this device begins with capturing the geometry of the bondwires and the relative positions of the MOS capacitors and conductive blocks within the package. A single wire in each array was selected to represent all the other wires. Slight variations due to manufacturing tolerances in the geometrical profiles were assumed to be negligible. In total, twelve simulations were required to characterize the total device. An illustration outlining the various simulations and how they represent the entire packaged transistor is provided in Fig. 9.2. Once all of the simulations were completed, the results were incorporated into a linear circuit simulator. A schematic representing the final model is provided in Fig. 9.3.

Measurements of the device were performed and a comparison between measured and simulated S-parameters is presented in Fig. 9.4. Excellent agreement is seen between all measured and simulated results. These results demonstrate that the matching networks found within high-power microwave transistors can be accurately simulated using full-wave simulators.

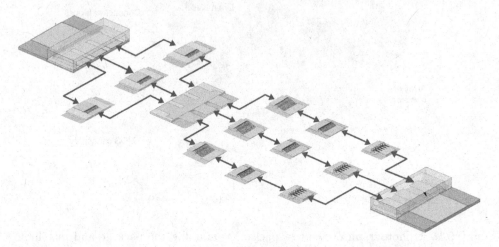

Fig. 9.2. An illustration of the individual components of the test-package and how the segments fit together to form the model of the entire test-package [6]. © 2006 IEEE. Reprinted with permission.

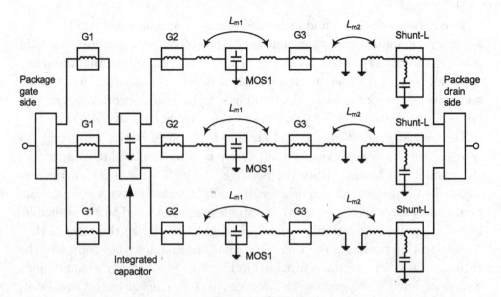

Fig. 9.3. The equivalent circuit representing the test-package illustrated in Fig. 9.1 [6]. © 2006 IEEE. Reprinted with permission.

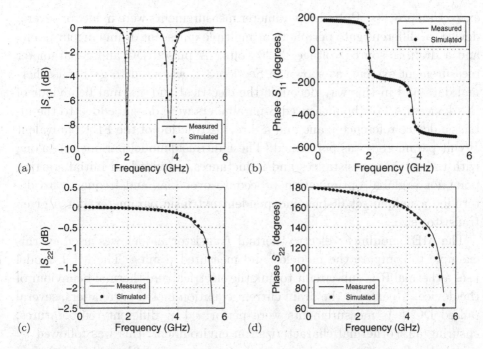

Fig. 9.4. The simulated and measured input and output reflection coefficients of the device illustrated in Fig. 9.1 [6]. © 2006 IEEE. Reprinted with permission.

9.4.2 Unmatched Silicon LDMOS Loadpull Model Example

This example focuses on the model validation of an unmatched 10.2 mm gate periphery LDMOS transistor. When modeling unmatched transistors, the most common measurement technique used to validate them is the automated loadpull. The flexibility and ability to measure accurately small-signal and large-signal figures of merit as a function of the input and output impedances makes the loadpull measurement system one of the most powerful tools, and perhaps the most commonly used measurement technique for model validation.

Before going into the details of the model validation procedure and a comparison of the modeled versus measured performance, we shall review the model extraction procedure used to generate the nonlinear transistor models. We have already highlighted the importance of obtaining unique and accurate extrinsic resistances and inductances to allow us to de-embed to the intrinsic transistor model. There are many techniques used to extract the extrinsic elements. A hybrid of the traditional cold-FET direct extraction modeling technique and a multi-bias hot-FET small-signal S-parameter optimization of the intrinsic model parameters was used during this model extraction process. Pulsed S-parameter measurements were made for several devices of different gate periphery at multiple quiescent drain current points and a drain-to-source voltage of 26 volts. A pulsed DC and S-parameter measurement system, as shown in Section 3.4, was used to gather isothermal data and in this way decouple the electrical and thermal behaviour of the devices. All of the measured S-parameters were de-embedded to the intrinsic device reference plane and a direct extraction of the FET equivalent circuit parameters was performed. These intrinsic model parameters, along with the extrinsic resistances and inductances, provided an initial starting point for a global optimization procedure over bias and frequency to determine a unique set of bias-independent extrinsic parameters for a given transistor size.

The MET nonlinear electro-thermal transistor model was used in this example to compare the modeled and measured results. The MET model uses a thermal R-C sub-circuit to link the electrical and thermal behaviour of the device. To extract the drain current equation model parameters, several pulsed DC I–V measurements were performed at different temperatures, ensuring an isothermal characterization environment, this was followed by a global optimization over bias and temperature of all the pulsed DC I–V data. From the pulsed S-parameter data, the capacitance versus voltage behaviour (C–V) can be obtained by fitting small-signal models over bias at each temperature. Then the MET model parameters that describe the C–V behaviour can be obtained.

The validation process involves measuring loadpull data under single-tone and two-tone stimuli on the same device that was used during the model extraction process, and in this way avoiding any issues related to device variability. The devices were measured with an automated loadpull system as described in Section 3.5.1. The input impedance was selected to conjugately-match the transistor and the load impedance was selected for a good trade-off between power-added efficiency and output power. Once the optimum input and output impedances were selected, the input power was swept from essentially small-signal to 3 dB gain compression, at which time gain, output power, efficiency, and intermodulation distortion (two-tone only) were measured. All loadpull measurements were performed at 2.14 GHz, and the two-tone loadpull measurements were performed with a tone separation of 100 kHz. The transistor was mounted in a 50 Ω test-fixture as shown in Section 3.3.3. The bias conditions were typical for wireless infrastructure transistors, that is, a drain-to-source voltage of 26 V, at several quiescent bias points yielding drain current densities ranging from 2 to 12 mA per mm of gate periphery.

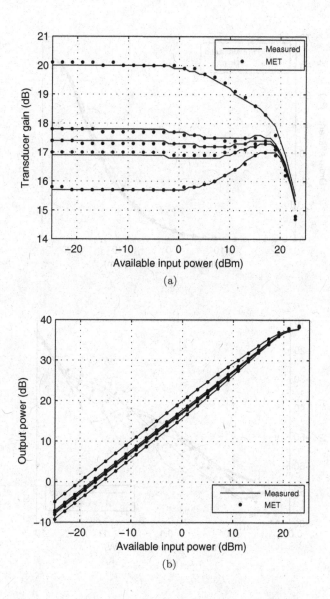

Fig. 9.5. Single-tone transducer gain (a) and output power (b) of MET LDMOS modeled versus measured data of 10.2 mm LDMOS FET [7]. © 2002 IEEE. Reprinted with permission.

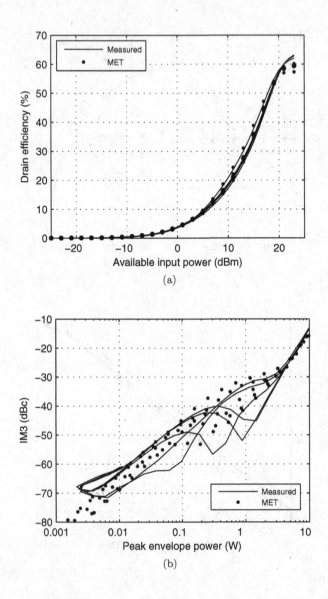

Fig. 9.6. Single-tone drain efficiency (a) and two-tone third order intermodulation distortion (b) of MET LDMOS modeled versus measured data of 10.2 mm LDMOS FET [7]. © 2002 IEEE. Reprinted with permission.

In this validation exercise, the measured loadpull data are compared with the simulated MET LDMOS model data under single-tone and two-tone excitations at five different current densities. Figures 9.5–9.6(a) show MET model predictions for transducer gain, output power, and drain efficiency versus available input power under one-tone stimuli. Figure 9.6(b) shows MET 3rd-order intermodulation distortion predictions versus peak envelope output power. These results show good agreement between measured and modeled data. The quality of the model reassures the model user of its predictive capabilities. The results show good agreement as the device moves into gain compression, as well as very good predictions of the transducer gain at multiple biases.

9.4.3 GaAs E-PHEMT Power Transistor Validation Example

In power amplifier design for wireless handset (and base-station) applications, the power amplifier should be as efficient as possible, while delivering as much power as possible. This has led to the adoption of higher efficiency amplifier modes such as Class AB, B, and so forth, which have higher drive requirements than a Class-A power amplifier at their optimum efficiency conditions, placing a premium on device gain, which has resulted in the use of technologies such as GaAs MESFET and PHEMT in wireless handset PA applications, even around 2 GHz.

In this example we shall describe the extraction and validation of a compact model of a GaAs enhancement-mode PHEMT transistor designed for use in an integrated circuit power amplifier for wireless handset applications. The power amplifier is biased in deep class AB and can deliver around 1–2 watts of RF output at 1.9 GHz; it has been adapted for CDMA and GSM modulation schemes.

The compact model must be capable of describing small- and large-signal device operation in Class AB or B bias conditions, so the threshold region of the FET characteristics must be modeled accurately. Typically, FET compact models use a hard switch at the threshold or pinch-off voltage V_T in the expression for the drain current characteristics. In addition to being inaccurate in the threshold region, a switch function can cause simulator difficulties unless the first derivative is continuous. In this compact model we have chosen to use the Chalmers University or 'Angelov' model for the drain current expression [8], because of its smooth transition from 'off' to 'on' in the threshold region.

The model must also be capable of describing accurately the high-order distortions at high and low (backed-off) drive conditions, in other words, it must be able to predict the ACLR performance of the power amplifier. A continuous or algebraic, continuously-differentiable expression to at least 3rd-order for the drain current model is required for this to be achieved at low drive. Further, Staudinger *et al.* [9] have demonstrated that a correctly formulated conservative charge model is required for accurate ACLR predictions. Additionally, the model must be able to predict accurately the RF output power and PAE under realistic load conditions, to be of value to the PA designer; therefore the dynamic electro-thermal coupling is included in the model description. The compact model follows the charge conservative approach outlined in Chapter 6, and includes the essential nonlinear features described therein.

The power transistor itself is a multi-finger device, whereas the transistor used for the model extraction is a two-finger device, of a total gate periphery small enough to be stable during S-parameter measurement, but of the same unit gate width as the power transistor. The extracted transistor model is also used in the design of the biasing and driver transistors in the PA IC. The scaling capability of the extracted model is therefore a necessary feature, in addition to the usual accuracy requirements. The extrinsic part of the model is based on the small-signal model reported in [10], and the values of the extrinsic components were found using cold-FET RF measurements [10] made on an array of transistors of varying unit gate width and number of gate fingers, to establish empirical scaling rules.

The model was extracted from on-wafer bias-dependent CW S-parameters taken over a wide V_{gs}–V_{ds} voltage space, at a constant chuck temperature, following the principles outlined in Chapter 6. The thermal resistance was optimized during the drain current parameter extraction, using an estimate of the GaAs thermal resistance as a starting value. The thermal time constant was defined as 1 microsecond in the dynamic electro-thermal equation in the model, to give an approximation to the dispersion of the drain current characteristic.

The drain current as a function of gate bias is shown in Fig. 9.7, illustrating the smooth, differentiable curve in the threshold region. The detailed shape of the curve is also well modeled, by adjustment of the ψ function in the Angelov drain current expression. This accurate, smooth curve is essential for the accurate prediction of distortions in Class AB/B bias conditions.

The large-signal model validation was performed using fundamental CW loadpull measurements at 1.9 GHz on the discrete power transistor, and by measurement of the distortion using ACLR measurements taken as a function of the drive-up power [11].

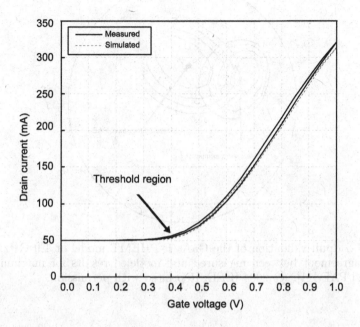

Fig. 9.7. Comparison between measured and modeled drain current versus gate voltage characteristics, showing good model accuracy in the threshold region [11]. © 2004 IEEE. Reprinted with permission.

Figure 9.8 shows that the compact model can predict the loadpull contours, and the optimum load impedances for both best power output and PAE. The accuracy of the predictions is excellent, with the maximum predicted output power is 26.26 dBm, in agreement with the loadpull measurement, and the maximum predicted PAE is 77.16%, compared with the measured value of 77.26%. The modeled and measured load impedances presented in each case are very close.

The adjacent (1st) and alternate (2nd) channel ACLR distortion predicted by the compact model demonstrated good agreement with measured data, as can be seen in Fig. 9.9, certainly down to the measurement noise floor. Since the large-signal measurements are made on the full-size power transistor, this is a validation of the scaling rules of the extrinsic parameters.

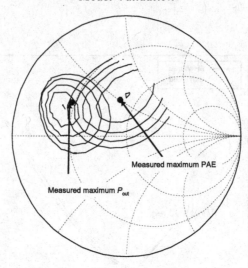

Fig. 9.8. Loadpull validation of the GaAs E-PHEMT model at 1.9 GHz, showing excellent agreement between measured and modeled results for maximum power output and PAE [11]. © 2004 IEEE. Reprinted with permission.

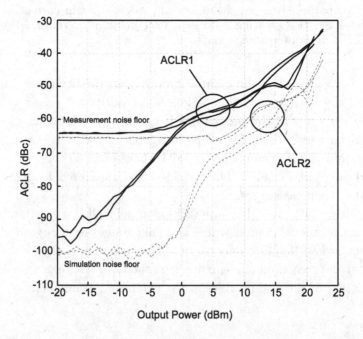

Fig. 9.9. Validation of the GaAs E-PHEMT model using large-signal distortion measurements of the 1st and 2nd channel ACLR [11]. © 2004. Reprinted with permission.

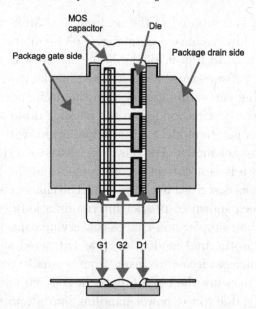

Fig. 9.10. Illustrations of the complex internal structure of the 900 MHz, 160 W power transistor are shown in cross-section and top-down views.

9.4.4 A 900 MHz, 160 W Silicon LDMOS Product Level Validation Example

All of the analysis and measurement techniques investigated so far have been developed with the objective of generating a model for the unmatched transistor and the matching networks found within high-power RF power transistors. To demonstrate the applicability of these techniques, we present a comparison of simulated and measured results of a high-power transistor designed and manufactured for cellular wireless infrastructure applications.

The selected transistor has been designed for N-CDMA, GSM and GSM EDGE base-station applications for the 860 and 960 MHz frequency bands. Additionally, the transistor is suitable for multi-carrier amplifier applications. In typical GSM applications, for $V_{ds} = 28$ V and $I_{DQ} = 1200$ mA, the packaged transistor is capable of 160 W when operating in continuous-wave mode (at its 1–dB compression point), with 20 dB gain and drain efficiency of 58%.

The intricate matching circuitry formed by the bondwires and MOS capacitors is illustrated in Fig. 9.10. Rows of parallel bondwires form arrays that interconnect the dies, MOS capacitors and the package. The packaged transistor contains three LDMOS die, each having a gate periphery of approximately 90 mm for a total gate periphery of around 270 mm. The matching network is composed of the package, one MOS capacitor, and 78 bondwires. All bondwires have a diameter of 63.5 μm and are made of aluminium. As indicated in Fig. 9.10, the MOS capacitor is a low value so an effective T-match network can be constructed on the gate side of the transistor. No matching capacitors are used on the drain side of the transistor so the dies are directly wire-bonded to the package drain lead.

The task of generating a model for this device begins with the extraction of the nonlinear transistor model. For this exercise we used the Root nonlinear model [12], because it is a correctly constructed and implemented charge-conserving model, as described in Chapter 6. The charge conservative model description has been shown to predict intermodulation distortion products more accurately than simpler non-charge-conserving capacitance models [9]. This model is an isothermal model, and was extracted at several heatsink temperatures from measurements taken from a smaller transistor, since it is not possible to measure the large (approximately 90 mm gate periphery) transistors directly, due to the power-handling limitations of the characterization equipment. After the model was extracted and scaled, the geometry of the bondwires and the relative positions of the MOS capacitors and die within the package were obtained with the use of an optical microscope.

Once the simulations of all of the constituent components were completed, the resulting models are incorporated into a large-signal circuit simulator. Large-signal single-tone and two-tone simulations were performed under pulsed conditions to provide an isothermal measurement environment. The measured versus simulated large-signal performance of the transistor model is shown in the following figures.

The output power and power added efficiency contours of the 900 MHz 160 W transistor are shown in Fig. 9.11, where the source has been conjugately-matched. These loadpull measurements and simulations were performed at 865 MHz under one-tone pulse conditions since CW measurements exceed the rated power limitation (< 100 W) of the passive tuners used in the loadpull system [13].

Measurements and simulations of the output power, 3rd-order intermodulation distortion (IM3), PAE, and transducer gain are shown in Figs. 9.12 and 9.13. These power sweep or drive-up measurements were performed while the packaged transistor was tuned at the output for maximum power-added efficiency, and maximum output power, respectively.

Overall, the measured and simulated results are in agreement over the test conditions presented. Often when generating a model for a packaged transistor of this complexity, there exists a difference between measured and simulated results and we turn to an optimizer to improve the fit. It is

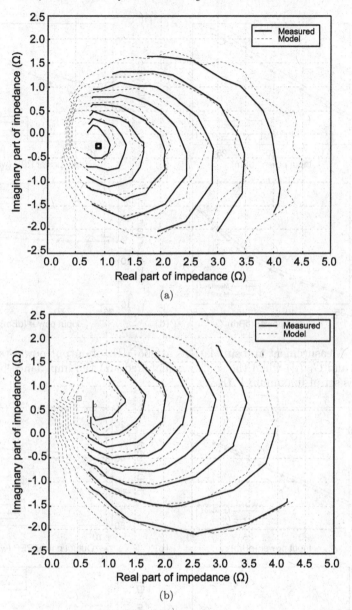

Fig. 9.11. Loadpull contours of the power transistor under single-tone pulsed excitation at 865 MHz. The output power contours are shown in (a) and the power-added efficiency contours are shown in (b).

often problematic when an optimizer is used to adjust the model parameters since, without extensive data, optimizing often forces the model to match the particular case at hand without consideration for the wider application. To

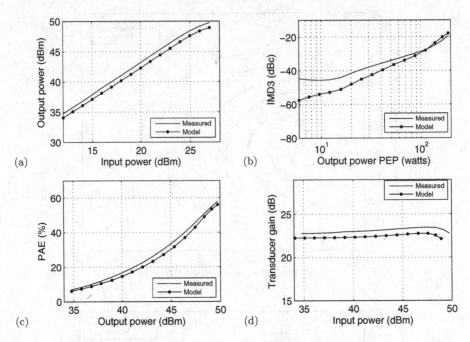

Fig. 9.12. Measurement and simulations at 865 MHz of output power (a), IM3 (b), PAE (c), and G_t (d) when the source is matched and the impedance presented to the load is set of maximum PAE.

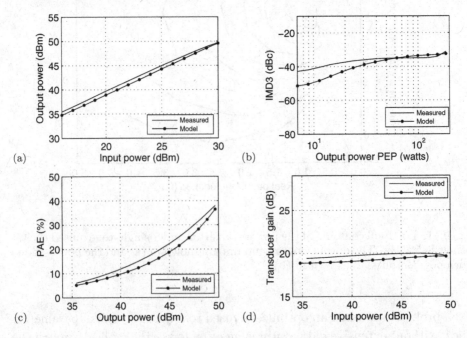

Fig. 9.13. Measurement and simulations at 865 MHz of output power (a), IM3 (b), PAE (c), and G_t (d) when the source is matched and the impedance presented to the load is set for maximum output power.

avoid this *ad hoc* fitting we adjust, if necessary, the model parameters around the physical parameters extracted during the model process. For example it is reasonable to adjust the bondwire inductances by 10% of the nominal value when using an inductive-only calculation, which is documented to be within its accuracy range [14]. Further adjustment is not warranted and will compromise the predictive capability of the model to other conditions that were not considered during the optimization process.

9.4.5 *Model Validation in the Time Domain*

All the model validation examples shown so far in this chapter are comparing measured versus simulated results in the frequency domain. For RF and microwave engineers, the analysis and understanding of circuits in the frequency domain is almost second nature. The S-parameters are complex numbers, providing magnitude and phase information at the device terminals. Under large-signal conditions the most common frequency-domain measurements used during the validation of large-signal models are input power sweeps at fixed impedances or loadpull measurements, which measure all the important large-signal figures of merit of the transistor as a function of the device terminal impedances. An inherent limitation of this large-signal characterization procedure is that the quantities being measured are scalar in nature, even though it is acknowledged that the input and output spectrum of the device at its terminals are complex in nature. Therefore, a model validation procedure that only uses large-signal figures of merit that are scalar in nature is only partially exercising the model, and will not address the richness of the device behaviour.

Recently, with the development and adoption of large-signal network analyzers, we can capture the magnitude and phase information of the input and output spectrum of the device. Another way to view this complex spectrum is by transforming the data into the time domain, which will allow us to reconstruct the time-domain waveforms of the transistor while operating under large-signal conditions. Because of this capability the LSNA often finds application in the extraction and development of behavioural models [15–17]. For compact model development the unique capability of extracting the time-domain waveforms at microwave frequencies provides an alternative method for validating large-signal transistor models.

By measuring magnitude and relative phase between the fundamental frequency and all harmonic content, the LSNA determines the time-domain waveforms for the specific impedance terminations applied to the transistor by the external matching networks. If we couple the LSNA with harmonic

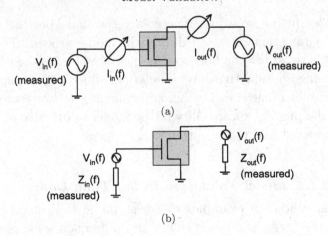

Fig. 9.14. Schematic of the harmonic balance simulation from measured time-domain large-signal data is shown in (a) [19]. The corresponding schematic using impedances specifying the test environment is illustrated in (b). © 2001 IEEE. Reprinted with permission.

loadpull capabilities we can then determine the performance of the transistor under many different operating conditions. The method is ideal for the validation of transistor models operating in amplifier classes for which having access to the time-domain waveforms is highly beneficial, for example, class F [18]. While this type of measurement system is often used for *waveform engineering* of the power amplifier, it offers the modeling engineer a powerful alternative approach to transistor model validation. The approach and results summarized here are an abbreviated summary of the work performed by Gaddi *et al.* [19].

A key aspect of any model validation is to be able to replicate the measurement environment within the circuit simulator. Depending on the type of measurement being performed, this task can be very cumbersome. The simulation of large-signal time-domain waveforms is possible with conventional circuit simulation packages offering harmonic balance simulation capabilities and fast Fourier transforms.

When performing a validation with time-domain measurement data, the issue of replicating the bench environment is avoided since the model is excited with the measured large-signal voltage waveform data. Since the large-signal data contain all of the information related to the terminations of the transistor, the measurement environment is completely specified within the circuit simulator and a comparison of measured versus simulated results is straightforward [17]. Figure 9.14 shows a schematic that implements the DUT extraction with the large-signal time-domain waveform data.

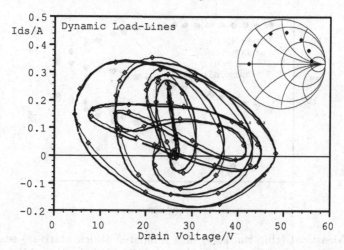

Fig. 9.15. Measured (markers) and simulated (solid lines) dynamic load-lines using rotating phase fundamental loads shown in inset [19]. © 2001 IEEE. Reprinted with permission.

In contrast, performing a loadpull frequency-domain validation requires that in addition to de-embedding the fixture at the fundamental frequency, knowledge of the terminations applied to the transistor over the entire frequency spectrum (fundamental, sub-harmonic, and harmonics frequencies) is required. The determination of these impedances for the entire spectrum is non-trivial and laborious [20].

The device used to illustrate the time-domain measurements is an on-wafer Si LDMOS transistor of 2.4 mm total gate periphery. All the measurements were performed with ground-signal-ground probes. The MET model was extracted for the same device used in this validation exercise. The model parameters describing the nonlinear DC drain current were extracted from pulsed DC-IV measurements at different heatsink temperatures. The nonlinear capacitance functions were fitted to the extracted capacitance versus voltage data obtained from small-signal S-parameter measurements.

Figure 9.15 shows the measured and simulated dynamic load-lines, that is, the plot of the measured drain current versus the measured drain-to-source voltage at each point in time of the waveform, for different output terminations as shown in the inset in the figure. Significant differences can be observed in the shape of the dynamic load-line under different output transistor terminations. The time-domain data can also be analyzed and displayed in the more traditional frequency-domain scalar representation.

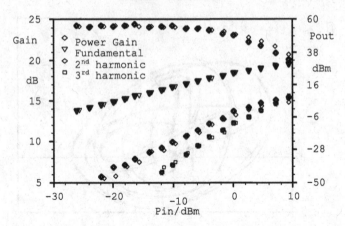

Fig. 9.16. Measured (thin markers) and simulated (thick markers) scalar output powers at the first three harmonics and power gain from the class AB measurement data [19]. © 2001 IEEE. Reprinted with permission.)

Figure 9.16 shows the measured versus simulated gain compression curves for the fundamental tone and two additional harmonic components.

The fundamental purpose of the model validation process is to increase the confidence level of the model user, and to provide guidance on the expectation to its accuracy. Time-domain measurements create a perfect complement to the more common model validation measurement obtained from traditional scalar loadpull measurement systems.

References

[1] O. Balci, "Verification, validation and accreditation of simulation models," in *Proc. of the Winter Simulation Conf.*, Atlanta, GA, 1997, pp. 135–141.

[2] R. G. Sargent, "Verification and validation of simulation models," in *Proc. of the Winter Simulation Conf.*, San Diego, CA, 1998, pp. 121–130.

[3] J. S. Carson, II, "Model verification and validation," in *Proc. of the Winter Simulation Conf.*, Washington, DC, 2002, pp. 52–58.

[4] J. F. Sevic, C. McGuire, G. R. Simpson, and J. A. Plá, "Data-based load pull simulation for large-signal transistor model validation," *Microwave Journal*, 40, no. 3, Mar. 1997.

[5] P. H. Aaen, J. A. Plá, and C. A. Balanis, "On the development of CAD techniques suitable for the design of high-power RF transistors," *IEEE Trans. Microwave Theory Tech.*, 53, no. 10, pp. 3067–3074, Oct. 2005.

[6] P. H. Aaen, J. A. Plá, and C. A. Balanis, "Modeling techniques suitable for CAD based design of internal matching networks of high-power RF/microwave

transistors," *IEEE Trans. Microwave Theory Tech.*, 54, no. 7, pp. 3052–3059, July 2006.

[7] J. A. Plá and D. Bridges, "A robust high voltage Si LDMOS model extraction process to achieve first pass linear RFIC amplifier design success," in *IEEE MTT-S Int. Microwave Symp. Dig.*, Seattle, WA, June 2002, pp. 263–266.

[8] I. Angelov, H. Zirath, and N. Rorsman, "A new empirical nonlinear model for HEMT and MESFET devices," *IEEE Trans. Microwave Theory Tech.*, 40, no. 12, pp. 2258–2266, Dec. 1992.

[9] J. Staudinger, M. C. de Baca, and R. Vaitkus, "An examination of several large-signal capacitance models to predict GaAs HEMT linear power amplifier performance," in *Proc. IEEE Radio and Wireless Conf. (RAWCON)*, Colorado Springs, CO, Aug. 1998, pp. 343–346.

[10] J. Wood and D. E. Root, "Bias-dependent linear scalable millimeter-wave FET model," *IEEE Trans. Microwave Theory Tech.*, 48, no. 12, pp. 2352–2360, Dec. 2000.

[11] J. Wood and D. E. Root, "A symmetric and thermally-de-embedded nonlinear FET model forwireless and microwave applications," in *IEEE MTT-S Int. Microwave Symp. Dig.*, Fort Worth, TX, June 2004, pp. 35–38.

[12] *IC-CAP Parameter Extraction and Device Modeling Software: Agilent Root MOSFET model*, Agilent Technologies, Agilent EEsof EDA.

[13] B. Noori, P. Hart, J. Wood, P. H. Aaen, M. Guyonnet, M. LeFevre, J. A. Plá, and J. Jones, "Load-pull measurements using modulated signals," in *Proc. 36th European Microwave Conf.*, Manchester, UK, Sept. 2006.

[14] K. Mouthaan, R. Tinti, M. de Kok, H. C. de Graaff, J. L. Tauritz, and J. Slotboom, "Microwave modeling and measurement of the self- and mutual inductances of coupled bondwires," in *Proc. IEEE Bipolar/BiCMOS Circuit and Technology Meeting*, Minneapolis, MN, 1997, pp. 166–169.

[15] D. E. Root, J. Verspecht, D. Sharrit, J. Wood, and A. Cognata, "Broadband poly-harmonic distortion (PHD) behavioral models from fast automated simulations and large-signal vectorial network measurements," *IEEE Trans. Microwave Theory Tech.*, 53, no. 11, pp. 3656–3664, Nov. 2005.

[16] J. Wood and D. E. Root, *Fundamentals of Nonlinear Behavioral Modeling for RF and Microwave Design.* Norwood, MA: Artech House, 2005.

[17] D. M. M.-P. Schreurs, J. Verspecht, S. Vandenberghe, and E. Vandamme, "Straightforward and accurate nonlinear device model parameter-estimation method based on vectorial large-signal measurements," *IEEE Trans. Microwave Theory Tech.*, 50, no. 10, pp. 2315–2319, Oct. 2002.

[18] S. C. Cripps, *RF Power Amplifiers for Wireless Communications*, 2nd edn. Norwood, MA: Artech House, 2006.

[19] R. Gaddi, J. A. Plá, J. Benedikt, and P. J. Tasker, "LDMOS electro-thermal model validation from large-signal time-domain measurements," in *IEEE MTT-S Int. Microwave Symp. Dig.*, Phoenix, AZ, May 2001, pp. 399–402.

[20] P. Aaen, J. Plá, D. Bridges, and E. Shumate, "A wideband method for the

rigorous low-impedance loadpull measurement of high-power transistors suitable for large-signal model validation," in *56th ARFTG Conference Digest*, Broomfield, CO, Dec. 2000.

About the Authors

Peter H. Aaen

Received the B.A.Sc. degree in Engineering Science and the M.A.Sc. degree in Electrical Engineering, both from the University of Toronto, Toronto, ON., Canada, and the Ph.D. degree in Electrical Engineering from Arizona State University, Tempe, AZ., USA, in 1995, 1997, and 2005, respectively. He is the Manager of the RF Modeling team of the RF Division of Freescale Semiconductor, Inc, Tempe, AZ, USA; a company which he joined in 1997 (then Motorola Inc. Semiconductor Product Sector). His areas of expertise include the development of passive and active compact models for the design and development of power transistors and ICs. Prior to his role of manager, his main focus was on the development of efficient electromagnetic simulation methodologies for complex packaged environments. His current work focuses on the development and validation of microwave transistor models and passive components. His technical interests include techniques for electromagnetic optimization, calibration techniques for microwave measurements and the development of package modeling techniques.

He has presented at various workshops at IMS and RWS in recent years; and has been a member of the IMS Technical Program Committee for the past two years. He is author or co-author of over a dozen papers, articles, and workshops in the fields of electromagnetic simulation, package modeling, and microwave device modeling and characterization. He is a Member of the IEEE and a member of the Microwave Theory and Techniques Society.

Jaime A. Plá

Received the B.S. degree in Electrical Engineering from the University of Puerto Rico, Mayagüez, in 1991, and the M.S. degree in Microwave Engineering from the University of Massachusetts at Amherst, in 1993. He is

353

currently the Manager of the Design Organization of the RF Division of Freescale Semiconductor, Inc, Tempe, AZ, USA; a company which he joined in 1995 (then Motorola Inc. Semiconductor Product Sector). Prior to his current design manager role, he was the Manager of the LDMOS Modeling Team where his work centered on the development of high-power RF electro-thermal device models for LDMOS devices. Other areas of focus were the development of package modeling techniques and modeling of passive components, as well as techniques for the measurement of electrical and thermal transistor characteristics related to small- and large-signal modeling extraction and validation. In 1991, he joined the Microwave Semiconductor Laboratory, Research Division, Raytheon, Lexington, MA. While with Raytheon, he was primarily involved with the development of microwave measurement techniques and linear and nonlinear models for monolithic-microwave integrated-circuit (MMIC) semiconductor devices such as GaAs MESFETs PHEMTs, and HBTs.

He is author or co-author of over two dozen papers and articles, in the fields of microwave device and package modeling and characterization, and microwave measurement techniques. He is a Member of the IEEE, and a member of the Microwave Theory and Techniques Society.

John Wood

Received B.Sc. and Ph.D. degrees in Electrical and Electronic Engineering from the University of Leeds, UK, in 1976 and 1980, respectively. He is currently a Senior Technical Contributor responsible for RF CAD & Modeling in the RF Division of Freescale Semiconductor, Inc, Tempe, AZ, USA. His areas of expertise include the development of compact device models and behavioral models for RF power transistors and ICs. To enable and support these modeling requirements, he has been involved in the specification of high power pulsed $I–V–RF$ test systems, for connectorized and on-wafer applications, and in the development of large-signal network analyzer (LSNA), loadpull, and envelope measurement techniques. From 1997–2005 he worked in the Microwave Technology Center of Agilent Technologies (then Hewlett Packard) in Santa Rosa, CA, USA, where his research work has included the investigation, characterization, and development of large-signal and bias-dependent linear FET models for millimetre-wave applications, and nonlinear behavioural modeling using LSNA measurements and nonlinear system identification techniques. Between 1983 and 1997 he was a Professor in the Department of Electronics at the University of York,

UK, where his research and teaching interests covered semiconductor devices, RF and microwave circuits, IC design, and device modeling. Prior to this time in academia, he was a Senior Research Engineer at STL in the UK, responsible for design and development of GaAs ICs and their fabrication processes.

He has organized and co-organized, and presented at workshops at IMS and RWS in recent years; he was on the Steering Committee for IMS 2006, and has been a member of the ARFTG Technical Program Committee for the past two years. He is author or co-author of over 80 papers and articles, in the fields of microwave device and system modeling and characterization, and microwave device technology. He is the co-editor of *Fundamentals of Nonlinear Behavioral Modeling for RF and Microwave Design* (Artech House, 2005). He is a Fellow of the IEEE, and a member of the Microwave Theory and Techniques, and Electron Devices Societies.

Index